Design of Structural Elements

OTHER TITLES FROM E & FN SPON

Building Services Engineering
D. V. Chadderton

Computer Methods in Structural Analysis
J. L. Meek

Concrete Masonry Designers's Handbook
J. J. Roberts, A. K. Tovey, W. B. Cranston and A. W. Beeby

Construction Contracts
J. R. Murdoch and W. Hughes

Construction Materials – Their Nature and Behaviour
Edited by J. M. Illston

Construction Methods and Planning
J. R. Illingworth

Design of Prestressed Concrete
R. I. Gilbert and N. C. Mickleborough

Design Strategies in Architecture
G. H. Baker

Examples of the Design of Reinforced Concrete Buildings to BS8110
C. E. Reynolds and J. C. Steedman

Fire from First Principles
P. Stollard and J. Abrahams

Geology for Civil Engineers
A. McLean and C. Gribble

Hydraulics in Civil and Environmental Engineering
A. J. Chadwick and J. C. Morfett

Introduction To Construction Economics
J. Manser

Limit States Design of Structural Steelwork
D. A. Nethercot

Project Management Demystified
G. Reiss

Reinforced Concrete Design Theory and Examples
T. J. MacGinley and B. S. Choo

Soil Mechanics
R. F. Craig

Structural Analysis
A. Ghali and A. M. Neville

The Behaviour and Design of Steel Structures
N. S. Trahair and M. A. Bradford

The Idea of Building
S. Groák

The Way We Build Now
A. Orton

Timber Engineering
E. M. Carmichael

For more information about these and other titles, please contact:

The Promotion Department, E & FN Spon, 2–6 Boundary Row, London SE1 8HN. Telephone 071 522 9966

Design of Structural Elements

Concrete, steelwork, masonry and timber design to British Standards and Eurocodes

Chanakya Arya

School of Architecture and Civil Engineering
South Bank University
London
UK

Chapters 1, 4 and 9 co-authored
by Peter R. Wright
South Bank University

E & FN SPON

An Imprint of Chapman & Hall

London · Glasgow · New York · Tokyo · Melbourne · Madras

**Published by E & FN Spon, an imprint of Chapman & Hall,
2–6 Boundary Row, London SE1 8HN**

Chapman & Hall, 2–6 Boundary Row, London SE1 8HN, UK

Blackie Academic & Professional, Wester Cleddens Road,
Bishopbriggs, Glasgow G64 2NZ, UK

Chapman & Hall Inc., One Penn Plaza, 41st Floor, New York NY
10119, USA

Chapman & Hall Japan, Thomson Publishing Japan,
Hirakawacho Nemoto Building, 6F,
1-7-11 Hirakawa-cho, Chiyoda-ku, Tokyo 102, Japan

Chapman & Hall Australia, Thomas Nelson Australia,
102 Dodds Street, South Melbourne, Victoria 3205, Australia

Chapman & Hall India, R. Seshadri,
32 Second Main Road, CIT East, Madras 600 035, India

First edition 1994

© 1994 Chanakya Arya

Typeset in Plantin by ROM-Data Corporation Ltd.

Printed in Great Britain by The Alden Press, Oxford

ISBN 0 419 17620 9

A catalogue record for this book is available from the British Library

Library of Congress Cataloging-in-Publication data

Arya, C.
    Design of structural elements : concrete, steelwork, masonry, and
timber design to British standards and Eurocodes / C. Arya. – 1st ed.
        p. cm.
    Includes index.
    ISBN 0–419–17620–9 (alk. paper)
    1. Structural design–Standards–Great Britain, 2. Structural
design–Standards–Europe.   I. Title
TA658.A79   1993
624.1'771'021841–dc20   93–32219
                                                CIP

In the memory of Biji

Contents

Preface

Structural design is a key element of all degree and diploma courses in civil and structural engineering. It involves the study of principles and procedures contained in the latest codes of practice for structural design for a range of materials, including concrete, steel, masonry and timber.

Most textbooks on structural design consider only one construction material and, therefore, the student may end up buying several books on the subject. This is undesirable both from the viewpoint of cost but also because it makes it difficult for the student to unify principles of structural design, because of differing presentation approaches adopted by the authors.

There are a number of combined textbooks which include sections on several materials. However, these tend to concentrate on application of the codes and give little explanation of the structural principles involved or, indeed, an awareness of material properties and their design implications. Moreover, none of the books refer to the new Eurocodes for structural design, some of which are scheduled to replace the equivalent British Standards around 1998.

The purpose of this book, then, is to describe the background to the principles and procedures contained in the latest British Standards and Eurocodes on the structural use of concrete, steelwork, masonry and timber. It is primarily aimed at students on civil and structural engineering degree and diploma courses. Allied professionals such as architects, builders and surveyors will also find it appropriate. In so far as it includes five chapters on the structural Eurocodes it will be of considerable interest to practising engineers too.

The subject matter is divided into 11 chapters and 3 parts:

Part 1   contains two chapters and explains the principles and philosophy of structural design, focusing on the limit state approach. It also explains how the overall loading on a structure and individual elements can be assessed, thereby enabling the designer to size the element.

Part 2   contains four chapters covering the design and detailing of a number of structural elements, e.g. floors, beams, walls, columns, connections and foundations to the latest British codes of practice for concrete, steelwork, masonry and timber design.

Part 3   contains five chapters on the Eurocodes for these materials. The first of these describes the purpose, scope and problems associated with drafting the Eurocodes. The remaining chapters describe the layout and contents of EC2, EC3, EC5 and EC6 for design in concrete, steelwork, timber and masonry respectively.

At the end of Chapters 1–6 a number of design problems have been included for the student to attempt, typical answers for which will subsequently be given in a companion to this book.

Although most of the tables and figures from the British Standards referred to in the text have been reproduced, it is expected that the reader will have either the full Standard or the publication *Extracts from British Standards for Students of Structural Design*, obtainable from BSI Sales, Linford Wood, Milton Keynes, Bucks, in order to gain the most from this book.

C. Arya
London
UK

Acknowledgements

The following people have made a significant contribution to the writing of this book, and their assistance is acknowledged with grateful thanks: Professor Laurence Wood, Head of the School of Architecture and Civil Engineering, South Bank University, for making available all necessary facilities; Fred Lambert formerly South Bank University, for reviewing Chapters 1–6; Tony Fewell, Head of Timber Structures, Building Research Establishment and John Moran, South Bank University, for reviewing Chapter 6; David Smith, South Bank University, for frequent discussions; Tony Threlfall, British Cement Association, for reviewing Chapters 3, 7 and 8; Colin Taylor (member of EC3's Editorial Board), Steel Construction Institute, for reviewing Chapters 7 and 9; John Steer (member of EC5's Editorial Board), Steer Consulting Engineers, for reviewing Chapter 11 and Peter Watt (member of EC6's Editorial Board), Brick Development Association, for reviewing Chapters 5 and 10.

Thanks are also due to Tony Fawcett for tracing the figures and the British Standards Institute, British Cement Association and British Steel PLC for permission to use extracts from their publications.

And finally to Hana, my friend, and to my family goes my appreciation for their support during a very difficult period of my life and without whom this book might never have been completed.

List of worked examples

xv

PART ONE

INTRODUCTION TO STRUCTURAL DESIGN

The primary aim of all structural design is to ensure that the structure will perform satisfactorily during its design life. Specifically, the designer must check that the structure is capable of carrying the loads safely and that it will not deform excessively due to the applied loads. This requires the designer to make realistic estimates of the strengths of the materials composing the structure and the loading to which it may be subject during its design life. Furthermore, the designer will need a basic understanding of structural behaviour.

The work that follows has two objectives:

1. to describe the philosophy of structural design;
2. to introduce various aspects of structural and material behaviour.

Towards the first objective, *Chapter 1* discusses the three main philosophies of structural design, emphasizing the limit state philosophy which forms the bases of design in many of the modern codes of practice. *Chapter 2* then outlines a method of assessing the design loading acting on individual elements of a structure and how this information can be used, together with the material properties, to size elements.

Chapter 1

Philosophy of design

This chapter is concerned with the philosophy of structural design. The chapter describes the overall aims of design and the many inputs into the design process. The primary aim of design is seen as the need to ensure that at no point in the structure do the design loads exceed the design strengths of the materials. This can be achieved by using the permissible stress or load factor philosophies of design. However, both suffer from drawbacks and it is more common to design according to limit state principles which involve considering all the mechanisms by which a structure could become unfit for its intended purpose during its design life.

1.1 Introduction

The task of the structural engineer is to design a structure which satisfies the needs of the client and the user. Specifically the structure should be safe, economical to build and maintain, and aesthetically pleasing. But what does the design process involve?

Design is a word that means different things to different people. In dictionaries the word is described as a mental plan, preliminary sketch, pattern, construction, plot or invention. Even among those closely involved with the built environment there are considerable differences in interpretation. Architects, for example, may interpret design as being the production of drawings and models to show what a new building will actually look like. To civil and structural engineers, however, design is taken to mean the entire planning process for a new building structure, bridge, tunnel, road, etc., from outline concepts and feasibility studies through mathematical calculations to working drawings which could show every last nut and bolt in the project. Together with the drawings there will be bills of quantities, a specification and a contract, which will form the necessary legal and organizational framework within which a contractor, under the supervision of engineers and architects, can construct the scheme.

There are many inputs into the engineering design process as illustrated by *Fig. 1.1* including:

1. client brief
2. experience
3. imagination
4. a site investigation
5. model and laboratory tests
6. economic factors
7. environmental factors.

The starting-point for the designer is normally a conceptual brief from the client, who may be a private developer or perhaps a government body. The conceptual brief may simply consist of some sketches prepared by the client or perhaps a detailed set of architect's drawings. Experience is crucially important, and a client will always demand that the firm he is employing to do the design has previous experience designing similar structures.

Although imagination is thought by some to be entirely the domain of the architect, this is not so. For engineers and technicians an imagination of how elements of structure interrelate in three dimensions is essential, as is an appreciation of the loadings to which structures might be subject in certain circumstances. In addition, imaginative solutions to engineering problems are often required to save money, time, or to improve safety or quality.

A site investigation is essential to determine the strength and other characteristics of the ground on which the structure will be founded. If the structure is unusual in any way, or subject to abnormal loadings, model or laboratory tests may also be used to help determine how the structure will behave.

In today's economic climate a structural designer must be constantly aware of the cost implications of his or her design. On the one hand design should aim to achieve economy of materials in the structure, but over-refinement can lead to an excessive number of different sizes and components in the structure, and labour costs will rise. In addition the actual cost

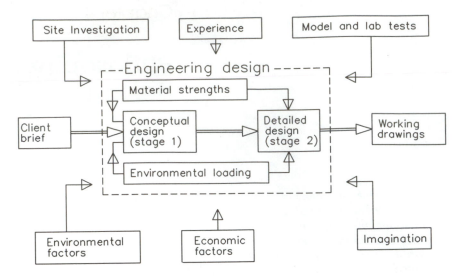

Fig. 1.1 *Inputs into the design process.*

of the designer's time should not be excessive, or this will undermine the employer's competitiveness. The idea is to produce a workable design achieving reasonable economy of materials, while keeping manufacturing and construction costs down, and avoiding unnecessary design and research expenditure. Attention to detailing and buildability of structures cannot be overemphasized in design. Most failures are as a result of poor detailing rather than incorrect analysis.

Designers must also understand how the structure will fit into the environment for which it is designed. Today many proposals for engineering structures stand or fall on this basis, so it is part of the designer's job to try to anticipate and reconcile the environmental priorities of the public and government.

The engineering design process can often be divided into two stages: (1) a feasibility study involving a comparison of the alternative forms of structure and selection of the most suitable type and (2) a detailed design of the chosen structure. The success of stage 1, the conceptual design, relies to a large extent on engineering judgement and instinct, both of which are the outcome of many years' experience of designing structures. Stage 2, the detailed design, also requires these attributes but is usually more dependent upon a thorough understanding of the codes of practice for structural design, e.g. BS 8110 and BS 5950. These documents are based on the amassed experience of many generations of engineers, and the results of research. They help to ensure safety and economy of construction, and that mistakes are not repeated. For instance, after the

infamous disaster at the Ronan Point block of flats in Newham, London, when a gas explosion caused a serious partial collapse, research work was carried out, and codes of practice were amended so that such structures could survive a gas explosion, with damage being confined to one level.

The aim of this book is to look at the procedures associated with the detailed design of structural elements such as beams, columns and slabs. *Chapter 2* will help the reader to revise some basic theories of structural behaviour. *Chapters 3–6* deal with design to British Standard (BS) codes of practice for the structural use of concrete (BS 8110), structural steelwork (BS 5950), masonry (BS 5628) and timber (BS 5268). *Chapter 7* introduces the new Eurocodes (EC) for structural design and *Chapters 8–11* then describe the layout and design principles in EC2, EC3, EC6 and EC5 for concrete, steelwork, masonry and timber respectively.

1.2 Basis of design

Table 1.1 illustrates some risk factors that are associated with activities in which people engage. It can be seen that some degree of risk is associated with air and road travel. However, people normally accept that the benefits of mobility outweigh the risks. Staying in buildings, however, has always been regarded as fairly safe. The risk of death or injury due to structural failure is extremely low, but as we spend most of our life in buildings this is perhaps just as well.

Table 1.1 Comparative death risk per 10^8 persons exposed

Mountaineering (international)	2700
Air travel (international)	120
Deep water trawling	59
Car travel	56
Coal mining	21
Construction sites	8
Manufacturing	2
Accidents at home	2
Fire at home	0.1
Structural failures	0.002

As far as the design of structures for safety is concerned, it is seen as the process of ensuring that stresses due to loading at all critical points in a structure have a very low chance of exceeding the strength of materials used at these critical points. *Figure 1.2* illustrates this in statistical terms.

In design there exist within the structure a number of critical points (e.g. beam mid-spans) where the design process is concentrated. The normal distribution curve on the left of *Fig. 1.2* represents the actual maximum material stresses at these critical points due to the loading. Because loading varies according to occupancy and environmental conditions, and because design is an imperfect process, the material stresses will vary about a modal value – the peak of the curve. Similarly the normal distribution curve on the right represents material strengths at these critical points, which are also not constant due to the variability of manufacturing conditions.

The overlap between the two curves represents a possibility that failure may take place at one of the critical points, as stress due to loading exceeds the strength of the material. In order for the structure to be safe the overlapping area must be kept to a minimum. The degree of overlap between the two curves can be minimized by using one of three distinct design philosophies, namely:

1. permissible stress design
2. load factor method
3. limit state design.

1.2.1 PERMISSIBLE STRESS DESIGN

In permissible stress design, sometimes referred to as modular ratio or elastic design, the stresses in the structure at working loads are not allowed to exceed a certain proportion of the yield stress of the construction material, i.e. the stress levels are limited to the elastic range. By assuming that the stress–strain relationship over this range is linear, it is possible to calculate the actual stresses in the material concerned. Such an approach formed the basis of the design methods used in CP 114 (the forerunner of BS 8110) and BS 449 (the forerunner of BS 5950).

However, although it modelled real building performance under actual conditions, this philosophy had two major drawbacks. Firstly, permissible design methods sometimes tended to overcomplicate the design process and also led to conservative solutions. Secondly, as the quality of materials increased and the safety margins decreased, the assumption that stress and strain are directly proportional became unjustifiable for materials such as concrete, making it impossible to estimate the true factors of safety.

1.2.2 LOAD FACTOR DESIGN

Load factor or plastic design was developed to take account of the behaviour of the structure once the yield point of the construction material had been reached. This approach involved calculating the collapse load of the structure. The working load was derived by dividing the collapse load by a load factor. This approach simplified methods of analysis and allowed actual factors of safety to be calculated. It was in fact permitted in CP 114 and BS 449 but was slow in gaining acceptance and was eventually superseded by the more comprehensive limit state approach.

The reader is referred to *Appendix A* for an example illustrating the differences between the permissible stress and load factor approaches to design.

1.2.3 LIMIT STATE DESIGN

Originally formulated in the former Soviet Union in the 1930s and developed in Europe in the 1960s, limit state design can perhaps be seen as a compromise between the permissible and load factor methods. It is in fact a more comprehensive approach which takes into account both methods in appropriate ways. Most modern structural codes of

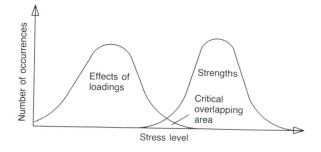

Fig. 1.2 *Relationship between stress and strength.*

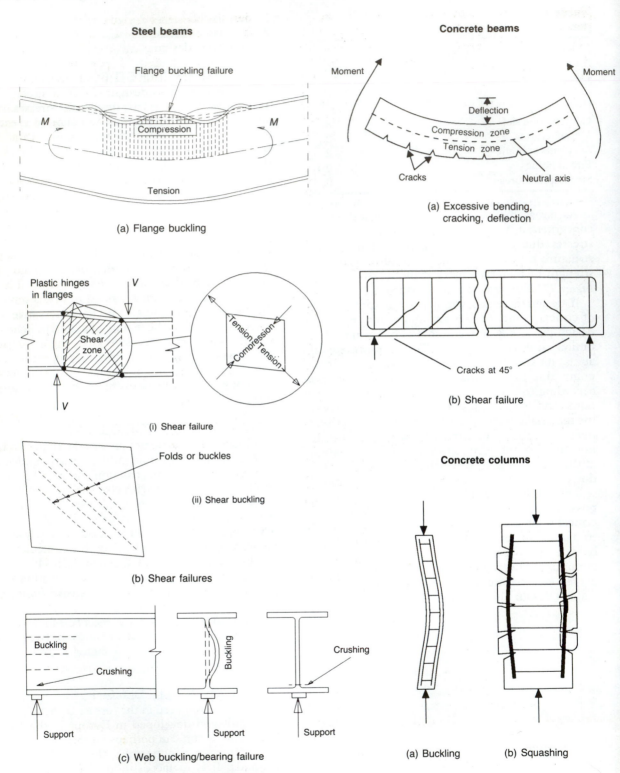

Steel beams

Flange buckling failure

M   Compression   M

Tension

(a) Flange buckling

Concrete beams

Moment   Moment

Deflection
Compression zone
Tension zone

Cracks   Neutral axis

(a) Excessive bending,
cracking, deflection

Plastic hinges
in flanges   V

Shear
zone

Tension   Compression
Compression   Tension

V

(i) Shear failure

Cracks at 45°

(b) Shear failure

Folds or buckles

(ii) Shear buckling

(b) Shear failures

Concrete columns

Buckling

Crushing

Buckling

Crushing

Support   Support   Support

(c) Web buckling/bearing failure

(a) Buckling   (b) Squashing

Fig. 1.3 *Typical modes of failure for beams and columns.*

practice are now based on the limit state approach. BS 8110 for concrete, BS 5950 for structural steel-work, BS 5400 for bridges and BS 5628 for masonry are all limit state codes. The principal exceptions are the code of practice for design in timber, BS 5268, and the old (but still current) structural steelwork code, BS 449, both of which are permissible stress codes. It should be noted, however, that the Eurocode for timber (EC5), which is expected to replace BS 5268 around 1999, is based on limit state principles.

As limit state philosophy forms the basis of the design methods in most modern codes of practice for structural design, it is essential that the design methodology is fully understood. This then is the purpose of the following subsections.

1.2.3.1 Ultimate and serviceability limit states

The aim of limit state design is to achieve acceptable probabilities that a structure will not become unfit for its intended use during its design life, that is, the structure will not reach a limit state. There are many ways in which a structure could become unfit for use, including excessive conditions of bending, shear, compression, deflection and cracking (*Fig. 1.3*). Each of these mechanisms is a limit state whose effect on the structure must be individually assessed.

Some of the above limit states, e.g. deflection and cracking, principally affect the appearance of the structure. Others, e.g. bending, shear and compression, may lead to partial or complete collapse of the structure. Those limit states which can cause failure of the structure are termed ultimate limit states. The others are categorized as serviceability limit states. The ultimate limit states enable the designer to calculate the strength of the structure. Serviceability limit states model the behaviour of the structure at working loads. In addition, there may be other limit states which may adversely affect the performance of the structure, e.g. durability and fire resistance, and which must therefore also be considered in design.

It is a matter of experience to be able to judge which limit states should be considered in the design of particular structures. Nevertheless, once this has been done, it is normal practice to base the design on the most critical limit state and then check for the remaining limit states. For example, for reinforced concrete beams the ultimate limit states of bending and shear are used to size the beam. The design is then checked for the remaining limit states, e.g. deflection and cracking. On the other hand, the serviceability limit state of deflection is normally critical in the design of concrete slabs. Again, once

the designer has determined a suitable depth of slab, he/she must then make sure that the design satisfies the limit states of bending, shear and cracking.

In assessing the effect of a particular limit state on the structure, the designer will need to assume certain values for the loading on the structure and the strength of the materials composing the structure. This requires an understanding of the concepts of characteristic and design values which are discussed below.

1.2.3.2 Characteristic and design values

As stated at the outset, when checking whether a particular member is safe, the designer cannot be certain about either the strength of the material composing the member or, indeed, the load which the member must carry. The material strength may be less than intended (a) because of its variable composition, and (b) because of the variability of manufacturing conditions during construction, and other effects such as corrosion. Similarly the load in the member may be greater than anticipated (a) because of the variability of the occupancy or environmental loading, and (b) because of unforeseen circumstances which may lead to an increase in the general level of loading, errors in the analysis, errors during construction, etc.

In each case, item (a) is allowed for by using a **characteristic** value. The characteristic strength is the value **below** which the strength lies in only a small number of cases. Similarly the characteristic load is the value **above** which the load lies in only a small percentage of cases. In the case of strength the characteristic value is determined from test results using statistical principles, and is normally defined as the value below which not more than 5% of the test results fall. However, at this stage there are insufficient data available to apply statistical principles to loads. Therefore the characteristic loads are normally taken to be the design loads from other codes of practice, e.g. BS 648 and BS 6399.

The overall effect of items under (b) is allowed for using a partial safety factor: γ_m for strength and γ_f for load. The design strength is obtained by dividing the characteristic strength by the partial safety factor for strength:

$$\text{Design strength} = \frac{\text{characteristic strength}}{\gamma_m} \quad (1.1)$$

The design load is obtained by multiplying the characteristic load by the partial safety factor for the load:

$$\text{Design load} = \text{characteristic load} \times \gamma_f \quad (1.2)$$

The value of γ_m will depend upon the properties of the actual construction material being used. Values for γ_f depend on other factors which will be discussed more fully in *Chapter 2*.

In general, once a preliminary assessment of the design loads has been made it is then possible to calculate the maximum bending moments, shear forces and deflections in the structure (*Chapter 2*). The construction material must be capable of withstanding these forces otherwise failure of the structure may occur, i.e.

$$\text{Design strength} \geqslant \text{design load} \qquad (1.3)$$

Simplified procedures for calculating the moment, shear and axial load capacities of structural elements together with acceptable deflection limits are described in the appropriate codes of practice. These allow the designer to rapidly assess the suitability of the proposed design. However, before discussing these procedures in detail, *Chapter 2* describes in general terms how the design loads acting on the structure are estimated and used to size individual elements of the structure.

1.3 Summary

This chapter has examined the bases of three philosophies of structural design: permissible stress, load factor and limit state. The chapter has concentrated on limit state design since it forms the basis of the design methods given in the codes of practice for concrete (BS 8110), structural steelwork (BS 5950) and masonry (BS 5628). The aim of limit state design is to ensure that a structure will not become unfit for its intended use, that is, it will not reach a limit state during its design life. Two categories of limit states are examined in design: ultimate and serviceability. The former is concerned with overall stability and determining the collapse load of the structure; the latter examines its behaviour under working loads. Structural design principally involves ensuring that the loads acting on the structure do not exceed its strength and the first step in the design process then is to estimate the loads acting on the structure.

Questions

1. Explain the difference between conceptual design and detailed design.
2. What is a code of practice and what is its purpose in structural design?
3. What are the primary variables in structural design? Discuss the difficulties associated with estimating representative values of these variables and how such difficulties can be overcome.
4. The characteristic strengths and design strengths are related via the partial safety factor for materials. The partial safety factor for concrete is higher than for steel reinforcement. Discuss why this should be so.
5. Describe in general terms the ways in which a beam and column could become unfit for use.

Chapter 2

Basic structural concepts and material properties

This chapter is concerned with general methods of sizing beams and columns in structures. The chapter describes how the characteristic and design loads acting on structures and on the individual elements are determined. Methods of calculating the bending moments, shear forces and deflections in beams are outlined. Finally, the chapter describes general approaches to sizing beams according to elastic and plastic criteria and sizing columns subject to axial loading.

2.1 Introduction

All structures are composed of a number of interconnected elements such as slabs, beams, columns, walls and foundations. Collectively, they enable the internal and external loads acting on the structure to be safely transmitted down to the ground. The actual way that this is achieved is difficult to model and many simplifying, but conservative, assumptions have to be made. For example, the degree of fixity at column and beam ends is usually uncertain but, nevertheless, must be estimated as it significantly affects the internal forces in the element. Furthermore,

it is usually assumed that the reaction from one element is a load on the next and that the sequence of load transfer between elements occurs in the order: ceiling/floor loads to beams to columns to foundations to ground (*Fig. 2.1*).

At the outset, the designer must make an assessment of the future likely level of loading, including self-weight, to which the structure may be subject during its design life. Using computer methods or hand calculations the design loads acting on individual elements can then be evaluated. The design loads are used to calculate the bending moments, shear forces and deflections at critical points along the elements. Finally, suitable dimensions for the element can be determined. This aspect requires an understanding of the elementary theory of bending and the behaviour of elements subject to compressive loading. These steps are summarized in *Fig. 2.2* and the following sections describe the procedures associated with each step.

2.2 Design loads acting on structures

The loads acting on a structure are divided into three basic types: dead, imposed and wind. For each type of loading there will be characteristic and design values, as discussed in *Chapter 1*, which must be estimated. In addition, the designer will have to determine the particular combination of loading which is likely to produce the most adverse effect on the structure in terms of bending moments, shear forces and deflections.

2.2.1 DEAD LOADS, G_k, g_k
Dead loads are all the permanent loads acting on the structure including self-weight, finishes, fixtures and partitions. The characteristic dead loads can be estimated using the schedule of weights of building materials given in BS 648 (*Table 2.1*) or from

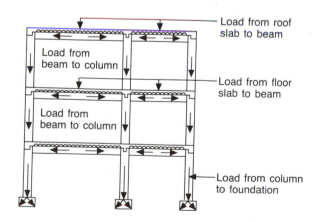

Fig. 2.1 *Sequence of load transfer between elements of a structure.*

Load from roof slab to beam

Load from beam to column

Load from floor slab to beam

Load from beam to column

Load from column to foundation

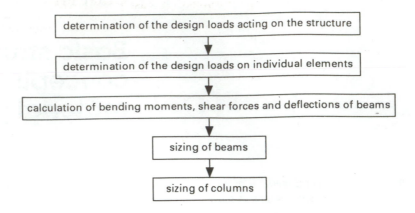

Fig. 2.2 *Design process.*

manufacturers' literature. The symbols G_k and g_k are normally used to denote the total and uniformly distributed characteristic dead loads respectively.

Estimation of the self-weight of an element tends to be a cyclic process since its value can only be assessed once the element has been designed which requires prior knowledge of the self-weight of the element. Generally, the self-weight of the element is likely to be small in comparison with other dead and live loads and any error in estimation will tend to have a minimal effect on the overall design.

residential dwellings, educational institutions, hospitals, and parts of the same structure, e.g. balconies, corridors and toilet rooms (*Table 2.2*).

2.2.3 WIND LOADS

Wind pressure can either add to the other gravitational forces acting on the structure or, equally well, exert suction or negative pressures on the structure. Under particular situations, the latter may well lead to critical conditions and must be considered in design. The characteristic wind loads acting on a

---

Example 2.1 Self–weight of a reinforced concrete beam

Calculate the self-weight of a reinforced concrete beam of breadth 300 mm, depth 600 mm and length 6000 mm.

From *Table 2.1*, unit mass of reinforced concrete is 2400 kg m^{-3}. Assuming that the gravitational constant is 10 m s^{-2} (strictly 9.807 m s^{-2}), the unit weight of reinforced concrete, ρ, is

$$\rho = 2400 \times 10 = 24\,000 \text{ N m}^{-3} = 24 \text{ kN m}^{-3}$$

Hence, the self-weight of beam, SW, is

$$SW = \text{volume} \times \text{unit weight}$$
$$= (0.3 \times 0.6 \times 6)24 = 25.92 \text{ kN}$$

2.2.2 IMPOSED LOADS Q_k, q_k

Imposed load, sometimes also referred to as live load, represents the load due to the proposed occupancy and includes the weights of the occupants, furniture and roof loads including snow. Since imposed loads tend to be much more variable than dead loads they are more difficult to predict. BS 6399: Part 1: 1984: *Code of Practice for Dead and Imposed Loads* gives typical characteristic imposed floor loads for different classes of structure, e.g.

structure can be assessed in accordance with the recommendations given in CP 3: Chapter V: Part 2: 1972 *Wind Load*. This code will eventually be superseded by Part 2 of BS 6399.

Wind loading is important in the design of masonry panel walls (*Chapter 5*). However beyond that, wind loading is not considered further since the emphasis in this book is on the design of elements rather than structures, which generally involves investigating the effects of dead and imposed loads only.

Table 2.1 Schedule of unit masses of building materials (based on BS 648)

Asphalt
Roofing 2 layers, 19 mm thick   42 kg m^{-2}
Damp-proofing, 19 mm thick   41 kg m^{-2}
Road and footpaths, 19 mm thick   44 kg m^{-2}

Bitumen roofing felts
Mineral surfaced bitumen   3.5 kg m^{-2}

Blockwork
Solid per 25 mm thick, stone   55 kg m^{-2}
  aggregate
Aerated per 25 mm thick   15 kg m^{-2}

Board
Blockboard per 25 mm thick   12.5 kg m^{-2}

Brickwork
Clay, solid per 25 mm thick   55 kg m^{-2}
  medium density
Concrete, solid per 25 mm thick   59 kg m^{-2}

Cast stone   2250 kg m^{-3}

Concrete
Natural aggregates   2400 kg m^{-3}
Lightweight aggregates (structural)   1760 + 240/- 160 kg m^{-3}

Flagstones
Concrete, 50 mm thick   120 kg m^{-2}

Glass fibre
Slab, per 25 mm thick   2.0–5.0 kg m^{-2}

Gypsum panels and partitions
Building panels 75 mm thick   44 kg m^{-2}

Lead
Sheet, 2.5 mm thick   30 kg m^{-2}

Linoleum
3 mm thick   6 kg m^{-2}

Plaster
Two coats gypsum, 13 mm thick   22 kg m^{-2}

Plastics sheeting (corrugated)   4.5 kg m^{-2}

Plywood
per mm thick   0.7 kg m^{-2}

Reinforced concrete   2400 kg m^{-3}

Rendering
Cement: sand (1:3), 13 mm thick   30 kg m^{-2}

Screeding
Cement: sand (1:3), 13 mm thick   30 kg m^{-2}

Slate tiles
(depending upon thickness and   24–78 kg m^{-3}
source)

Steel
Solid (mild)   7850 kg m^{-3}
Corrugated roofing sheets, per mm   10 kg m^{-2}
  thick

Tarmacadam
25 mm thick   60 kg m^{-2}

Terrazzo
25 mm thick   54 kg m^{-2}

Tiling, roof
Clay   70 kg m^{-2}

Timber
Softwood   590 kg m^{-3}
Hardwood   1250 kg m^{-3}

Water   1000 kg m^{-3}

Woodwool
Slabs, 25 mm thick   15 kg m^{-2}

Table 2.2 Imposed loads for residential occupancy class

Floor area usage	Intensity of distributed load $kN\,m^{-2}$	Concentrated load kN
Type 1. Self-contained dwelling units		
All	1.5	1.4
Type 2. Apartment houses, boarding houses, lodging houses, guest houses, hostels, residential clubs and communal areas in blocks of flats		
Boiler rooms, motor rooms, fan rooms and the like including the weight of machinery	7.5	4.5
Communal kitchens, laundries	3.0	4.5
Dining rooms, lounges, billiard rooms	2.0	2.7
Toilet rooms	2.0	–
Bedrooms, dormitories	1.5	1.8
Corridors, hallways, stairs, landings, footbridges, etc.	3.0	4.5
Balconies	Same as rooms to which they give access but with a minimum of 3.0	1.5 per metre run concentrated at the outer edge
Cat walks	–	1.0 at 1 m centres
Type 3. Hotels and motels		
Boiler rooms, motor rooms, fan rooms and the like, including the weight of machinery	7.5	4.5
Assembly areas without fixed seating, dance halls	5.0	3.6
Bars	5.0	–
Assembly areas with fixed seating [a]	4.0	–
Corridors, hallways, stairs, landings, footbridges, etc.	4.0	4.5
Kitchens, laundries	3.0	4.5
Dining rooms, lounges, billiard rooms	2.0	2.7
Bedrooms	2.0	1.8
Toilet rooms	2.0	–
Balconies	Same as rooms to which they give access but with a minimum of 4.0	1.5 per metre run concentrated at the outer edge
Cat walks	–	1.0 at 1 m centres

[a] Fixed seating is seating where its removal and the use of the space for other purposes are improbable.

2.2.4 LOAD COMBINATIONS AND DESIGN LOADS

The design loads are obtained by multiplying the characteristic loads by the partial safety factor for loads, γ_f (*Chapter 1*). The value for γ_f depends on several factors including the limit state under consideration, i.e. ultimate or serviceability, the accuracy of predicting the load and the particular combination of loading which will produce the worst possible effect on the structure in terms of bending moments, shear forces and deflections.

In most of the simple structures which will be considered in this book, the worst possible combination will arise due to the maximum dead and maximum imposed loads acting on the structure together. In such cases, the partial safety factors

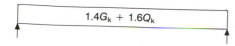

Fig. 2.3

for dead and imposed loads are 1.4 and 1.6 respectively (*Fig. 2.3*) and hence the design load is given by

$$\text{Design load} = 1.4G_k + 1.6Q_k$$

However, it should be appreciated that theoretically the design dead loads can vary between the characteristic and ultimate values, i.e. $1.0G_k$ and $1.4G_k$. Similarly, the design imposed loads can vary between zero and the ultimate value, i.e. $0.0Q_k$ and $1.6Q_k$. Thus for a simply supported beam with an overhang (*Fig. 2.4(a)*) the load cases shown in *Figs 2.4(b)–(d)* will need to be considered in order to determine the design bending moments and shear forces in the beam.

2.3 Design loads acting on elements

Once the design loads acting on the structure have been estimated it is then possible to calculate the design loads acting on individual elements. As was pointed out at the beginning of this chapter, this usually requires the designer to make assumptions regarding the support conditions and how the loads will eventually be transmitted down to the ground. *Figures 2.5 (a)* and *(b)* illustrate some of the more commonly assumed support conditions at the ends of beams and columns respectively.

In design it is common to assume that all the joints in the structure are pinned and that the sequence of load transfer occurs in the order: ceiling/floor loads to beams to columns to foundations to ground. These assumptions will considerably simplify calculations and lead to conservative estimates of the design loads acting on individual elements of the structure. The actual calculations to determine the forces acting on the elements are best illustrated by a number of worked examples as follows.

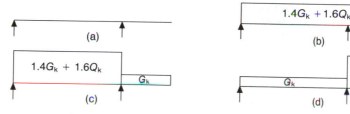

Fig. 2.4

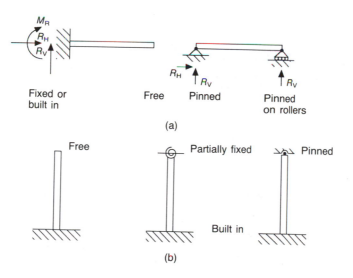

Fig. 2.5 *Typical beams and column support conditions.*

Example 2.2 Design loads on a floor beam

A composite floor consisting of a 150 mm thick reinforced concrete slab supported on steel beams spanning 5 m and spaced at 3 m centres is to be designed to carry an imposed load of 3.5 kN m^{-2}. Assuming that the unit mass of the steel beams is 50 kg m^{-1} run, calculate the design loads on a typical internal beam.

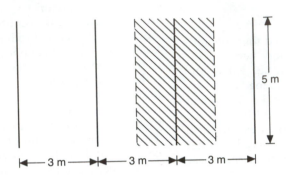

UNIT WEIGHTS OF MATERIALS

Reinforced concrete

From *Table 2.1*, unit mass of reinforced concrete is 2400 kg m^{-3}. Assuming the gravitational constant is 10 m s^{-2}, the unit weight of reinforced concrete is

$$2400 \times 10 = 24\,000 \text{ N m}^{-3} = 24 \text{ kN m}^{-3}$$

Steel beams

Unit mass of beam $= 50$ kg m^{-1} run
Unit weight of beam $= 50 \times 10 = 500$ N m^{-1} run $= 0.5$ kN m^{-1} run

LOADING

Slab

Slab dead load (g_k) $=$ self-weight $= 0.15 \times 24 = 3.6$ kN m^{-2}
Slab imposed load (q_k) $= 3.5$ kN m^{-2}
Slab ultimate load $= 1.4 g_k + 1.6 q_k = 1.4 \times 3.6 + 1.6 \times 3.5$
                        $= 10.64$ kN m^{-2}

Beam

Beam dead load (g_k) $=$ self-weight $= 0.5$ kN m^{-1} run
Beam ultimate load $= 1.4 g_k$ $= 1.4 \times 0.5 = 0.7$ kN m^{-1} run

DESIGN LOAD

Each internal beam supports a uniformly distributed load from a 3 m width of slab (hatched \\\\\\\\) plus self-weight. Hence

$$\text{Design load on beam} = \text{slab load} + \text{self-weight of beam}$$
$$= 10.64 \times 5 \times 3 + 0.7 \times 5$$
$$= 159.6 + 3.5 = 163.1 \text{ kN}$$

Example 2.3 Design loads on floor beams and columns

The floor shown below with an overall depth of 225 mm is to be designed to carry an imposed load of 3 kN m^{-2} plus floor finishes and ceiling loads of 1 kN m^{-2}. Calculate the design loads acting on beams B1–C1, B2–C2 and B1–B3 and columns B1 and C1. Assume that all the column heights are 3 m and that the beam and column weights are 70 and 60 kg m^{-1} run respectively.

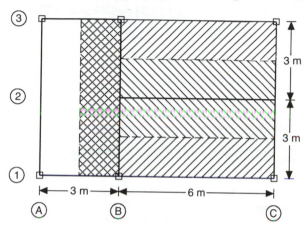

UNIT WEIGHTS OF MATERIALS

Reinforced concrete

From *Table 2.1*, unit mass of reinforced concrete is 2400 kg m^{-3}. Assuming the gravitational constant is 10 m s^{-2}, the unit weight of reinforced concrete is

$$2400 \times 10 = 24\,000 \text{ N m}^{-3} = 24 \text{ kN m}^{-3}$$

Steel beams

Unit mass of beam = 70 kg m^{-1} run
Unit weight of beam = 70×10 = 700 N m^{-1} run = 0.7 kN m^{-1} run

Steel columns

Unit mass of column = 60 kg m^{-1} run
Unit weight of column = 60×10 = 600 N m^{-1} run = 0.6 kN m^{-1} run

LOADING

Slab

Slab dead load (g_k) = self-weight + finishes
= 0.225× 24 + 1 = 6.4 kN m^{-2}
Slab imposed load (q_k) = 3 kN m^{-2}
Slab ultimate load = 1.4 g_k + 1.6 q_k
= 1.4 ×6.4 + 1.6×3 = 13.76 kN m^{-2}

Beam

Beam dead load (g_k) = self-weight = 0.7 kN m^{-1} run
Beam ultimate load = 1.4 g_k = 1.4×0.7 = 0.98 kN m^{-1} run

Column

Column dead load (g_k) = 0.6 kN m^{-1} run
Column ultimate load = 1.4 g_k = 1.4×0.6 = 0.84 kN m^{-1} run

DESIGN LOADS

Beam B1–C1

Assuming that the slab is simply supported, beam B1–C1 supports a uniformly distributed load from a 1.5 m width of slab (hatched //////) plus self-weight of beam. Hence

$$Design\ load\ on\ beam\ B1{-}C1 = slab\ load + self\text{-}weight\ of\ beam$$
$$= 13.76 \times 6 \times 1.5 + 0.98 \times 6$$
$$= 123.84 + 5.88 = 129.72\ kN$$

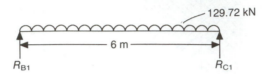

Since the beam is symmetrically loaded,

$$R_{B1} = R_{C1} = 129.72/2 = 64.86\ kN$$

Beam B2–C2

Assuming that the slab is simply supported, beam B2–C2 supports a uniformly distributed load from a 3 m width of slab (hatched \\\\\\) plus its self-weight. Hence

$$Design\ load\ on\ beam\ B2{-}C2 = slab\ load + self\text{-}weight\ of\ beam$$
$$= 13.76 \times 6 \times 3 + 0.98 \times 6$$
$$= 247.68 + 5.88 = 253.56\ kN$$

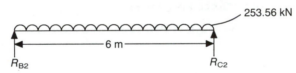

Since the beam is symmetrically loaded, R_{B2} and R_{C2} are the same and equal to 253.56/2 = 126.78 kN.

Beam B1–B3

Assuming that the slab is simply supported, beam B1–B3 supports a uniformly distributed load from a 1.5 m width of slab (shown cross-hatched) plus the self-weight of the beam and the reaction transmitted from beam B2–C2 which acts as a point load at mid-span. Hence

$$Design\ load\ on\ beam\ B1{-}B3 = uniformly\ distributed\ load\ from\ slab\ plus\ self\text{-}weight\ of\ beam$$
$$+ point\ load\ from\ reaction\ R_{B2}$$
$$= (13.76 \times 1.5 \times 6 + 0.98 \times 6) + 126.78$$
$$= 129.72 + 126.78 = 256.5\ kN$$

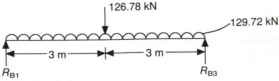

Since the beam is symmetrically loaded,

$$R_{B1} = R_{B3} = 256.5/2 = 128.25\ kN$$

Column B1

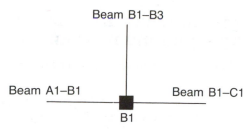

Column B1 supports the reactions from beams A1–B1, B1–C1 and B1–B3 and its self-weight. From the above, the reaction at B1 due to beam B1–C1 is 64.86 kN and from beam B1–B3 is 128.25 kN. Beam A1–B1 supports only its self-weight = 0.98×3 = 2.94 kN. Hence reaction at B1 due to A1–B1 is 2.94/2 = 1.47 kN. Since the column height is 3 m, self-weight of column = 0.84×3 = 2.52 kN. Hence

$$\text{Design load on column B1} = 64.86 + 128.25 + 1.47 + 2.52$$
$$= 197.1 \text{ kN}$$

Column C1

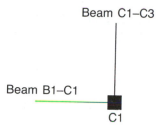

Column C1 supports the reactions from beams B1–C1 and C1–C3 and its self-weight. From the above, the reaction at C1 due to beam B1–C1 is 64.86 kN. Beam C1–C3 supports the reactions from B2–C2 (= 126.78 kN) and its self-weight (= 0.98×6) = 5.88 kN. Hence the reaction at C1 is (126.78 + 5.88)/2 = 66.33 kN. Since the column height is 3 m, self-weight = 0.84×3 = 2.52 kN. Hence

$$\text{Design load on column C1} = 64.86 + 66.33 + 2.52 = 133.71 \text{ kN}$$

2.4 Structural analysis

The design axial loads can be used directly to size columns. Column design will be discussed more fully in *section 2.5*. However, before flexural members such as beams can be sized, the design bending moments and shear forces must be evaluated. Such calculations can be performed by a variety of methods as noted below, depending upon the complexity of the loading and support conditions:

1. equilibrium equations
2. formulae
3. computer methods.

Hand calculations are suitable for analysing statically determinate structures such as simply supported beams and slabs (*section 2.4.1*). For various standard load cases, formulae for calculating the maximum bending moments, shear forces and deflections are available which can be used to rapidly analyse beams, as will be discussed in *section 2.4.2*. Alternatively, the designer may resort to using various commercially available computer packages, e.g. SAND. Their use is not considered in this book.

2.4.1 EQUILIBRIUM EQUATIONS

It can be demonstrated that if a body is in equilibrium under the action of a system of external forces, all parts of the body must also be in equilibrium. This principle can be used to determine the bending moments and shear forces along a beam. The actual procedure simply involves making fictitious 'cuts' at intervals along the beam and applying the equilibrium equations given below to the cut portions of the beam.

$$\Sigma \text{ moments } (M) = 0 \qquad (2.1)$$

$$\Sigma \text{ vertical forces } (V) = 0 \qquad (2.2)$$

Example 2.4 Design moments and shear forces in beams using equilibrium equations

Calculate the design bending moments and shear forces in beams B2–C2 and B1–B3 of *Example 2.3*.

BEAM B2–C2

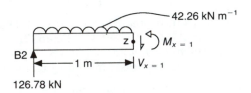

253.56 kN $\equiv$ 42.26 kN m^{-1}

B2 ⟩⟩⟩⟩⟩⟩⟩⟩⟩⟩ C2

6 m

$R_{B2} = 126.78$ kN   126.78 kN $= R_{C2}$

Let the longitudinal centroidal axis of the beam be the x axis and $x = 0$ at support B2.

x = 0

By inspection,

$$\text{Moment at } x = 0 \ (M_{x=0}) = 0$$
$$\text{Shear force at } x = 0 \ (V_{x=0}) = R_{B2} = 126.78 \text{ kN}$$

x = 1

Assuming that the beam is cut 1 m from support B2, i.e. $x = 1$ m, the moments and shear forces acting on the cut portion of the beam will be those shown in the free body diagram below:

42.26 kN m^{-1}

z) $M_{x=1}$

B2

1 m $V_{x=1}$

126.78 kN

From equation 2.1, taking moments about Z gives

$$126.78 \times 1 - 42.26 \times 1 \times 0.5 - M_{x=1} = 0$$

Hence

$$M_{x=1} = 105.65 \text{ kN m}$$

From equation 2.2, summing the vertical forces gives

$$126.78 - 42.26 \times 1 - V_{x=1} = 0$$

Hence

$$V_{x=1} = 84.52 \text{ kN}$$

x = 2

The free body diagram for the beam, assuming that it has been cut 2 m from support B2, is shown below:

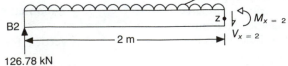

42.26 kN m^{-1}

z) $M_{x=2}$

B2

2 m $V_{x=2}$

126.78 kN

From equation 2.1, taking moments about Z gives

$$126.78 \times 2 - 42.26 \times 2 \times 1 - M_{x=2} = 0$$

Hence

$$M_{x=2} = 169.04 \text{ kN m}$$

From equation 2.2, summing the vertical forces gives

$$126.78 - 42.26 \times 2 - V_{x=2} = 0$$

Hence

$$V_{x=2} = 42.26 \text{ kN}$$

If this process is repeated for values of x equal to 3, 4, 5 and 6 m, the following values of the moments and shear forces in the beam will result:

x(m)	0	1	2	3	4	5	6
M(kN m)	0	105.65	169.04	190.17	169.04	105.65	0
V(kN)	126.78	84.52	42.26	0	- 42.26	- 84.52	- 126.78

This information is better presented diagrammatically as shown below:

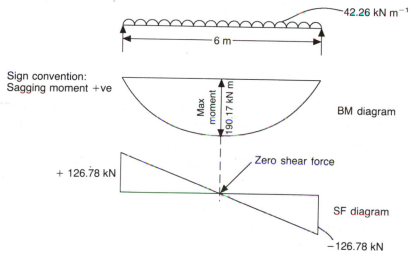

Hence, the design moment (M) is 190.17 kN m. Note that this occurs at mid-span and coincides with the point of zero shear force. The design shear force (V) is 126.78 kN and occurs at the supports.

BEAM B1–B3

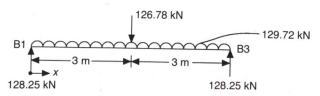

Again, let the longitudinal centroidal axis of the beam be the x axis of the beam and set $x = 0$ at support B1. The steps outlined earlier can be used to determine the bending moments and shear forces at $x = 0$, 1 and 2 m and since the beam is symmetrically loaded and supported, these values of bending moment and shear forces will apply at $x = 4$, 5 and 6 m respectively.

The bending moment at $x = 3$ m can be calculated by considering all the loading immediately to the left of the point load as shown below:

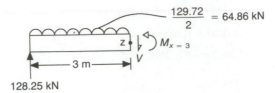

From equation 2.1, taking moments about Z gives

$$128.25 \times 3 - 64.86 \times 3/2 - M_{x = 3} = 0$$

Hence

$$M_{x = 3} = 287.46 \text{ kN m}$$

In order to determine the shear force at $x = 3$ the following two load cases need to be considered:

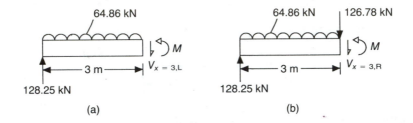

In (a) it is assumed that the beam is cut immediately to the left of the point load. In (b) the beam is cut immediately to the right of the point load. In (a), from equation 2.2, the shear force to the left of the cut, $V_{x = 3, L}$, is given by

$$128.25 - 64.86 - V_{x = 3, L} = 0$$

Hence

$$V_{x = 3, L} = 63.39 \text{ kN}$$

In (b), from equation 2.2, the shear force to the right of the cut, $V_{x = 3,R}$, is given by

$$128.25 - 64.86 - 126.78 - V_{x = 3,R} = 0$$

Hence

$$V_{x = 3,R} = -63.39 \text{ kN}$$

Summarizing the results in tabular and graphical form gives

x(m)	0	1	2	3	4	5	6
M(kN m)	0	117.44	213.26	287.46	213.26	117.44	0
V(kN)	128.25	106.63	85.01	63.39 | -63.39	-85.01	-106.63	-128.25

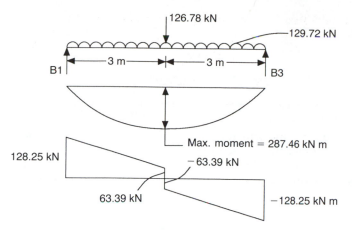

Max. moment = 287.46 kN m

Hence, the design moment for beam B1–B3 is 287.46 kN m and occurs at mid-span and the design shear force is 128.25 kN and occurs at the supports.

2.4.2 FORMULAE

An alternative method of determining the design bending moments and shear forces in beams involves using the formulae quoted in *Table 2.3*. The table also includes formulae for calculating the maximum deflections for each load case. The formulae can be derived using a variety of methods for analysing statically indeterminate structures, e.g. slope deflection and virtual work, and the interested reader is referred to any standard work on this subject for background information. There will be many instances in practical design where the use of standard formulae is not convenient and equilibrium methods are preferable.

Example 2.5 Design moments and shear forces in beams using formulae

Repeat *Example 2.4* using the formulae given in *Table 2.3*.

BEAM B2–C2

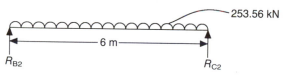

By inspection, maximum moment and maximum shear force occur at beam mid-span and supports respectively. From *Table 2.3*, design moment, *M*, is given by

$$M = \frac{Wl}{8} = \frac{253.56 \times 6}{8} = 190.17 \text{ kN m}$$

and design shear force, *V*, is given by

$$V = \frac{W}{2} = \frac{253.56}{2} = 126.78 \text{ kN}$$

BEAM B1–B3

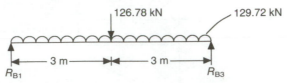

This load case can be solved using the principle of superposition which can be stated in general terms as follows: 'The effect of several actions taking place simultaneously can be reproduced exactly by adding the effects of each case separately.' Thus, the loading on beam B1–B3 can be considered to be the sum of a uniformly distributed load (W_{udl}) of 129.72 kN and a point load (W_{pl}) at mid-span of 126.78 kN.

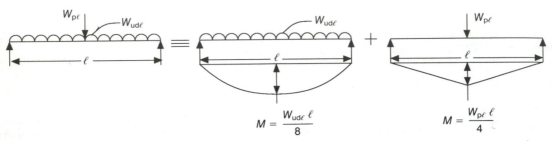

$$M = \frac{W_{ud\ell}\,\ell}{8} \qquad M = \frac{W_{p\ell}\,\ell}{4}$$

By inspection, the maximum bending moment and shear force for both load cases occur at beam mid-span and supports respectively. Thus, the design moment, M, is given by

$$M = \frac{W_{udl}l}{8} + \frac{W_{pl}l}{4} = \frac{129.72 \times 6}{8} + \frac{126.78 \times 6}{4} = 287.46 \text{ kN m}$$

and the design shear force, V, is given by

$$V = \frac{W_{udl}}{2} + \frac{W_{pl}}{2} = \frac{129.72}{2} + \frac{126.78}{2} = 128.25 \text{ kN}$$

2.5 Beam design

Having calculated the design bending moment and shear force, all that now remains to be done is to assess the size and strength of beam required. Generally, the ultimate limit state of bending will be critical for medium-span beams which are moderately loaded and shear for short-span beams which are heavily loaded. For long-span beams the serviceability limit state of deflection may well be critical. Irrespective of the actual critical limit state, once a preliminary assessment of the size and strength of beam needed has been made, it must be checked for the remaining limit states that may influence its long-term integrity.

The processes involved in such a selection will depend on whether the construction material behaves (i) elastically or (ii) plastically. If the material is elastic, it obeys Hooke's law, that is, the stress in the material due to the applied load is directly proportional to its strain (*Fig. 2.6*) where

$$\text{Stress } (\sigma) = \frac{\text{force}}{\text{area}}$$

and Strain $(\varepsilon) = \frac{\text{change in length}}{\text{original length}}$

The slope of the graph of stress vs. strain (*Fig. 2.6*) is therefore constant, and this gradient is normally referred to as the elastic or Young's modulus and is denoted by the letter E. It is given by

$$\text{Young's modulus } (E) = \frac{\text{stress}}{\text{strain}}$$

Note that strain is dimensionless but that both stress and Young's modulus are usually expressed in N mm^{-2}.

Table 2.3: Bending moments, shear forces and deflections for various standard load cases

Loading	Maximum bending moment	Maximum shearing force	Maximum deflection
W at midspan, $L/2 + L/2$	$\dfrac{WL}{4}$	$\dfrac{W}{2}$	$\dfrac{WL^3}{48EI}$
$W/2$, $W/2$ at third points, $L/3 + L/3 + L/3$	$\dfrac{WL}{6}$	$\dfrac{W}{2}$	$\dfrac{23WL^3}{1296EI}$
$W/2$, $W/2$, $L/4 + L/2 + L/4$	$\dfrac{WL}{8}$	$\dfrac{W}{2}$	$\dfrac{11WL^3}{768EI}$
$ud\ell = W$, L	$\dfrac{WL}{8}$	$\dfrac{W}{2}$	$\dfrac{5WL^3}{384EI}$
W triangular, L	$\dfrac{WL}{6}$	$\dfrac{W}{2}$	$\dfrac{WL^3}{60EI}$
W fixed ends, $L/2 + L/2$	$\dfrac{WL}{8}$ (at supports and at midspan)	$\dfrac{W}{2}$	$\dfrac{WL^3}{192EI}$
$ud\ell = W$ fixed ends, L	$\dfrac{WL}{12}$ at supports $\dfrac{WL}{24}$ at midspan	$\dfrac{W}{2}$	$\dfrac{WL^3}{384EI}$
W cantilever, L	WL	W	$\dfrac{WL^3}{3EI}$
$ud\ell = W$ cantilever, L	$\dfrac{WL}{2}$	W	$\dfrac{WL^3}{8EI}$

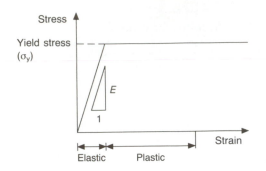

Fig. 2.6 *Stress–strain plot for steel.*

A material is said to be plastic if it strains without a change in stress. Plasticine and clay are plastic materials but so is steel beyond its yield point (*Fig. 2.6*). As will be seen in *Chapter 3*, reinforced concrete design also assumes that the material behaves plastically.

The structural implications of elastic and plastic behaviour are best illustrated by considering how bending is resisted by the simplest of beams – a rectangular section b wide and d deep.

2.5.1 ELASTIC CRITERIA

When any beam is subject to load it bends as shown in *Fig. 2.7(a)*. The top half of the beam is put into compression and the bottom half into tension. In the middle, there is neither tension nor compression. This axis is normally termed the neutral axis.

If the beam is elastic and stress and strain are directly proportional for the material, the variation in strain and stress from the top to middle to bottom is linear (*Figs 2.7 (b)* and *(c)*). The maximum stress in compression and tension is σ_y. The average stress in compression and tension is $\sigma_y/2$. Hence the compressive

force, F_c, and tensile force, F_t, acting on the section are equal and are given by

$$F_c = F_t = F = \text{stress} \times \text{area}$$
$$= \frac{\sigma_y}{2} \frac{bd}{2} = \frac{bd\sigma_y}{4}$$

The tensile and compression forces are separated by a distance s whose value is equal to $2d/3$ (*Fig. 2.7(a)*). Together they make up a couple, or moment, which acts in the opposite sense to the design moment. The value of this moment of resistance, M_r, is given by

$$M_r = Fs = \frac{bd\sigma_y}{4} \frac{2d}{3} = \frac{bd^2\sigma_y}{6} \qquad (2.3)$$

At equilibrium, the design moment in the beam will equal the moment of resistance, i.e.

$$M = M_r \qquad (2.4)$$

Provided that the yield strength of the material, i.e. σ_y is known, equations 2.3 and 2.4 can be used to calculate suitable dimensions for the beam needed to resist a particular design moment. Alternatively if b and d are known, the required material strength can be evaluated.

Equation 2.3 is more usually written as

$$M = \sigma Z \qquad (2.5)$$

or

$$M = \frac{\sigma I}{y} \qquad (2.6)$$

where Z is the elastic section modulus and is equal to $bd^2/6$ for a rectangular beam, I is the moment of inertia or, more correctly, the second moment of area of the section and y the distance from the

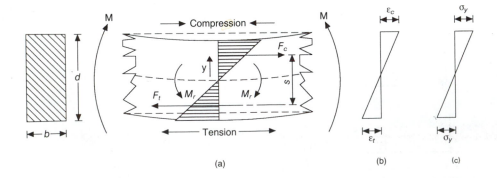

Fig. 2.7 *Strain and stress in an elastic beam.*

Table 2.4 Second moments of area

Shape	Second moment of area
Rectangle	$I_{xx} = bd^3/12 \quad I_{aa} = \dfrac{bd^3}{3}$ $I_{yy} = db^3/12$
Triangle	$I_{xx} = bh^3/36$ $I_{aa} = bh^3/12$
Disk	$I_{xx} = \pi R^4/4$ $I_{polar} = 2\,I_{xx}$
Annular disk	$I_{xx} = \pi(R_1{}^4 - R_2{}^4)/4$
Annular disk (approx.)	$I_{xx} = \pi R^3 t \text{ (for } t/R < 1/3)$ $R = \text{mean radius}$
Hollow rectangle	$I_{xx} = \dfrac{BD^3 - bd^3}{12}$ $I_{yy} = \dfrac{DB^3 - bd^3}{12}$
I-section	$I_{xx} = \dfrac{BD^3 - bd^3}{12}$ $I_{yy} = \dfrac{2TB^3}{12} + \dfrac{dt^3}{12}$

neutral axis (*Fig. 2.7(a)*). The elastic section modulus can be regarded as an index of the strength of the beam in bending. The second moment of area about an axis x–x in the plane of the section is defined by

$$I_{xx} = \int y^2 \, \mathrm{d}A \qquad (2.7)$$

Second moments of area of some common shapes are given in *Table 2.4*.

2.5.2 PLASTIC CRITERIA

While the above approach would be suitable for design involving the use of materials which have a linear elastic behaviour, materials such as reinforced concrete and steel have a substantial plastic performance. In practice this means that on reaching an elastic yield point the material continues to deform

but with little or no change in maximum stress.

Figure 2.8 shows what this means in terms of stresses in the beam. As the loading on the beam increases extreme fibre stresses reach the yield point, σ_y, and remain constant as the beam continues to

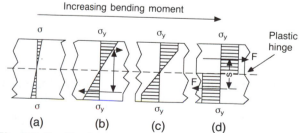

Fig. 2.8 *Bending failure of a beam (a) below yield (elastic); (b) at yield point (elastic); (c) beyond yield (partially plastic); (d) beyond yield (fully plastic).*

bend. A zone of plastic yielding begins to penetrate into the interior of the beam, until a point is reached immediately prior to complete failure, when practically all the cross-section has yielded plastically. The average stress at failure is σ_y, rather than $\sigma_y/2$ as was found to be the case when the material was assumed to have a linear-elastic behaviour. The moment of resistance assuming plastic behaviour is given by

$$M_r = Fs = \frac{bd\sigma_y}{2}\frac{d}{2} = \frac{bd^2\sigma_y}{4} = S\sigma_y \qquad (2.8)$$

where S is the plastic section modulus and is equal to $bd^2/4$ for rectangular beams. By setting the design moment equal to the moment of resistance of the beam its size and strength can be calculated according to plastic criteria.

Load capacity = design stress × area of column cross-section (2.10)

The design stress is related to the crushing strength of the material(s) concerned. However, not all columns fail in this mode. Some fail due to a combination of buckling followed by crushing of the material(s) (*Fig. 1.3*). These columns will tend to have lower load-carrying capacities, a fact which is taken into account by reducing the design stresses in the column. The design stresses are related to the slenderness ratio of the column, which is found to be a function of the following factors:

1. geometric properties of the column cross-section, e.g. lateral dimension of column, radius of gyration;
2. length of column;

Example 2.6 Elastic and plastic moments of resistance of a beam section

Calculate the moment of resistance of a beam 50 mm wide by 100 mm deep with $\sigma_y = 20 \text{ N mm}^{-2}$ according to (i) elastic criteria and (ii) plastic criteria.

ELASTIC CRITERIA
From equation 2.3,

$$M_{r,el} = \frac{bd^2}{6}\sigma_y = \frac{50 \times 100^2 \times 20}{6} = 1.67 \times 10^6 \text{ N mm}$$

PLASTIC CRITERIA
From equation 2.8,

$$M_{r,pl} = \frac{bd^2\sigma_y}{4} = \frac{50 \times 100^2 \times 20}{4} = 2.5 \times 10^6 \text{ N mm}$$

Hence it can be seen that the plastic moment of resistance of the section is greater than the maximum elastic moment of resistance. This will always be the case but the actual difference between the two moments will depend upon the shape of the section.

2.6   Column design

Generally, column design is relatively straightforward. The design process simply involves making sure that the design load does not exceed the load capacity of the column, i.e.

Load capacity ≥ design axial load (2.9)

Where the column is required to resist a predominantly axial load, its load capacity is given by

3. support conditions (*Fig. 2.5*).

The radius of gyration, r, is a sectional property which provides a measure of the column's ability to resist buckling. It is given by

$$r = (I/A)^{1/2} \qquad (2.11)$$

Generally, the higher the slenderness ratio, the greater the tendency for buckling and hence the lower the load capacity of the column. Most practical reinforced concrete columns are designed to fail by crushing and the design equations for this medium are based on equation 2.10 (*Chapter 3*). Steel columns, on the other hand, are designed to fail by a combination of buckling and crushing. Empirical relations have been derived to predict the design stress in steel columns in terms of the slenderness ratio (*Chapter 4*).

In most real situations, however, the loads will be applied eccentric to the axes of the column. Columns will therefore be required to resist axial loads and bending moments about one or both axes. In some cases, the bending moments may be small and can be accounted for in design simply by reducing the allowable design stresses in the material(s). Where the moments are large, for instance columns along the outside edges of buildings, the analysis procedures become more complex and the designer may have to resort to the use of design charts. However, as different codes of practice treat this in different ways, the details will be discussed separately in the chapters which follow.

2.7 Summary

This chapter has presented a simple but conservative method for assessing the design loads acting on individual members of building structures. This assumes that all the joints in the structure are pin-ended and that the sequence of load transfer occurs in the order: ceiling/floor loads to beams to columns to foundations to ground. The design loads are used to calculate the design bending moments, shear forces, axial loads and deflections in the member. Based on these design strengths, the sizes of the structural members are determined.

Example 2.7 Analysis of column section

Determine whether the reinforced concrete column cross-section shown below would be suitable to resist an axial load of 1500 kN. Assume that the design compressive strengths of the concrete and steel reinforcement are 14 and 345 N mm^{-2} respectively.

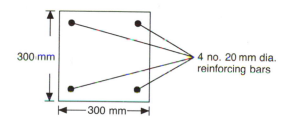

$$\text{Area of steel bars} = 4 \times (\pi 20^2/4) = 1256 \text{ mm}^2$$
$$\text{Net area of concrete} = 300 \times 300 - 1256 = 88\,744 \text{ mm}^2$$

Load capacity of column = force in concrete + force in steel
$$= 14 \times 88\,744 + 345 \times 1256$$
$$= 1675\,736 \text{ N} = 1675.7 \text{ kN}$$

Hence the column cross-section would be suitable to resist the design load of 1500 kN.

Questions

1. For the beams shown, calculate and sketch the bending moment and shear force diagrams.

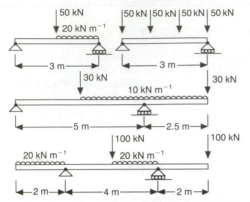

2. For the three load cases shown in *Fig. 2.4*, sketch the bending moment and shear force diagrams and hence determine the design bending moments and shear forces for the beam.

3. (a) Calculate the area, moment of inertia, plastic and elastic moduli and radius of gyration about the x–x axis of the rectangular beam section shown below.

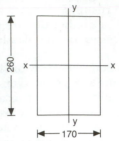

(b) Assuming $\sigma_y = 100$ N mm^{-2}, calculate the moment of resistance of the beam according to (i) elastic and (ii) plastic criteria. Determine the maximum uniformly distributed load the beam can carry in both cases assuming that it is 6 m long and is simply supported at its ends.

4. What are the most common ways in which columns can fail? Discuss the factors which influence their behaviour under axial load.

PART TWO

STRUCTURAL DESIGN TO BRITISH STANDARDS

Part One has discussed the principles of limit state design and outlined general approaches towards assessing the sizes of beams and columns in building structures. Since limit state design forms the basis of the design methods in most modern codes of practice on structural design, there is considerable overlap in the design procedures presented in these codes.

The aim of this part of the book is to give detailed guidance on the design of a number of structural elements in the four media: concrete, steel, masonry and timber to the current British Standards. The work has been divided into four chapters as follows:

1. *Chapter 3* discusses the design procedures in BS 8110: *Code of Practice for the Structural Use of Concrete* relating to beams, slabs, pad foundations, retaining walls and columns.
2. *Chapter 4* discusses the design procedures in BS 5950: *Code of Practice for the Structural Use of Steelwork in Buildings* relating to beams, columns and connections.
3. *Chapter 5* discusses the design procedures in BS 5628: *Code of Practice for Use of Masonry* relating to unreinforced loadbearing and panel walls.
4. *Chapter 6* discusses the design procedures in BS 5268: *Code of Practice for Structural Use of Timber* relating to beams, columns and stud walling.

Chapter 3

Design in reinforced concrete to BS 8110

This chapter is concerned with the detailed design of reinforced concrete elements to BS 8110. A general discussion of the different types of commonly occurring beams, slabs, walls, foundations and columns is given together with a number of fully worked examples covering the design of the following elements: singly and doubly reinforced beams, one-way spanning solid slab, pad foundation, cantilever retaining wall and short braced columns supporting axial loads and uniaxial or biaxial bending. The section which deals with singly reinforced beams is, perhaps, the most important since it introduces the design procedures and equations which are common to the design of the other elements mentioned above, with the exception of columns.

3.1 Introduction

Reinforced concrete is one of the principal materials used in structural design. It is a composite material, consisting of steel reinforcing bars embedded in concrete. These two materials have complementary properties. Concrete, on the one hand, has high compressive strength but low tensile strength. Steel bars, on the other, can resist high tensile stresses but will buckle when subjected to comparatively low compressive stresses. Steel is much more expensive than concrete. By providing steel bars predominantly in those zones within a concrete member which will be subjected to tensile stresses, an economical structural material can be produced which is both strong in compression and strong in tension. In addition, the concrete provides corrosion protection and fire resistance to the more vulnerable embedded steel reinforcing bars.

Reinforced concrete is used in many civil engineering applications such as the construction of structural frames, foundations, retaining walls, water-retaining structures, highways and bridges. They are normally designed in accordance with

the recommendations given in various documents including BS 5400: *Code of Practice for the Design of Steel, Concrete and Composite Bridges*, BS 8007: *Code of Practice for the Design of Concrete Structures for Retaining Aqueous Liquids* and BS 8110: *Code of Practice for the Structural Use of Concrete*. Since the primary aim of this book is to give guidance on the design of structural elements, this is best illustrated by considering the contents of BS 8110.

BS 8110 is divided into the following three parts:

Part 1: *Code of Practice for Design and Construction.*
Part 2: *Code of Practice for Special Circumstances.*
Part 3: *Design Charts for Singly Reinforced Beams, Doubly Reinforced Beams and Rectangular Columns.*

Part 1 covers most of the material required for everyday design. Since most of this chapter is concerned with the contents of Part 1, it should be assumed that all references to BS 8110 refer to Part 1 exclusively. Part 2 covers subjects such as torsional resistance, calculation of deflections and estimation of elastic deformations. These aspects of design are beyond the scope of this book and Part 2, therefore, is not discussed here. Part 3 of BS 8110 contains charts for use in the design of singly reinforced beams, doubly reinforced beams and rectangular columns. A number of design examples illustrating the use of these charts are included in the relevant sections of this chapter.

3.2 Objectives and scope

All reinforced concrete building structures are composed of various categories of elements including slabs, beams, columns, walls and foundations (*Fig. 3.1*). Within each category is a range of element types. The aim of this chapter is to describe the element types and, for selected elements, to give guidance on their design.

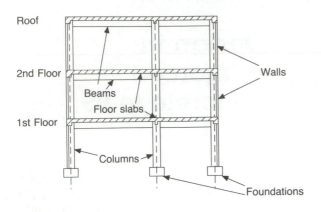

Fig. 3.1 *Some elements of a structure.*

A great deal of emphasis has been placed in the text to highlight the similarities in structural behaviour and, hence, design of the various categories of elements. Thus, certain slabs can be regarded for design purposes as a series of transversely connected beams. Columns may support slabs and beams but columns may also be supported by slabs and beams, in which case the latter are more commonly referred to as foundations. Cantilever retaining walls are usually designed as if they consist of three cantilever beams as shown in *Fig. 3.2*. Columns are different in that they are primarily compression members rather than beams and slabs which predominantly resist bending. Therefore columns are dealt with separately at the end of the chapter.

Irrespective of the element being designed, the designer will need an understanding of the following aspects which are discussed next:

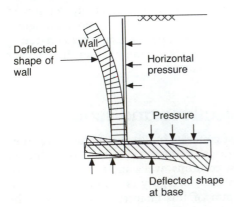

Fig. 3.2 *Cantilever retaining wall.*

1. symbols
2. basis of design
3. material properties
4. loading
5. stress–strain relationships
6. durability and fire resistance.

The detailed design of beams, slabs, foundations, retaining walls and columns will be discussed in *sections 3.9, 3.10, 3.11, 3.12* and *3.13* respectively.

3.3 Symbols

For the purpose of this book, the following symbols have been used. These have largely been taken from BS 8110. Note that in one or two cases the same symbol is differently defined. Where this occurs the reader should use the definition most appropriate to the element being designed.

GEOMETRIC PROPERTIES

b	width of section
d	effective depth of the tension reinforcement
h	overall depth of section
x	depth to neutral axis
z	lever arm
d	depth to the compression reinforcement
l	effective span
c	nominal cover to reinforcement

BENDING

F_k	characteristic load
g_k, G_k	characteristic dead load
q_k, Q_k	characteristic imposed load
w_k, W_k	characteristic wind load
f_k	characteristic strength
f_{cu}	characteristic compressive cube strength of concrete
f_y	characteristic tensile strength of reinforcement
γ_f	partial safety factor for load
γ_m	partial safety factor for material strengths
K	coefficient given by $M/f_{cu}bd^2$
K'	coefficient given by $M_u/f_{cu}bd^2 = 0.156$ when redistribution does not exceed 10%
M	design ultimate moment
M_u	design ultimate moment of resistance
A_s	area of tension reinforcement
A_s'	area of compression reinforcement
Φ	diameter of main steel
Φ'	diameter of links

SHEAR

f_{yv}	characteristic strength of links
s_v	spacing of links along the member
V	design shear force due to ultimate loads
v	design shear stress
v_c	design concrete shear stress
A_{sv}	total cross-sectional area of shear reinforcement

COMPRESSION

b	width of column
h	depth of column
l_0	clear height between end restraints
l_e	effective height
l_{ex}	effective height in respect of x–x axis
l_{ey}	effective height in respect of y–y axis
N	design ultimate axial load
A_c	net cross-sectional area of concrete in a column
A_{sc}	area of longitudinal reinforcement

3.4 Basis of design

The design of reinforced concrete elements to BS 8110 is based on the limit state method. As discussed in *Chapter 1*, the two principal categories of limit states normally considered in design are (i) ultimate limit state and (ii) serviceability limit state. The ultimate limit state models the behaviour of the element at failure due to a variety of mechanisms including excessive bending, shear and compression or tension. The serviceability limit state models the behaviour of the member at working loads and in the context of reinforced concrete design is principally concerned with the limit states of deflection and cracking.

Having identified the relevant limit states, the design process simply involves basing the design on the most critical one and then checking for the remaining limit states. This requires an understanding of (1) material properties and (2) loadings.

3.5 Material properties

The two materials whose properties must be known are concrete and steel reinforcement. In the case of concrete, the property with which the designer is primarily concerned is its compressive strength. For steel, however, it is its tensile strength capacity which is important.

3.5.1 CHARACTERISTIC COMPRESSIVE STRENGTH OF CONCRETE, f_{cu}

Concrete is a mixture of water, coarse and fine aggregate and a cementitious binder (normally Portland cement) which hardens to a stone-like mass. As can be appreciated, it is difficult to produce a homogeneous material from these components. Furthermore, its strength and other properties may vary considerably due to operations such as transportation, compaction and curing.

The compressive strength of concrete is usually determined by carrying out compression tests on 28-day-old cubes which have been prepared using a standard procedure laid down in BS 1881. If a large number of compression tests were carried out on concrete cubes made from the same mix, it would be found that a plot of crushing strength against frequency of occurrence would approximate to a normal distribution (*Fig. 3.3*).

For design purposes it is necessary to assume a unique value for the strength of the mix. However, choosing too high a value will result in a high probability that most of the structure will be constructed with concrete having a strength below this value. Conversely, too low a value will result in inefficient use of the material. As a compromise between economy and safety, BS 8110 refers to the characteristic strength (f_{cu}) which is defined as the value below which not more than 5% of the test results fall.

The characteristic and mean strength (f_m) of a sample are related by the expression

$$f_{cu} = f_m - 1.64 \text{ s.d.}$$

where s.d. is the standard deviation. Thus assuming that the mean strength is 35 N mm^{-2} and standard deviation is 3 N mm^{-2}, the characteristic strength of the mix is $35 - 1.64 \times 3 = 30$ N mm^{-2}.

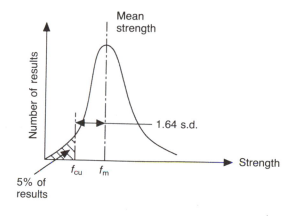

Fig. 3.3 *Normal frequency distribution of strengths.*

Table 3.1 Concrete compressive strengths

Concrete grade	Characteristic strength, f_{cu} ($N\ mm^{-2}$)
C25	25
C30	30
C35	35
C40	40
C45	45
C50	50

Concrete of a given strength can be identified by its 'grade'. For example a grade 30 concrete (C30) has a characteristic strength of 30 N mm^{-2}. For reinforced concrete made with normal aggregates, BS 8110 recommends that the lowest grade of concrete should be C25 although, in practice, a C30 mix is invariably necessary because of durability considerations (*section 3.8*). *Table 3.1* shows the characteristic strengths of various grades of concrete normally specified in reinforced concrete design.

3.5.2 CHARACTERISTIC STRENGTH OF REINFORCEMENT, f_y

Concrete is strong in compression but weak in tension. Because of this it is normal practice to provide steel reinforcement in those areas where tensile stresses in the concrete are most likely to develop. Consequently, it is the tensile strength of the reinforcement which most concerns the designer.

The tensile strength of reinforcement can be determined using the procedure laid down in BS EN 10 002: Part 1. The tensile strength will also vary 'normally' with specimens of the same composition. Using the same reasoning as above, BS 8110 recommends that design should be based on the characteristic strength of the reinforcement (f_y) and gives typical values for mild steel and high yield steel, the two reinforcement types used in UK practice, of 250 and 460 N mm^{-2} respectively (*Table 3.2*).

Table 3.2 Strength of reinforcement (Table 3.1, BS 8110)

Reinforcement type	Characteristic strength, f_y ($N\ mm^{-2}$)
Hot rolled mild steel	250
High yield steel (hot rolled or cold worked)	460

Table 3.3 Values of γ_m for the ultimate limit state (Table 2.2, BS 8110)

Material/stress type	Partial safety factor, γ_m
Reinforcement	1.15
Concrete in flexure or axial load	1.50
Concrete shear strength without shear reinforcement	1.25
Concrete bond strength	1.40
Concrete, others (e.g. bearing stress)	≥ 1.50

3.5.3 DESIGN STRENGTH

Tests to determine the characteristic strengths of concrete and steel reinforcement are carried out on near-perfect specimens which have been prepared under laboratory conditions. Such conditions will seldom exist in practice. Therefore it is undesirable to use characteristic strengths to size members.

To take account of differences between actual and laboratory values, local weaknesses and inaccuracies in assessment of the resistance of sections, the characteristic strengths (f_k) are divided by appropriate partial safety factor for strengths (γ_m), obtained from *Table 3.3*. The resulting values are termed design strengths and it is the design strengths which are used to size members.

$$\text{Design strength} = \frac{f_k}{\gamma_m} \qquad (3.1)$$

It should be noted that for the ultimate limit state the partial safety factor for reinforcement (γ_{ms}) is always 1.15, but for concrete (γ_{mc}) assumes different values depending upon the stress type under consideration. Furthermore, the partial safety factors for concrete are all greater than that for reinforcement since concrete quality is less controllable.

3.6 Loading

In addition to the material properties, the designer needs to know the type and magnitude of the loading to which the structure may be subject during its design life.

The loads acting on a structure are divided into three basic types: dead, imposed and wind (*section 2.2*). Associated with each type of loading there are characteristic and design values which must be assessed before the individual elements of the structure can be designed. These aspects are discussed next.

3.6.1 CHARACTERISTIC LOAD

As noted in *Chapter 2*, it is not possible to apply statistical principles to determine characteristic dead

Table 3.4 Values of γ_f for various load combinations (based on Table 2.1, BS 8110)

Load combination	Load type				
	Dead, G_k		Imposed, Q_k		Wind, W_k
	Adverse	Beneficial	Adverse	Beneficial	
1. Dead and imposed	1.4	1.0	1.6	0	—
2. Dead and wind	1.4	1.0	—	—	1.4
3. Dead and wind and imposed	1.2	1.2	1.2	1.2	1.2

(G_k), imposed (Q_k) and wind (W_k) loads simply because there are insufficient data. Therefore, the characteristic loads are taken to be those given in the following documents:

1. BS 648: *Schedule of Weights for Building Materials.*
2. BS 6399: *Design Loadings for Buildings*, Part 1: *Code of Practice for Dead and Imposed Loads.*
3. CP 3: Chapter V: *Wind Loads.* This code will eventually be superseded by Part 2 of BS 6399.

3.6.2 DESIGN LOAD
Variations in the characteristic loads may arise due to a number of reasons such as errors in the analysis and design of the structure, constructional inaccuracies and possible unusual load increases. In order to take account of these effects, the characteristic loads (F_k) are multiplied by the appropriate partial safety factor for loads (γ_f), taken from *Table 3.4*, to give the design loads acting on the structure:

$$\text{Design load} = \gamma_f F_k \qquad (3.2)$$

Generally, the 'adverse' factors will be used to derive the design loads acting on the structure. For example, for single-span beams subject to only dead and imposed loads the appropriate values of γ_f are generally 1.4 and 1.6 respectively (*Fig. 3.4(a)*). However, for continuous beams, load cases must be analysed which should include the maximum design load on all spans and maximum and minimum design loads on alternate spans (*Fig. 3.4(b)*).

The design loads are used to calculate the distribution of bending moments and shear forces in the structure usually using elastic analysis methods as discussed in *Chapter 2*. At no point should they exceed the corresponding design strengths of the member, otherwise failure of the structure may arise.

The design strength is a function of the distribution of stresses in the member. Thus, for the simple case of a steel bar in direct tension the design strength is equal to the cross-sectional area of the bar multiplied by the average stress at failure (*Fig. 3.5*). The distribution of stresses in reinforced concrete

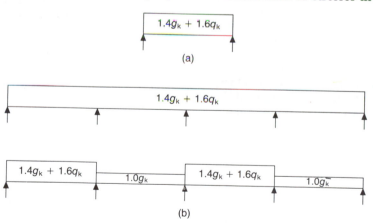

(a)

(b)

Fig. 3.4 Ultimate design loads: (a) single span beam; (b) continuous beam.

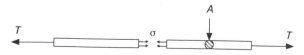

Fig. 3.5 Design strength of the bar. Design strength $T = \sigma A$, where σ is the average stress at failure and A the cross-sectional area of the bar.

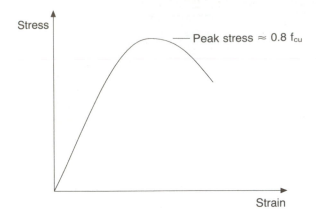

Fig. 3.6 *Actual stress–strain curve for concrete in compression.*

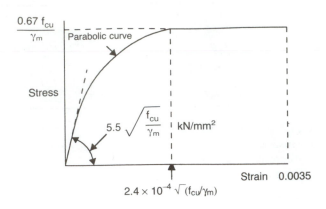

Fig. 3.7 *Design stress–strain curve for concrete in compression (Fig. 2.1, BS 8110).*

members is usually more complicated, but can be estimated once the stress–strain behaviour of the concrete and steel reinforcement is known. This aspect is discussed next.

3.7 Stress–strain curves

3.7.1 STRESS–STRAIN CURVE FOR CONCRETE

Figure 3.6 shows a typical stress–strain curve for a concrete cylinder under uniaxial compression. Note that the stress–strain behaviour is never truly linear and that the maximum compressive stress at failure is approximately $0.8 \times$ characteristic cube strength (i.e. $0.8f_{cu}$).

However, the actual behaviour is rather complicated to model mathematically and, therefore, BS 8110 uses the modified stress–strain curve shown in *Fig. 3.7* for design. This assumes that the peak stress is only 0.67 (rather than 0.8) times the characteristic strength and hence the design stress for concrete is given by

$$\text{Design compressive stress for concrete} = \frac{0.67f_{cu}}{\gamma_{mc}}$$

In other words, the failure stress assumed in design is approximately $0.45/0.8 = 56\%$ of the actual stress at failure when near-perfect specimens are tested.

3.7.2 STRESS–STRAIN CURVE FOR STEEL REINFORCEMENT

A typical tensile stress–strain curve for steel reinforcement is shown in *Fig. 3.8*. It can be divided into two regions: (i) an elastic region where strain is

proportional to stress, and (ii) a plastic region where small increases in stress produce large increases in strain. The change from elastic to plastic behaviour occurs at the yield stress and is significant since it defines the characteristic strength of reinforcement (f_y).

Once again, the actual material behaviour is rather complicated to model mathematically and therefore BS 8110 modifies it to the form shown in *Fig. 3.9* which also includes the idealized stress–strain relationship for reinforcement in compression. The maximum design stress for reinforcement in tension and compression is given by

$$\text{Design stress for reinforcement} = \frac{f_y}{\gamma_{ms}} \quad (3.4)$$

From the foregoing it is possible to determine the distribution of stresses at a section and hence calculate the design strength of the member. The latter is normally carried out using the equations given in BS

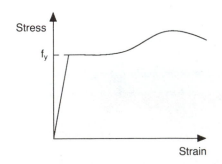

Fig. 3.8 *Actual stress–strain curve for reinforcement.*

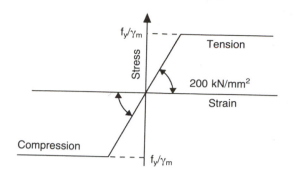

Fig. 3.9 Design stress–strain curve for reinforcement (Fig. 2.2, BS 8110).

8110. However, before considering these in detail, it is useful to pause for a moment in order to introduce BS 8110's requirements in respect of durability and fire resistance since these requirements are common to several of the elements which will be subsequently discussed.

3.8 Durability and fire resistance

Apart from the need to ensure that the design is structurally sound, the designer must also verify the proper performance of the structure in service. Principally this will involve the two limit states of (i) durability and (ii) fire resistance.

3.8.1 DURABILITY

Many concrete structures are showing signs of severe deterioration after only a few years of service. Repair of these structures is both difficult and very costly. Therefore, over recent years, much effort has been directed towards improving the durability requirements, particularly with respect to the protection of embedded reinforcing steel from corrosion. This has been largely achieved by specifying limits for:

1. nominal depth of cover to the reinforcement;
2. minimum cement content;
3. maximum water/cement ratios;
4. maximum allowable crack widths.

Table 3.5 gives the nominal (minimum plus 5mm) depths of concrete cover to all reinforcement for specified exposure conditions, minimum cement contents and maximum water/cement ratios. The limits on water/cement ratio and cement content will automatically be assured by specifying the minimum grades of concrete indicated in the table.

BS 8110 further recommends that the maximum crack width should not exceed 0.3 mm in order to avoid corrosion of the reinforcing bars. This requirement will generally be satisfied by observing the detailing rules given in BS 8110 with regard to (1) minimum reinforcement areas and (2) maximum clear spacings between reinforcing bars. These requirements will be discussed individually for beams, slabs and columns in *sections 3.9.1.6, 3.10.2.4* and *3.13.6* respectively.

Table 3.5 Exposure conditions and nominal covers to all reinforcement (including shear reinforcement) to meet durability requirements (based on Tables 3.2 and 3.4, BS 8110)

Conditions of exposure	Nominal cover (mm)				
Mild: for example, protected against weather or aggressive condition	25	20	20	20	20
Moderate: for example, sheltered from severe rain and freezing while wet; subject to condensation or continuously under water; in contact with non-aggressive soil	–	35	30	25	20
Severe: for example, exposed to severe rain; alternate wetting and drying; occasional freezing or severe condensation	–	–	40	30	25
Very severe: for example, exposed to sea water spray, de-icing salts, corrosive fumes or severe wet freezing	–	–	50	40	30
Extreme: for example, exposed to abrasive action (sea water and solids, flowing acid water, machinery or vehicles)	–	–	–	60	50
Maximum free water/cement ratio	0.65	0.60	0.55	0.50	0.45
Minimum cement content (kg m^{-3})	275	300	325	350	400
Lowest concrete grade	C30	C35	C40	C45	C50

Table 3.6 Nominal cover to all reinforcement to meet specified periods of fire resistance (based on Tables 3.5, BS 8110)

Fire resistance (hours)	Nominal cover (mm)				
	Beams		Floors		Columns
	Simply supported	Continuous	Simply supported	Continuous	
0.5	20	20	20	20	20
1.0	20	20	20	20	20
1.5	20	20	25	20	20
2.0	40	30	35	25	25
3.0	60	40	45	35	25
4.0	70	50	55	45	25

3.8.2 FIRE PROTECTION

Fire protection of reinforced concrete members is largely achieved by specifying limits for (1) nominal thickness of cover to the reinforcement and (2) minimum dimensions of members.

Table 3.6 gives the actual values of the nominal depths of concrete covers to all reinforcement for specified periods of fire resistance and member types. The covers in the table may need to be increased because of durability considerations. The minimum dimensions of members for fire resistance are shown in *Fig. 3.10*.

Having discussed these more general aspects relating to structural design, the detailed design of beams is considered in the following section.

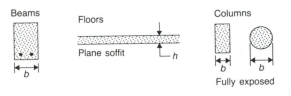

Fire resistance hours	Minimum dimension		
	Beam width (b)	Floor thickness (h)	Fully exposed column width (b)
0.5	200	75	150
1.0	200	95	200
1.5	200	110	250
2.0	200	125	300
3.0	240	150	400
4.0	280	170	450

Fig. 3.10 *Minimum dimensions of reinforced concrete members for fire resistance (based on Fig. 3.2, BS 8110).*

3.9 Beams

Beams in reinforced concrete structures can be defined according to:

1. cross-section
2. position of reinforcement
3. support conditions.

Some common beam sections are shown in *Fig. 3.11*. Beams reinforced with tension steel only are referred to as singly reinforced. Beams reinforced with tension and compression steel are termed doubly reinforced. Inclusion of compression steel will increase the moment capacity of the beam and hence allow more slender sections to be used. Thus, doubly reinforced beams are used in preference to singly reinforced beams when there is some restriction on the construction depth of the section.

Under certain conditions, T- and L-beams are more economical than rectangular beams since some of the concrete below the dotted line (neutral axis),

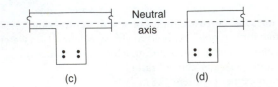

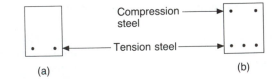

Fig. 3.11 *Beam sections: (a) singly reinforced; (b) doubly reinforced; (c) T-section; (d) L-section.*

Fig. 3.12 *Support conditions: (a) simply supported; (b) continuous.*

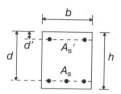

Fig. 3.13 *Notation.*

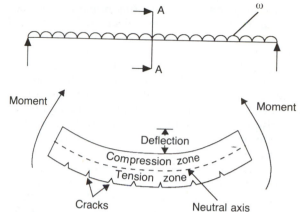

Fig. 3.14

which serves only to contain the tension steel, is removed resulting in a reduced unit weight of beam. Furthermore, beams may be simply supported at their ends or continuous, as illustrated in *Fig. 3.12*.

Figure 3.13 illustrates some of the notation used in beam design. Here b is the width of the beam, h the overall depth of section, d the effective depth of tension reinforcement, d' the depth of compression reinforcement, A_s the area of tension reinforcement and A_s' the area of compression reinforcement.

The following subsections consider the design of:

1. singly reinforced beams
2. doubly reinforced beams
3. continuous, L- and T-beams.

3.9.1 SINGLY REINFORCED BEAM DESIGN

All beams may fail due to excessive bending or shear. In addition, excessive deflection of beams must be avoided otherwise the efficiency or appearance of the structure may become impaired. As discussed in

section 3.4, bending and shear are ultimate states while deflection is a serviceability state. Generally, structural design of concrete beams primarily involves consideration of the following aspects which are discussed next:

1. bending
2. shear
3. deflection.

3.9.1.1 Bending (clause 3.4.4.4, BS 8110)

Consider the case of a simply supported, singly reinforced rectangular beam subject to a uniformly distributed load ω as shown in *Figs 3.14 and 3.15*. The load causes the beam to deflect downwards, putting the top portion of the beam into compression and the bottom portion into tension. At some distance x below the compression face, the section is neither in compression nor tension and therefore the strain at this level is zero. This axis is normally referred to as the neutral axis.

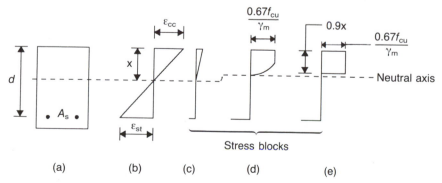

Fig. 3.15 *Stress and strain distributions at section A–A; (a) section; (b) strains; (c) triangular (low strain); (d) rectangular parabolic (large strain); (e) equivalent rectangular.*

Assuming that plane sections remain plane, the strain distribution will be triangular (*Fig. 3.15(b)*). The stress distribution in the concrete above the neutral axis is initially triangular (*Fig. 3.15(c)*), for low values of strain, because stress and strain are directly proportional (*Fig. 3.7*). However, the stress in the concrete below the neutral axis is zero since it is assumed that the concrete is cracked, being unable to resist any tensile stress. All the tensile stresses in the member are assumed to be resisted by the steel reinforcement and this is reflected in a peak in the tensile stress at the level of the reinforcement.

As the intensity of loading on the beam increases, the mid-span moment increases and the distribution of stresses changes from that shown in *Fig. 3.15(c)* to *3.15(d)*. The stress in the reinforcement increases linearly with strain up to the yield point. Thereafter it remains at a constant value (*Fig. 3.9*). However, as the strain in the concrete increases, the stress distribution is assumed to follow the parabolic form of the stress–strain relationship for concrete under compression (*Fig. 3.7*).

The actual stress distribution at a given section and the mode of failure of the beam will depend upon whether the section is (1) under-reinforced or (2) over-reinforced. If the section is over-reinforced the steel does not yield and the failure mechanism will be crushing of the concrete due to its compressive capacity being exceeded. Steel is expensive and, therefore, over-reinforcing will lead to uneconomical design. Furthermore, with this type of failure there may be no external warning signs; just sudden, catastrophic collapse.

If the section is under-reinforced, the steel yields and failure will again occur due to crushing of the concrete. However, the beam will show considerable deflection which will be accompanied by severe cracking and spalling from the tension face, thus providing ample warning signs of failure. Moreover, this form of design is more economical since a greater proportion of the steel strength is utilized. Therefore, it is normal practice to design sections which are under-reinforced rather than over-reinforced.

In an under-reinforced section, since the reinforcement will have yielded, the tensile force in the steel (F_{st}) at the ultimate limit state can be readily calculated using the following:

$$F_{st} = \text{design stress} \times \text{area}$$
$$= \frac{f_y A_s}{\gamma_{ms}} \quad \text{(using equation 3.4)} \quad (3.5)$$

where f_y is the yield stress, A_s the area of reinforcement and γ_{ms} the factor of safety for reinforcement (= 1.15).

However, it is not an easy matter to calculate the compressive force in the concrete because of the complicated pattern of stresses in the concrete. To simplify the situation, BS 8110 replaces the rectangular–parabolic stress distribution with an equivalent rectangular stress distribution (*Fig. 3.15(e)*). And it is the rectangular stress distribution which is used in order to develop the design formulae for rectangular beams given in clause 3.4.4.4 of BS 8110. Specifically, the code gives formulae for the following design parameters which are derived below:

1. ultimate moment of resistance
2. area of tension reinforcement
3. lever arm.

Ultimate moment of resistance, M_u. Consider the singly reinforced beam shown in *Fig. 3.16*. The loading on the beam gives rise to an ultimate design moment (M) at mid-span. The resulting curvature of the beam produces a compression force in the concrete (F_{cc}) and a tensile force in the reinforcement (F_{st}). Since there is no resultant axial force on the beam, the force in the concrete must equal the force in the reinforcement:

$$F_{cc} = F_{st} \quad (3.6)$$

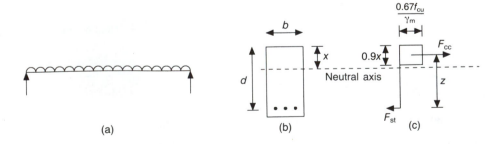

Fig. 3.16 *Ultimate moment of resistance for singly reinforced section.*

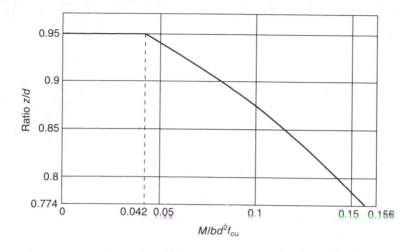

Fig. 3.17 Lever-arm curve.

These two forces are separated by a distance z, the moment of which forms a couple (M_u) which opposes the design moment. For structural stability $M_u \geqslant M$ where

$$M_u = F_{cc}z = F_{st}z \qquad (3.7)$$

From the stress block shown in *Fig. 3.16(c)*

$$F_{cc} = \text{stress} \times \text{area} = \frac{0.67f_{cu}}{\gamma_{mc}}\,0.9xb \qquad (3.8)$$

and

$$z = d - 0.9x/2 \qquad (3.9)$$

In order to ensure that the section is under-reinforced, BS 8110 limits the depth of the neutral axis (x) to a maximum of $0.5d$, where d is the effective depth (*Fig. 3.16(b)*). Hence

$$x \leqslant 0.5d \qquad (3.10)$$

By combining equations 3.7–3.10 and putting $\gamma_{mc} = 1.5$ (*Table 3.3*) it can be shown that the ultimate moment of resistance is given by

$$M_u = 0.156f_{cu}bd^2 \qquad (3.11)$$

Note that M_u depends only on the properties of the concrete and not the steel reinforcement. Provided that the design moment does not exceed M_u (i.e. $M \not> M_u$), a beam whose section is singly reinforced will be sufficient to resist the design moment. The following section derives the equation necessary to calculate the area of reinforcement needed for such a case.

Area of tension reinforcement, A_s. At the limiting condition $M_u = M$, equation 3.7 becomes

$$M = F_{st}\,z = \frac{f_yA_s}{\gamma_{ms}}\,z \text{ (from equation 3.5)}$$

Rearranging and putting $\gamma_{ms} = 1.15$ (*Table 3.3*) gives

$$A_s = \frac{M}{0.87f_yz} \qquad (3.12)$$

Solution of this equation requires an expression for z which can either be obtained graphically (*Fig. 3.17*) or by calculation as discussed below.

Lever arm, z. At the limiting condition $M_u = M$, equation 3.7 becomes

$$M = F_{cc}\,z = \frac{0.67f_{cu}}{\gamma_{mc}}\,0.9bxz \quad \text{(from equation 3.8)}$$

$$= 0.4f_{cu}\,bzx \quad \text{(putting } \gamma_{mc} = 1.5\text{)}$$

$$= 0.4f_{cu}\,bz2\frac{(d-z)}{0.9} \quad \text{(from equation 3.9)}$$

$$= \frac{8}{9}f_{cu}bz\,(d-z)$$

$$\frac{M}{f_{cu}\,bd^2} = \frac{8}{9}\,(z/d)\,(1-z/d) \quad \begin{array}{l}\text{(dividing both sides by}\\ f_{cu}\,bd^2)\end{array}$$

Substituting $K = \dfrac{M}{f_{cu}\,bd^2}$ and putting $z_o = \dfrac{z}{d}$ gives

41

$$K = \frac{8}{9}z_0(1-z_0) \text{ or } 0 = z_0^2 - z_0 + 9K/8$$

This is a quadratic and can be solved to give

$$z_0 = z/d = 0.5 + \sqrt{(0.25 - K/0.9)}$$

This equation is used to draw the lever arm curve shown in *Fig. 3.17*, and is usually expressed in the following form

$$z = d[0.5 + \sqrt{(0.25 - K/0.9)}] \qquad (3.13)$$

Once z has been determined, the area of tension reinforcement A_s can be calculated using equation 3.12. In Clause 3.4.4.1 of BS 8110 it is noted that z should not exceed $0.95d$ in order to give a reasonable concrete area in compression. Moreover it should be

remembered that equation 3.12 can only be used to determine A_s provided that $M \not> M_u$ or $K \not> K'$ where

$$K = \frac{M}{f_{cu}bd^2} \text{ and } K' = \frac{M_u}{f_{cu}bd^2}$$

To summarize, design for bending requires the calculation of the maximum design moment (M) and corresponding ultimate moment of resistance of the section (M_u). Provided $M \not> M_u$, or $K \not> K'$, only tension reinforcement is needed and the area of steel can be calculated using equation 3.12 via equation 3.13. Where $M > M_u$ the designer has the option to either increase the section sizes (i.e. $M \leqslant M_u$) or design as a doubly reinforced section. The latter option is discussed more fully in *section 3.9.2*.

Example 3.1 Design of bending reinforcement for a singly reinforced beam

A simply supported rectangular beam of 7 m span carries characteristic dead (including self-weight of beam), g_k, and imposed, q_k, loads of 12 and 8 kN m^{-1} respectively (*Fig. 3.18*). The beam dimensions are breadth, b, 275 mm and effective depth, d, 450 mm. Assuming the following material strengths, calculate the area of reinforcement required.

$$f_{cu} = 30 \text{ N mm}^{-2} \qquad f_y = 460 \text{ N mm}^{-2}$$

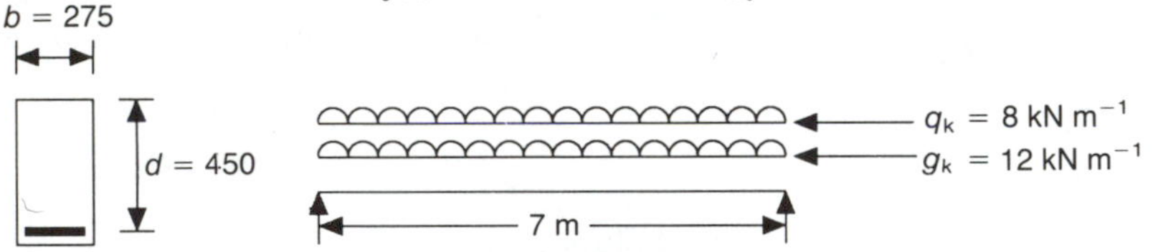

Fig. 3.18

$$\text{Ultimate load } (w) = 1.4g_k + 1.6q_k$$
$$= 1.4 \times 12 + 1.6 \times 8 = 29.6 \text{ kN m}^{-1}$$

Design moment $(M) = \dfrac{wl^2}{8} = \dfrac{29.6 \times 7^2}{8} = 181.3 \text{ kN m}$

Ultimate moment of resistance, M_u, is $M_u = 0.156f_{cu}bd^2 = 0.156 \times 30 \times 275 \times 450^2$
$$= 260.6 \times 10^6 \text{ N mm} = 260.6 \text{ kN m}$$

Since $M_u > M$ design as a singly reinforced beam.

$$K = \frac{M}{f_{cu}bd^2} = \frac{181.3 \times 10^6}{30 \times 275 \times 450^2} = 0.1085$$

$$z = d[0.5 + \sqrt{(0.25 - K/0.9)}]$$
$$= 450[0.5 + \sqrt{(0.25 - 0.1085/0.9)}]$$
$$= 386.9 \text{ mm} \not> 0.95d (= 427.5 \text{ mm}) \quad \text{OK}$$

$$A_s = \frac{M}{0.87f_yz} = \frac{181.3 \times 10^6}{0.87 \times 460 \times 386.9} = 1171 \text{ mm}^2$$

For detailing purposes this area of steel has to be transposed into a certain number of bars of a given diameter. This is usually achieved using steel area tables similar to that shown in *Table 3.7*. Thus it can be seen that four 20 mm diameter bars have a total cross-sectional area of 1260 mm^2 and would therefore be suitable. Hence provide 4T20. (N.B. T refers to high yield steel bars ($f_y = 460$ N mm^{-2}); R refers to mild steel bars ($f_y = 250$ N mm^{-2}).)

Table 3.7   Cross-sectional areas of groups of bars (mm^2)

Bar size (mm)	1	2	3	4	Number of bars 5	6	7	8	9	10
6	28.3	56.6	84.9	113	142	170	198	226	255	283
8	50.3	101	151	201	252	302	352	402	453	503
10	78.5	157	236	314	393	471	550	628	707	785
12	113	226	339	452	566	679	792	905	1020	1130
16	201	402	603	804	1010	1210	1410	1610	1810	2010
20	314	628	943	1260	1570	1890	2200	2510	2830	3140
25	491	982	1470	1960	2450	2950	3440	3930	4420	4910
32	804	1610	2410	3220	4020	4830	5630	6430	7240	8040
40	1260	2510	3770	5030	6280	7540	8800	10100	11300	12600

3.9.1.2 Design charts

An alternative method of determining the area of tensile steel required in singly reinforced rectangular beams is by using the design charts given in Part 3 of BS 8110. Chart 2 is reproduced here as *Fig. 3.19* and is appropriate for use with grade 460 reinforcement.

The chart is based on the rectangular parabolic stress distribution shown in *Fig. 3.15(d)* rather than the simplified rectangular distribution in *Fig. 3.15(e)*, used to develop the design formulae in the previous section. Generally, design charts will result in slightly greater estimates of steel areas than those obtained by using the design formulae.

The procedure involves the following steps:

1. Check $M \ngtr M_u$.
2. Select appropriate chart from Part 3 of BS 8110 based on the grade of tensile reinforcement.
3. Calculate M/bd^2.
4. Plot M/bd^2 ratio on chart and read off corresponding $100A_s/bd$ value using curve appropriate to grade of concrete selected for design.
5. Calculate A_s.

Using the figures given in *Example 3.1*,

$$M = 181.3 \text{ kN m} \ngtr M_u = 260.6 \text{ kN m}$$

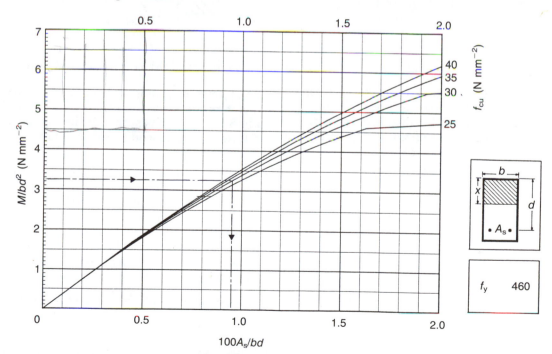

Fig. 3.19 *Design chart for singly reinforced beam (chart no. 2, BS 8110: Part 3).*

$$\frac{M}{bd^2} = \frac{181.3 \times 10^6}{275 \times 450^2} = 3.26$$

From *Fig. 3.19*, using the $f_{cu} = 30$ N mm^{-2} curve,

$$\frac{100A_s}{bd} = 0.95$$

Hence,

$$A_s = 1176 \text{ mm}^2$$

Therefore provide 4T20 ($A_s = 1260$ mm^2).

3.9.1.3 Shear (clause 3.4.5, BS 8110)

Another way in which failure of a beam may arise is due to its shear capacity being exceeded. Shear failure may arise in several ways, but the two principal failure mechanisms are shown in *Fig. 3.20*. With reference to *Fig. 3.20(a)*, as the loading increases, an inclined crack rapidly develops between the edge of the support and the load point, resulting in splitting of the beam into two pieces. This is normally termed diagonal tension failure and can be prevented by providing shear reinforcement.

The second failure mode, termed diagonal compression failure (*Fig. 3.20(b)*), occurs under the action of large shear forces acting near the support, resulting in crushing of the concrete. This type of failure is avoided by limiting the maximum shear stress to 5 N mm^{-2} or $0.8\sqrt{f_{cu}}$, whichever is the lesser.

The design shear stress, v, at any cross-section can be calculated from

$$v = V/bd \qquad (3.14)$$

where V is the design shear force due to ultimate loads, b the breadth of section and d the effective depth of section. In order to determine whether shear reinforcement is required, it is necessary to calculate the shear resistance, or using BS 8110 terminology the design concrete shear stress, at critical sections along the beam. The design concrete shear stress, v_c, is found to be composed of three major components, namely:

1. concrete in the compression zone;
2. aggregate interlock across the crack zone;
3. dowel action of the tension reinforcement.

The design concrete shear stress can be determined using *Table 3.8*. The values are in terms of the percentage area of longitudinal tension reinforcement ($100A_s/bd$) and effective depth of the section (d). The table assumes that the grade of concrete is 25. For other grades of concrete, up to a maximum of 40, the design shear stresses can be determined by multiplying the values in the table by the factor $(f_{cu}/25)^{1/3}$.

Generally, where the design shear stress exceeds the design concrete shear stress, shear reinforcement will be needed. This is normally done by providing (1) vertical shear reinforcement commonly referred to as 'links' or (2) a combination of vertical and inclined (or bent-up) bars as shown below.

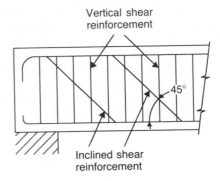

Beam with vertical and inclined shear reinforcement

The former is the most widely used method and will therefore be the only one discussed here. The following section derives the design equations for calculating the area and spacing of links.

Shear resistance of links. Consider a reinforced concrete beam with links uniformly spaced at a distance s_v, under the action of a shear force V. The resulting failure plane is assumed to be inclined approximately 45° to the horizontal as shown in *Fig. 3.21*. The number of links intersecting the potential crack is equal to d/s_v, and it follows therefore that the shear resistance of these links, V_{link}, is given by

$$V_{link} = \text{number of links} \times \text{total cross-sectional area}$$
$$\text{of link } (Fig.\ 3.22) \times \text{design stress}$$
$$= (d/s_v) \times A_{sv} \times 0.87 \, f_{yv}$$

The shear resistance of concrete, V_{conc}, can be calculated from

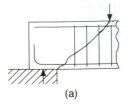

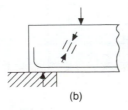

Fig. 3.20 *Types of shear failure: (a) diagonal tension; (b) diagonal compression.*

Table 3.8 Values of design concrete shear stress, v_c (N mm^{-2}) (Table 3.9, BS 8110)

$\dfrac{100 A_s}{bd}$	Effective depth (d) mm							
	125	150	175	200	225	250	300	≥400
≤0.15	0.45	0.43	0.41	0.40	0.39	0.38	0.36	0.34
0.25	0.53	0.51	0.49	0.47	0.46	0.45	0.43	0.40
0.50	0.67	0.64	0.62	0.60	0.58	0.56	0.54	0.50
0.75	0.77	0.73	0.71	0.68	0.66	0.65	0.62	0.57
1.00	0.84	0.81	0.78	0.75	0.73	0.71	0.68	0.63
1.50	0.97	0.92	0.89	0.86	0.83	0.81	0.78	0.72
2.00	1.06	1.02	0.98	0.95	0.92	0.89	0.86	0.80
≥3.00	1.22	1.16	1.12	1.08	1.05	1.02	0.98	0.91

$$V_{conc} = v_c bd \quad \text{(using equation 3.14)}$$

The design shear force due to ultimate loads, V, must be less than the sum of the shear resistance of the concrete (V_{conc}) plus the shear resistance of the links (V_{link}), otherwise failure of the beam may arise. Hence

$$V \leqslant V_{conc} + V_{link}$$
$$\leqslant v_c bd + (d/s_v) A_{sv} 0.87 f_{yv} \quad \text{(see above)}$$
$$V/bd \leqslant v_c + (1/bs_v) A_{sv} 0.87 f_{yv} \quad \text{(dividing both sides by } bd)$$
$$v \leqslant v_c + (1/bs_v) A_{sv} 0.87 f_{yv} \quad \text{(from equation 3.14)}$$

Rearranging gives

$$\frac{A_{sv}}{s_v} = \frac{b (v - v_c)}{0.87 f_{yv}} \quad (3.15)$$

Where $(v - v_c)$ is less than 0.4 N mm^{-2} then links should be provided according to

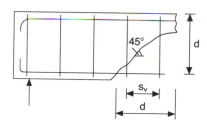

Fig. 3.21 Shear resistance of links.

$$\frac{A_{sv}}{s_v} = \frac{0.4b}{0.87 f_{yv}} \quad (3.16)$$

Equations 3.15 and 3.16 provide a basis for calculating the minimum area and spacing of links. The details are discussed next.

Form, area and spacing of links. Shear reinforcement should be provided in beams according to the criteria given in *Table 3.9*. Thus where the

Table 3.9 Form, area and spacing of links (Table 3.8, BS 8110)

Values of v (N mm^{-2})	Area of shear reinforcement to be provided
$v < 0.5 v_c$ throughout the beam	No links required but normal practice to provide nominal links in members of structural importance
$0.5 v_c < v < (v_c + 0.4)$	Nominal (or minimum) links for whole length of beam $A_{sv} \geq \dfrac{0.4 b s_v}{0.87 f_{yv}}$
$(v_c + 0.4) < v < 0.8\sqrt{f_{cu}}$ or 5 N mm^{-2}	Design links $A_{sv} \geq \dfrac{b s_v (v - v_c)}{0.87 f_{yv}}$

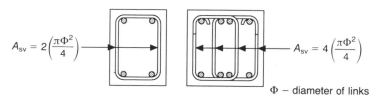

Φ – diameter of links

Fig. 3.22 A_{sv} for varying shear reinforcement arrangements.

design shear stress is less than half the design concrete shear stress (i.e. $v < 0.5v_C$), no shear reinforcement will be necessary although, in practice, it is normal to provide nominal links in all beams of structural importance. Where $0.5\, v_C < v < (v_C + 0.4)$ nominal links based on equation 3.16 should be provided. Where $v > v_C + 0.4$, design links based on equation 3.15 should be provided.

BS 8110 further recommends that the spacing of links in the direction of the span should not exceed $0.75d$. This will ensure that at least one link will cross the potential crack.

Example 3.2 Design of shear reinforcement for a beam

Design the shear reinforcement for the beam shown in *Fig. 3.23* using mild steel ($f_y = 250$ N mm^{-2}) links for the following load cases: (i) $q_k = 0$, (ii) $q_k = 10$ kN m^{-1}, (iii) $q_k = 27$ kN m^{-1} and (iv) $q_k = 40$ kN m^{-1}.

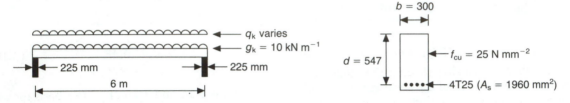

Fig. 3.23

DESIGN CONCRETE SHEAR STRESS, v_C

$$\frac{100A_s}{bd} = \frac{100 \times 1960}{300 \times 547} = 1.19$$

From *Table 3.8*, $v_C = 0.66$ N mm^{-2} (see below).

$\dfrac{100A_s}{bd}$	Effective depth (mm)	
	300	≥ 400
	N mm^{-2}	N mm^{-2}
1.00	0.68	0.63
1.19		0.66
1.50	0.78	0.72

$q_k = 0$

Design shear stress (v)

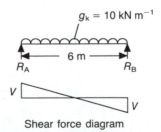

Shear force diagram

Total ultimate load, W, is

$$W = (1.4\, g_k + 1.6 q_k)\ \text{span} = (1.4 \times 10 + 1.6 \times 0)6 = 84\ \text{kN}$$

Since beam is symmetrically loaded

$$R_A = R_B = W/2 = 42\ \text{kN}$$

Ultimate shear force $(V) = 42$ kN and design shear stress, v, is

$$v = \frac{V}{bd} = \frac{42 \times 10^3}{300 \times 547} = 0.26 \text{ N mm}^{-2}$$

Diameter and spacing of links
By inspection

$$v < v_c/2$$

i.e. $0.26 \text{ N mm}^{-2} < 0.33 \text{ N mm}^{-2}$. Hence from *Table 3.9*, shear reinforcement may not be necessary.

$q_k = 10 \text{ kN m}^{-1}$

Design shear stress (v)

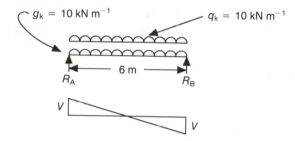

Total ultimate load, W, is

$$W = (1.4\, g_k + 1.6 q_k) \text{ span} = (1.4 \times 10 + 1.6 \times 10)6 = 180 \text{ kN}$$

Since beam is symmetrically loaded

$$R_A = R_B = W/2 = 90 \text{ kN}$$

Ultimate shear force $(V) = 90$ kN and design shear stress, v, is

$$v = \frac{V}{bd} = \frac{90 \times 10^3}{300 \times 547} = 0.55 \text{ N mm}^{-2}$$

Diameter and spacing of links
By inspection

$$v_c/2 < v < (v_c + 0.4)$$

i.e. $0.33 < 0.55 < 1.06$. Hence, from *Table 3.9*, provide nominal links for whole length of beam according to

$$\frac{A_{sv}}{s_v} = \frac{0.4b}{0.87 f_{yv}} = \frac{0.4 \times 300}{0.87 \times 250} = 0.55$$

This value has to be translated into a certain bar size and spacing of links and is usually achieved using shear reinforcement tables similar to *Table 3.10*. The spacing of links should not exceed 0.75 $d = 0.75 \times 547 = 410$ mm. From *Table 3.10* it can be seen that 10 mm diameter links spaced at 275 mm centres provide a A_{sv}/s_v ratio of 0.571 and would therefore be suitable. Hence provide R10 links at 275 mm centres for whole length of beam.

Table 3.10 Values of A_{sv}/s_v

Diameter of links (mm)	Spacing of links (mm)										
	85	90	100	125	150	175	200	225	250	275	300
8	1.183	1.118	1.006	0.805	0.671	0.575	0.503	0.447	0.402	0.336	0.335
10	1.847	1.744	1.57	1.256	1.047	0.897	0.785	0.698	0.628	0.571	0.523
12	2.659	2.511	2.26	1.808	1.507	1.291	1.13	1.004	0.904	0.822	0.753
16	4.729	4.467	4.02	3.216	2.68	2.297	2.01	1.787	1.608	1.462	1.34

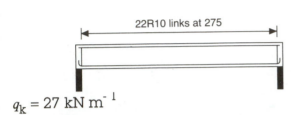

22R10 links at 275

$q_k = 27$ kN m^{-1}

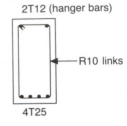

2T12 (hanger bars)

R10 links

4T25

Design shear stress (v)

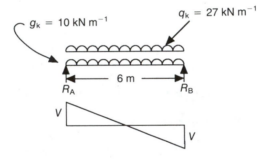

$g_k = 10$ kN m^{-1}

$q_k = 27$ kN m^{-1}

6 m

R_A R_B

V

V

Total ultimate load, W, is

$$W = (1.4\, g_k + 1.6\, q_k)\, \text{span} = (1.4 \times 10 + 1.6 \times 27)6 = 343.2 \text{ kN}$$

Since beam is symmetrically loaded

$$R_A = R_B = W/2 = 171.6 \text{ kN}$$

Ultimate shear force (V) = 171.6 kN and design shear stress, v, is

$$v = \frac{V}{bd} = \frac{171.6 \times 10^3}{300 \times 547} = 1.05 \text{ N mm}^{-2}$$

Diameter and spacing of links

By inspection

$$v_c/2 < v < (v_c + 0.4)$$

i.e. $0.33 < 1.05 < 1.06$. Hence from *Table 3.9*, provide nominal links for whole length of beam according to

$$\frac{A_{sv}}{s_v} = \frac{0.4b}{0.87 f_{yv}} = \frac{0.4 \times 300}{0.87 \times 250} = 0.55$$

Therefore as in case (ii) ($q_k = 10$ kN m^{-1}), provide R10 links at 275 mm centres.

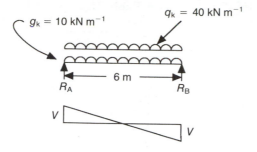

$q_k = 40 \text{ kN m}^{-1}$

Design shear stress (v)
Total ultimate load, W, is

$$W = (1.4g_k + 1.6q_k) \text{ span} = (1.4 \times 10 + 1.6 \times 40)6 = 468 \text{ kN}$$

Since beam is symmetrically loaded

$$R_A = R_B = W/2 = 234 \text{ kN}$$

Ultimate shear force $(V) = 234$ kN and design shear stress, v, is

$$v = \frac{V}{bd} = \frac{234 \times 10^3}{300 \times 547} = 1.43 \text{ N mm}^{-2} < \text{permissible} = 0.8 \sqrt{25} = 4 \text{ N mm}^{-2}$$

Diameter and spacing of links
Where $v < (v_c + 0.4) = 0.66 + 0.4 = 1.06 \text{ N mm}^{-2}$, nominal links are required according to

$$\frac{A_{sv}}{s_v} = \frac{0.4b}{0.87f_{yv}} = \frac{0.4 \times 300}{0.87 \times 250} = 0.55$$

Hence, from *Table 3.10*, provide R10 links at 275 mm centres where $v < 1.06 \text{ N mm}^{-2}$, i.e. 2.223 m either side of the mid-span of beam.

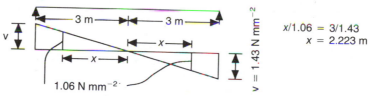

$x/1.06 = 3/1.43$
$x = 2.223$ m

Where $v > (v_c + 0.4) = 1.06 \text{ N mm}^{-2}$ design links required according to

$$\frac{A_{sv}}{s_v} = \frac{b(v-v_c)}{0.87f_{yv}} = \frac{300(1.43 - 0.66)}{0.87 \times 250} = 1.06$$

Hence from *Table 3.10*, provide R10 links at 125 mm centres ($A_{sv}/s_v = 1.26$), where $v > 1.06 \text{ N mm}^{-2}$, i.e. 0.777 m in from both supports.

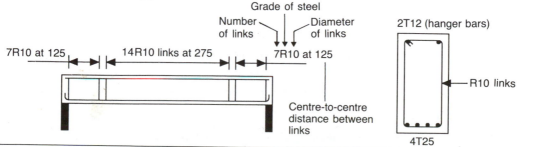

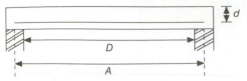

Fig. 3.24 Effective span of simply supported beam.

Table 3.11 Basic span/effective depth ratio for rectangular sections (based on Table 3.10, BS 8110)

Support conditions	Span/effective depth ratio (rectangular section)
Cantilever	7
Simply supported	20
Continuous	26

Table 3.12 Modification factors for compression reinforcement (based on Table 3.12, BS 8110)

$\dfrac{100A'_s{}^{,prov}}{bd}$	Factor
0.00	1.0
0.15	1.05
0.25	1.08
0.35	1.10
0.5	1.14
0.75	1.20
1	1.25
1.5	1.33
2.0	1.40
2.5	1.45
$\geqslant 3.0$	1.5

3.9.1.4 Deflection (Clause 3.4.6, BS 8110)

In addition to checking that failure of the member does not arise due to the ultimate limit states of bending and shear, the designer must ensure that the deflections under working loads do not adversely affect either the efficiency or appearance of the structure. BS 8110 describes the following criteria for ensuring the proper performance of rectangular beams:

1. Final deflection should not exceed span/250.
2. Deflection after construction of finishes and partitions should not exceed span/500 or 20 mm, whichever is the lesser, for spans up to 10 m.

However, it is rather difficult to make accurate predictions of the deflections that may arise in concrete members principally because the member may be cracked under working loads and the degree of restraint at the supports is uncertain. Therefore, BS 8110 uses an approximate method based on permissible ratios of the span/effective depth. Before discussing this method in detail it is worth clarifying what is meant by the span, or more correctly the effective span, of a beam.

Table 3.13 Modification factors for tension reinforcement (Table 3.11, BS 8110)

Service stress		M/bd^2								
		0.50	0.75	1.00	1.50	2.00	3.00	4.00	5.00	6.00
	100	2.00	2.00	2.00	1.86	1.63	1.36	1.19	1.08	1.01
	150	2.00	2.00	1.98	1.69	1.49	1.25	1.11	1.01	0.94
$(f_y = 250)$	156	2.00	2.00	1.96	1.66	1.47	1.24	1.10	1.00	0.94
	200	2.00	1.95	1.76	1.51	1.35	1.14	1.02	0.94	0.88
	250	1.90	1.70	1.55	1.34	1.20	1.04	0.94	0.87	0.82
$(f_y = 460)$	288	1.68	1.50	1.38	1.21	1.09	0.95	0.87	0.82	0.78
	300	1.60	1.44	1.33	1.16	1.06	0.93	0.85	0.80	0.76

Note 1. The values in the table derive from the equation:

$$\text{Modification factor} = 0.55 + \frac{(477 - f_s)}{120 \left(0.9 + \dfrac{M}{bd^2}\right)} \leq 2.0 \quad \text{(equation 7)}$$

where

f_s is the design service stress in the tension reinforcement

M is the design ultimate moment at the centre of the span or, for a cantilever, at the support.

Note 2. The design service stress in the tension reinforcement in a member may be estimated from the equation:

$$f_s = \frac{5f_y A_{s,\,req}}{8A_{s,\,prov}} \times \frac{1}{\beta_b} \quad \text{(equation 8)}$$

where β_b is the percentage moment redistribution, equal to 1 for simply supported beams.

Effective span (clause 3.4.1.2, BS 8110). All calculations relating to beam design should be based on the effective span of the beam. For a simply supported beam this should be taken as the lesser of (1) the distance between centres of bearings, A, or (2) the clear distance between supports, D, plus the effective depth, d, of the beam (*Fig. 3.24*).

Span/effective depth ratio. Generally, the deflection criteria in (1) and (2) will be satisfied provided that the span/effective depth ratio of the beam does not exceed the appropriate limiting values given in *Table 3.11*. The reader is referred to the *Handbook to BS 8110* which outlines the basis of this approach.

The span/effective depth ratios given in the table apply to spans up to 10 m long. Where the span exceeds 10 m, these ratios should be multiplied by 10/span (except for cantilevers). The basic ratios may be further modified by factors taken from *Tables 3.12* and *3.13*, depending upon the amount of compression and tension reinforcement respectively. Deflection is usually critical in the design of slabs rather than beams and, therefore, modification fac-

Table 3.14 Span/effective depth ratios for initial design

Support condition	Span/depth
Cantilever	6
Simply supported	12
Continuous	15

tors will be discussed more fully in the context of slab design (*section 3.10*).

3.9.1.5. Member sizing

The dual concepts of span/effective depth ratios and maximum design concrete shear stress can be used not only to assess the performance of members with respect to deflection and shear but also for preliminary sizing of members. *Table 3.14* gives modified span/depth ratios for estimating the effective depth of a concrete beam provided that its span is known. The width of the beam can then be determined by limiting the maximum design concrete shear stress to around (say) 1.2 N mm^{-2}.

Example 3.3 Sizing a concrete beam

A simply supported beam has an effective span of 8 m and supports characteristic dead (g_k) and live (q_k) loads of 15 kN m^{-1} and 10 kN m^{-1} respectively. Determine suitable dimensions for the effective depth and width of the beam.

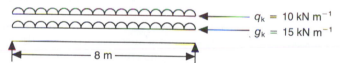

$$q_k = 10 \text{ kN m}^{-1}$$
$$g_k = 15 \text{ kN m}^{-1}$$
8 m

From *Table 3.14*, span/depth ratio for a simply supported beam is 12. Hence effective depth, d, is

$$d = \frac{\text{span}}{\text{ratio}} = \frac{8000}{12} \approx 670 \text{ mm}$$

Total ultimate load is =

$$(1.4g_k + 1.6q_k) \text{ span} = (1.4 \times 15 + 1.6 \times 10)8 = 296 \text{ kN}$$

Design shear force (V) = 296/2 = 148 kN and design shear stress, v, is

$$v = \frac{V}{bd} = \frac{148 \times 10^3}{670b}$$

Assuming v is equal to (say) 1.2 N mm^{-2} gives a width of beam, b, of

$$b = \frac{148 \times 10^3}{670 \times 1.2} \approx 185 \text{ mm}$$

Hence a beam of width 185 mm and effective depth 670 mm would be suitable to support the given design loads.

3.9.1.6 Reinforcement details (clause 3.12, BS 8110)

The previous sections have covered much of the theory required to design singly reinforced concrete beams. However, there are a number of code provisions with regard to:

1. maximum and minimum reinforcement areas
2. spacing of reinforcement
3. curtailment and anchorage of reinforcement
4. lapping of reinforcement.

These need to be taken into account since they may affect the final design.

Reinforcement areas (clauses 3.12.5.3 and 3.12.6.1, BS 8110). As pointed out in *section 3.8*, there is a need to control cracking of the concrete because of durability and aesthetic considerations. This is usually achieved by providing minimum areas of reinforcement in the member. However, too large an area of reinforcement should also be avoided since it will hinder proper placing and adequate compaction of the concrete around the reinforcement.

For rectangular beams with overall dimensions b and h, the area of tension reinforcement, A_s, should lie within the following limits:

$$0.24\% \; bh \leq A_s \leq 4\% \; bh \text{ when } f_y = 250 \text{ N mm}^{-2}$$

$$0.13\% \; bh \leq A_s \leq 4\% \; bh \text{ when } f_y = 460 \text{ N mm}^{-2}$$

Spacing of reinforcement (clause 3.12.11.1, BS 8110). BS 8110 specifies minimum and maximum distances between tension reinforcement. The actual limits vary, depending upon the grade of reinforcement. The minimum distance is based on the need to achieve good compaction of the concrete around the reinforcement. The limits on the maximum distance between bars arise from the need to ensure that the maximum crack width does not exceed 0.3 mm in order to prevent corrosion of embedded bars (*section 3.8*).

For singly reinforced simply supported beams the clear horizontal distance between tension bars, s_b, should lie within the following limits:

$$h_{agg} + 5 \text{ mm or bar size} \leq s_b \leq 300 \text{ mm} \quad \text{when } f_y = 250 \text{ N mm}^{-2}$$

$$h_{agg} + 5 \text{ mm or bar size} \leq s_b \leq 160 \text{ mm} \quad \text{when } f_y = 460 \text{ N mm}^{-2}$$

where h_{agg} is the maximum size of the coarse aggregate.

Curtailment and anchorage of bars (clause 3.12.9, BS 8110). The design process for simply supported beams, in particular the calculations re-

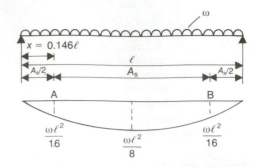

Fig. 3.25

lating to the design moment and area of bending reinforcement, is concentrated at mid-span. However, the bending moment decreases either side of the mid-span and it follows, therefore, that it should be possible to reduce the corresponding area of bending reinforcement by curtailing bars. For the beam shown in *Fig. 3.25*, theoretically 50% of the main steel can be curtailed at points A and B. However, in order to develop the design stress in the reinforcement (i.e. $0.87f_y$ at mid-span), these bars must be anchored into the concrete. Except at end supports, this is normally achieved by extending the bars beyond the point at which they are theoretically no longer required by a distance equal to the greater of (i) the effective depth of the member and (ii) 12 times the bar size.

Where a bar is stopped off in the tension zone, e.g. beam shown in *Fig. 3.25*, this distance should be increased to the full anchorage bond length in accordance with the values given in *Table 3.15*. However, simplified rules for the curtailment of bars are given in clause 3.12.10.2 of BS 8110. These are shown diagrammatically in *Fig. 3.26* for simply supported and cantilever beams.

The code also gives rules for the anchorage of bars at the supports. Thus, at a simply supported end

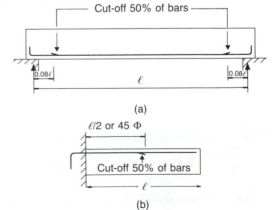

Fig. 3.26 *Simplified rules for curtailment of bars in beams: (a) simply supported; (b) cantilever beam.*

Table 3.15 Anchorage lengths as multiples of bar size (based on Table 3.29, BS 8110)

Reinforcement type	L_A			
	$f_{cu} = 25$	30	35	40 or more
Plain (250)				
Tension	39	36	33	31
Compression	32	29	27	25
Deformed type 1 (460)				
Tension	51	46	43	40
Compression	41	37	34	32
Deformed type 2 (460)				
Tension	41	37	34	32
Compression	32	29	27	26

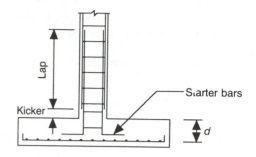

Fig. 3.29 *Lap lengths.*

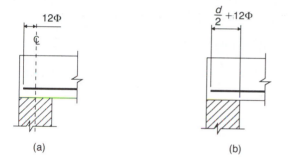

Fig. 3.27 *Anchorage requirements at simple supports.*

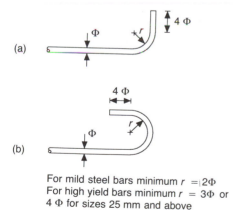

For mild steel bars minimum $r = 2\Phi$
For high yield bars minimum $r = 3\Phi$ or 4Φ for sizes 25 mm and above

Fig. 3.28 *Anchorage lengths for hooks and bends (a) anchorage length for 90° bend = 4r but not greater than 12 φ; (b) anchorage length for hook = 8r but not greater than 24 φ.*

each tension bar will be properly anchored provided the bar extends a length equal to one of the following: (a) 12 times the bar size beyond the centre line of the support, or (b) 12 times the bar size plus $d/2$ from the face of the support (*Fig. 3.27*). Sometimes it is not possible to use straight bars due to limitations of space and, in this case, anchorage must be provided by using hooks or bends in the reinforcement. The anchorage values of hooks and bends are shown in *Fig. 3.28*. Where hooks or bends are provided, BS 8110 states that they should not begin before the centre of the support for rule (a) nor before $d/2$ from the face of the support for rule (b).

Laps in reinforcement (clause 3.12.8, BS 8110). It is not possible nor, indeed, practicable to construct the reinforcement cage for an individual element or structure without joining some of the bars. This is normally achieved by lapping bars (*Fig. 3.29*). Bars which have been joined in this way must act as a single length of bar. This means that the lap length should be sufficiently long in order that stresses in one bar can be transferred to the other. The minimum lap length should not be less than 15 times the bar diameter or 300 mm. For **tension laps** it should normally be equal to the tension anchorage length, but will often need to be increased as outlined in clause 3.12.8.13 of BS 8110. The anchorage length (L) is calculated using

$$L = L_A \times \Phi \qquad (3.17)$$

where Φ is the diameter of the (smaller) bar and L_A is obtained from *Table 3.15* and depends upon the stress type, grade of concrete and reinforcement type. For **compression laps** the lap length should be at least 1.25 times the compression anchorage length.

Example 3.4 Design of a simply supported concrete beam

A reinforced concrete beam which is 300 mm wide and 600 mm deep is required to span 6.0 m between the centres of supporting piers 300 mm wide (*Fig. 3.30*). The beam carries dead and imposed loads of 25 and 19 kN m^{-1} respectively. Assuming $f_{cu} = 30$ N mm^{-2}, $f_y = 460$ N mm^{-2}, $f_{yv} = 250$ N mm^{-2} and the exposure condition is mild, design the beam.

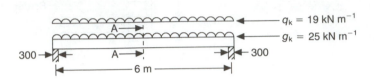

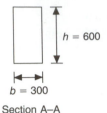

Section A–A

Fig. 3.30

DESIGN MOMENT, *M*

Loading

Dead

Self-weight of beam = $0.6 \times 0.3 \times 24 = 4.32$ kN m^{-1}
Total dead load (g_k) = $25 + 4.32 = 29.32$ kN m^{-1}

Imposed

Total imposed load (q_k) = 19 kN m^{-1}

Ultimate load

Total ultimate load (W) $= (1.4g_k + 1.6q_k)$ span
$= (1.4 \times 29.32 + 1.6 \times 19)\,6$
$= 428.7$ kN

Design moment

Maximum design moment (M) = $\dfrac{Wl}{8} = \dfrac{428.7 \times 6}{8}$
$= 321.5$ kN m

ULTIMATE MOMENT OF RESISTANCE, M_u

Effective depth, *d*

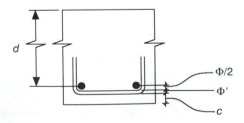

Assume diameter of main bars (Φ) = 25 mm
Assume diameter of links (Φ')　= 10 mm

From *Table 3.5*, cover for mild conditions of exposure (c) = 25 mm.

$$d = h - c - \Phi' - \Phi/2$$
$$= 600 - 25 - 10 - 25/2 = 552 \text{ mm}$$

Ultimate moment

$$M_u = 0.156 f_{cu} b d^2 = 0.156 \times 30 \times 300 \times 552^2$$
$$= 427.8 \times 10^6 \text{ N mm} = 427.8 \text{ kN m} > M$$

Since $M_u > M$ no compression reinforcement is required.

MAIN STEEL, A_s

$$K = \frac{M}{f_{cu}\,bd^2} = \frac{321.5 \times 10^6}{30 \times 300 \times 552^2} = 0.117$$

$$z = d[0.5 + \sqrt{(0.25 - K/0.9)}]$$
$$= 552[0.5 + \sqrt{(0.25 - 0.117/0.9)}] = 467 \text{ mm}$$

$$A_s = \frac{M}{0.87 f_y z} = \frac{321.5 \times 10^6}{0.87 \times 460 \times 467} = 1720 \text{ mm}^2$$

Hence from *Table 3.7*, provide 4T25 $(A_s = 1960 \text{ mm}^2)$.

SHEAR REINFORCEMENT

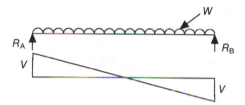

Ultimate design load, $W = 428.7$ kN

Shear stress, v
Since beam is symmetrically loaded

$$R_A = R_B = W/2 = 214.4 \text{ kN}$$

Ultimate shear force (V) = 214.4 kN and design shear stress, v, is

$$v = \frac{V}{bd} = \frac{214.4 \times 10^3}{300 \times 552} = 1.29 \text{ N mm}^{-2}$$

(Check v < permissible = $0.8 \sqrt{30} = 4.38 \text{ N mm}^{-2}$.)

Design concrete shear stress, v_c

$$100 A_s/bd = 100 \times 1960/300 \times 552 = 1.18$$

From *Table 3.8*,

$$v_c = (30/25)^{1/3} \times 0.66 \text{ N mm}^{-2} = 0.70 \text{ N mm}^{-2}$$

Diameter and spacing of links
Where v < $(v_c + 0.4) = 0.70 + 0.4 = 1.10 \text{ N mm}^{-2}$, nominal links are required according to

$$\frac{A_{sv}}{s_v} = \frac{0.4b}{0.87 f_{yv}} = \frac{0.4 \times 300}{0.87 \times 250} = 0.552$$

Hence, from *Table 3.10*, provide 10 mm links at 275 mm centres where $v < 1.10$ N mm^{-2}, i.e. 2.558 m either side of mid-span of beam.

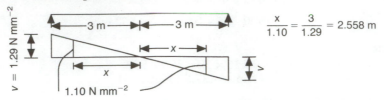

$$\frac{x}{1.10} = \frac{3}{1.29} = 2.558 \text{ m}$$

Where $v > (v_c + 0.4) = 1.10$ N mm^{-2} design links required according to

$$\frac{A_{sv}}{s_v} = \frac{b\,(v - v_c)}{0.87 f_{yv}} = \frac{300\,(1.29 - 0.7)}{0.87 \times 250} = 0.81$$

Maximum spacing of links is $0.75d = 0.75 \times 552 = 414$ mm. Hence from *Table 3.12*, provide 10 mm links at 175 mm centres ($A_{sv}/s_v = 0.897$) where $v > 1.10$ N mm^{-2}, i.e. 0.442 m in from both supports.

EFFECTIVE SPAN
The above calculations were based on an effective span of 6 m, but this needs to be confirmed. As stated in *section 3.9.1.4* the effective span is the lesser of (1) centre-to-centre distance between support, i.e. 6 m, and (2) clear distance between supports plus the effective depth, i.e. $5700 + 552 = 6252$ mm. Therefore assumed span length of 6 m is correct.

DEFLECTION
Actual span/depth ratio = 6000/552 = 10.8

$$\frac{M}{bd^2} = \frac{321.5 \times 10^6}{300 \times 552^2} = 3.5$$

and from equation 8 (*Table 3.13*)

$$f_s = 5 f_y \frac{A_{s,\,req}}{8 A_{s,\,prov}} = \frac{5 \times 460 \times 1720}{8 \times 1960} = 252 \text{ N mm}^{-2}$$

From *Table 3.11*, basic span/depth ratio for a simply supported beam is 20 and from *Table 3.13*, modification factor ≈ 0.99. Hence permissible span/depth ratio = $20 \times 0.99 = 19.8 >$ actual (= 10.8) and the beam therefore satisfies deflection criteria in BS 8110.

REINFORCEMENT DETAILS
The sketch below shows the main reinforcement requirements for the beam. For reasons of buildability, the actual reinforcement details may well be slightly different and the reader is referred to the following publications for further information on this point:

1. *Designed and Detailed* (BS 8110:1985), Higgins, J. B. and Rogers, B. R., British Cement Association, 1989.
2. *Standard Method of Detailing Structural Concrete*, the Concrete Society and the Institution of Structural Engineers, London, 1989.

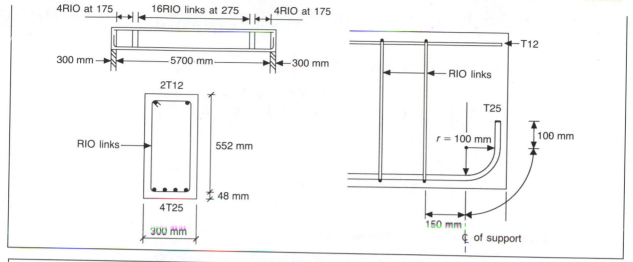

Example 3.5 Analysis of a singly reinforced concrete beam

A singly reinforced concrete beam in which $f_{cu} = 30$ N mm^{-2} and $f_y = 460$ N mm^{-2} contains 1960 mm^2 of tension reinforcement (*Fig. 3.31*). If the effective span is 7 m and the density of reinforced concrete is 24 kN m^{-3}, calculate the maximum imposed load that the beam can carry assuming that the load is (a) uniformly distributed and (b) occurs as a point load at mid-span.

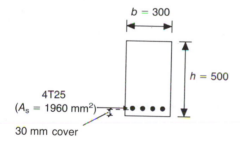

Fig. 3.31

MAXIMUM UNIFORMLY DISTRIBUTED IMPOSED LOAD, q_k

Moment capacity of section, M

Effective depth, d, is

$$d = h - \text{cover} - \phi/2 = 500 - 30 - 25/2 = 457 \text{ mm}$$

$$\frac{100 A_s}{bd} = \frac{100 \times 1960}{300 \times 457} = 1.43$$

From chart No. 2 of BS 8110 Part 3 (*Fig. 3.19*)

$$\frac{M}{bd^2} = 4.5 \Rightarrow M = 4.5 \times 300 \times 457^2 \times 10^{-6} = 281.9 \text{ kN m}$$

Maximum uniformly distributed imposed load, q_k

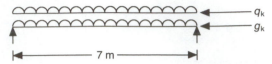

Dead load Self-weight of beam $(g_k) = 0.5 \times 0.3 \times 24 = 3.6 \text{ kN m}^{-1}$

Ultimate load Total ultimate load $(W) = (1.4g_k + 1.6q_k) \text{ span}$
$$= (1.4 \times 3.6 + 1.6q_k) \, 7$$

Imposed load Maximum design moment $(M) = \dfrac{Wl}{8} = \dfrac{(5.04 + 1.6q_k) \, 7^2}{8}$
$$= 281.9 \text{ kN m} \quad \text{(from above)}$$

Hence the maximum uniformly distributed imposed load the beam can support is

$$q_k = \frac{(281.9 \times 8) / 7^2 - 5.04}{1.6} = 25.6 \text{ kN m}^{-1}$$

MAXIMUM POINT LOAD AT MID-SPAN, Q_k

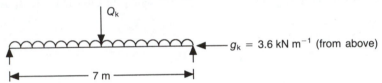

Loading

Ultimate load Ultimate dead load $(W_D) = 1.4g_k \times \text{span} = 1.4 \times 3.6 \times 7 = 35.3 \text{ kN}$
Ultimate imposed load (W_I) $\qquad = 1.6Q_k$

Imposed load Maximum design moment, M, is

$$M = \frac{W_D l}{8} + \frac{W_I l}{4} \quad (Example \ 2.5, \text{ beam B1–B3})$$

$$= \frac{35.3 \times 7}{8} + \frac{1.6Q_k \times 7}{4} = 281.9 \text{ kN m (from above)}$$

Hence the maximum point load which the beam can support at mid-span is

$$Q_k = \frac{(281.9 - 35.3 \times 7/8)4}{1.6 \times 7} = 89.6 \text{ kN}$$

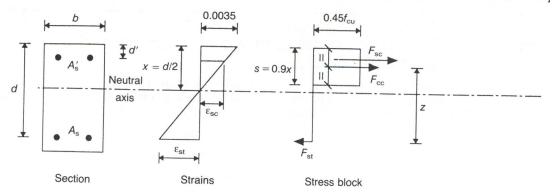

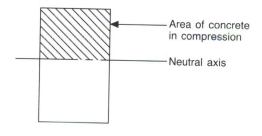

Fig 3.32 *Section with compression reinforcement.*

3.9.2 DOUBLY REINFORCED BEAM DESIGN

If the design moment is greater than the ultimate moment of resistance, i.e. $M > M_u$, or $K > K'$ where $K = M/f_{cu} bd^2$ and $K' = M_{cu}/f_{cu} bd^2$ the concrete will have insufficient strength in compression to generate this moment and maintain an under-reinforced mode of failure.

The required compressive strength can be achieved by increasing the proportions of the beam, particularly its overall depth. However, this may not always be possible due to limitations on the headroom in the structure, and in such cases it will be necessary to provide reinforcement in the compression face. The compression reinforcement will be designed to resist the moment in excess of M_u. This will ensure that the compressive stress in the concrete does not exceed the permissible value and ensure an under-reinforced failure mode.

Beams which contain tension and compression reinforcement are termed doubly reinforced. They are generally designed in the same way as singly reinforced beams except in respect of the calculations needed to

determine the areas of tension and compression reinforcement. This aspect is discussed below.

3.9.2.1 Compression and tensile steel areas
(clause 3.4.4.4, BS 8110)

The area of compression steel (A_s') is calculated from

$$A_s' = \frac{M - M_u}{0.87f_y (d-d')} \qquad (3.18)$$

where d' is the depth of the compression steel from the compression face (*Fig. 3.32*). The area of tension reinforcement is calculated from

$$A_s = \frac{M_u}{0.87f_{yz}} + A_s' \qquad (3.19)$$

where $z = d[0.5 + \sqrt{(0.25 - K'/0.9)}]$ and $K' = 0.156$

Equations 3.18 and 3.19 can be derived using the stress block shown in *Fig. 3.32*. This is basically the same stress block used in the analysis of a singly reinforced section (*Fig. 3.16*) except for the additional compression force (F_{sc}) in the steel.

In the derivation of equations 3.18 and 3.19 it is assumed that the compression steel has yielded (i.e. design stress = $0.87 f_y$) and this condition will be met only if

$$\frac{d'}{x} \not> 0.43 \quad \text{or} \quad \frac{d'}{d} \not> 0.215 \quad \text{where } x = \frac{d-z}{0.45}$$

If $d'/x > 0.43$, the compression steel will not have yielded and, therefore, the compressive stress will be less than $0.87f_y$. In such cases, the design stress can be obtained using *Fig. 3.9*.

Example 3.6 Design of bending reinforcement for a doubly reinforced beam

The reinforced concrete beam shown in *Fig. 3.33* has an effective span of 9 m and carries uniformly distributed dead (including self-weight of beam) and imposed loads of 4 and 5 kN m^{-1} respectively. Design the bending reinforcement assuming the following:

$$f_{cu} = 30 \text{ N mm}^{-2}$$

$$f_y = 460 \text{ N mm}^{-2}$$

Cover to *main* steel = 40 mm

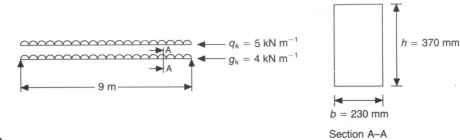

Fig. 3.33

DESIGN MOMENT, *M*

Loading

Ultimate load
Total ultimate load $(W) = (1.4g_k + 1.6q_k)$ span
$$= (1.4 \times 4 + 1.6 \times 5) \, 9 = 122.4 \text{ kN}$$

Design moment
Maximum design moment $(M) = \dfrac{Wl}{8} = \dfrac{122.4 \times 9}{8} = 137.7 \text{ kN m}$

ULTIMATE MOMENT OF RESISTANCE, M_u

Effective depth, *d*
Assume diameter of tension bars $(\phi) = 25$ mm:

$$d = h - \phi/2 - \text{cover} = 370 - 25/2 - 40 = 317 \text{ mm}$$

Ultimate moment
$M_u = 0.156f_{cu}bd^2 = 0.156 \times 30 \times 230 \times 317^2$
$\quad = 108.2 \times 10^6 \text{ N mm} = 108.2 \text{ kN m}$
Since $M_u < M$ compression reinforcement is required.

COMPRESSION REINFORCEMENT
Assume diameter of compression bars $(\phi) = 16$ mm. Hence

$$d' = \text{cover} + \phi/2 = 40 + 16/2 = 48 \text{ mm}$$

$$z = d[0.5 + \sqrt{(0.25 - K'/0.9)}]$$

$$= 317[0.5 + \sqrt{(0.25 - 0.156/0.9)}] = 246 \text{ mm}$$

$$x = (d - z)/0.45 = (317 - 246)/0.45 = 158 \text{ mm}$$

$$\frac{d'}{x} = \frac{48}{158} = 0.3 \not> 0.43, \text{ i.e. compression steel has yielded.}$$

$$A_s' = \frac{M-M_u}{0.87f_y(d-d')} = \frac{(137.7-108.2) \times 10^6}{0.87 \times 460 \times (317-48)} = 274 \text{ mm}^2$$

Hence from *Table 3.7*, provide 2T16 ($A_s' = 402$ mm^2).

TENSION REINFORCEMENT

$$A_s' = \frac{M_u}{0.87f_y z} + A_s' = \frac{(108.2 \times 10^6)}{0.87 \times 460 \times 246} + 274 = 1373 \text{ mm}^2$$

Hence provide 3T25 ($A_s = 1470$ mm^2).

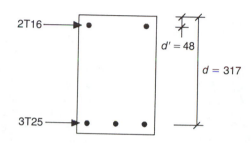

3.9.2.2 Design charts

Rather than solving equations 3.18 and 3.19 it is possible to determine the area of tension and compression reinforcement simply by using the design charts for doubly reinforced beams given in Part 3 of BS 8110. Such charts are available for design involving the use of reinforcement grade 460, concrete grades 25, 30, 35, 40, 45 and 50 and d'/d ratios of 0.1, 0.15 and 0.2. Chart No. 7 is reproduced here as *Fig. 3.34*.

The procedure involves the following steps:

1. Check $M_u < M$.
2. Calculate d'/d.

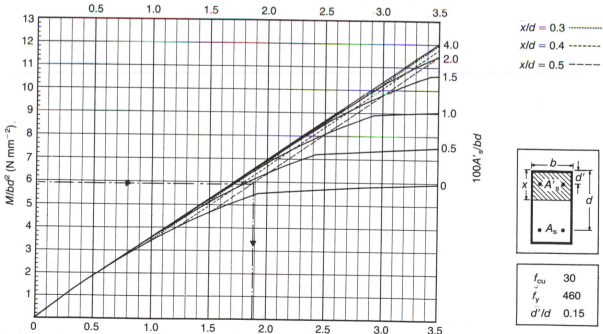

Fig. 3.34 *Design chart for doubly reinforced beams (chart no. 7, BS 8110: Part 3).*

3. Select appropriate chart from Part 3 of BS 8110 based on grade of concrete and d'/d ratio.
4. Calculate M/bd^2.
5. Plot M/bd^2 ratio on chart and read off corresponding $100A_s'/bd$ and $100A_s/bd$ values (Fig. 3.34).
6. Calculate A_s' and A_s.

Using the figures given in *Example 3.6*,

$M_u = 108.2$ kN m $< M = 137.7$ kN m

Since d'/d ($= 48/317$) $= 0.15$ and $f_{cu} = 30$ N mm^{-2}, chart No. 7 is appropriate. Furthermore, since the beam is simply supported, no redistribution of moments is possible, therefore, use $x/d = 0.5$ construction line in order to determine areas of reinforcement.

$$\frac{M}{bd^2} = \frac{137.7 \times 10^6}{230 \times 317^2} = 5.95$$

$$100A_s'/bd = 0.36 \Rightarrow A_s' = 263 \text{ mm}^2$$
$$100A_s/bd = 1.88 \Rightarrow A_s = 1370 \text{ mm}^2$$

Hence from *Table 3.7*, provide 2T16 compression steel and 3T25 tension steel.

3.9.3 CONTINUOUS, L- AND T-BEAMS

In most real situations, the beams in buildings are seldom single span but continuous over the supports, e.g. beams 1, 2, 3 and 4 in *Fig. 3.35(a)*. The

design process for such beams is similar to that outlined above for single-span beams. However, the main difference arises from the fact that with continuous beams, the designer will need to consider the various loading arrangements discussed in *section 3.6.2* in order to determine the design bending moments and shear forces in the beam. The analysis to calculate the bending moments and shear forces can be carried out by moment distribution or, provided the conditions in clause 3.4.3 of BS 8110 are satisfied, by using the coefficients given in Table 3.6 of BS 8110 (not reproduced here). Once this has been done, the beam can be sized and the areas of bending reinforcement calculated as discussed in *section 3.9.1* or *3.9.2*. At the internal supports, the bending moment is reversed and it should be remembered that the tensile reinforcement will occur in the top half of the beam and compression reinforcement in the bottom half of the beam.

Generally, beams and slabs are cast monolithically, that is, they are structurally tied. At mid-span, it is more economical in such cases to design the beam as an L- or T-section by including the adjacent areas of the slab (*Fig. 3.35(b)*). The actual width of slab which acts together with the beam is normally termed the effective flange. Its width can be determined in accordance with clause 3.4.1.5 of BS 8110. The depth of the neutral axis in relation to the depth of flange will influence the design process and must therefore be determined. The depth of the neutral axis, x, can be calculated using equation 3.9 derived in *section 3.9.1*, i.e.

$$x = \frac{d-z}{0.45}$$

Where the neutral axis lies within the flange, which will normally be the case in practice, the beam can be designed as being singly reinforced taking the breadth of the beam, b, equal to the effective flange width. At the supports of a continuous member, e.g. at columns B2, B3, C2 and C3, due to the moment reversal, b should be taken as the actual width of the beam.

3.9.4 SUMMARY FOR BEAM DESIGN

Figure 3.36 shows the basic steps that should be followed in order to design reinforced concrete beams.

3.10 Slabs

If a series of very wide, shallow rectangular beams were placed side by side and connected transversely such that it was possible to share the load between adjacent beams, the combination of beams would

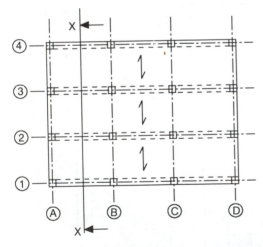

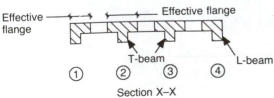

Section X–X

Fig. 3.35

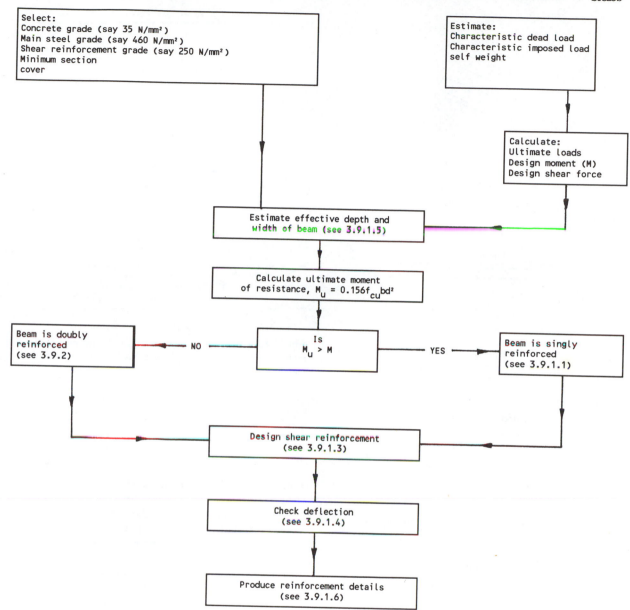

Fig. 3.36 *Beam design procedure.*

act as a slab (*Fig. 3.37*). Reinforced concrete slabs are used to form a variety of elements in building structures such as floors, roofs, staircases, foundations and some types of walls (*Fig. 3.38*). Since these elements can be modelled as a set of transversely connected beams, it follows that the design of slabs is similar, in principle, to that for beams. The major difference is that in slab design the serviceability limit state of deflection is normally critical, rather than the ultimate limit states of bending and shear.

3.10.1 TYPES OF SLABS

Slabs may be solid, ribbed (with hollow blocks or voids), flat or of composite construction. The latter type is not referred to in BS 8110 but has been included in this discussion for the sake of completeness. In practice, the choice of slab for a particular structure will largely depend upon the loading conditions and the length of the span. Thus for short-span slabs, generally less than 5 m, the most economical solution is to provide a solid slab of

Fig. 3.37 *Floor slab as a series of beams connected transversely.*

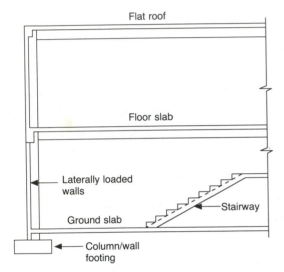

Fig. 3.38 *Various applications for slabs in reinforced concrete structures.*

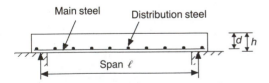

Fig. 3.39 *Elevation of solid slab.*

constant thickness over the complete span (*Fig. 3.39*).

With longer spans and light to moderate live loads, generally less than about 3 kN m^{-2}, it is more economical to provide a ribbed slab constructed using removable forms, hollow blocks or permanent or removable void formers (*Fig. 3.40*). Such slabs have a reduced self-weight compared to solid slabs since part of the concrete in the tension zone is omitted.

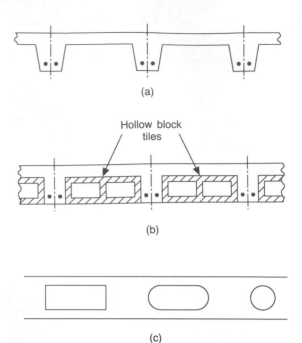

Fig. 3.40 *Ribbed slabs: (a) ribbed slab constructed with removal forms; (b) ribbed (hollow-block) slab; (c) ribbed (voided) slab.*

Fig. 3.41 *Example of flat slab construction.*

For some applications, flat slabs are superior to solid or ribbed slabs since they are generally easier and hence cheaper to construct. The ease of construction chiefly arises from the fact that the depth of the beams supporting the slab is the same as the depth of the slab. This avoids having to erect complicated shuttering, thereby reducing the overall cost of construction (*Fig. 3.41*). High live loads on flat slabs result in high shear stresses at the supports which may allow the columns to punch through the slab unless appropriate steps are taken. The shear stresses in the slab can be reduced by (i) using columns with large diameters and/or thicker slabs, (ii) enlarging column heads, i.e. by providing flared column heads or (iii) local thickening of the slab, i.e. by providing drop panels (*Fig. 3.42*).

However, all these methods have drawbacks and research effort has therefore been directed at finding alternative solutions. The work to date has resulted in the development of a structural component called

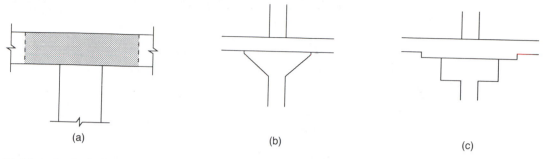

Fig. 3.42 *Methods of reducing shear stresses in flat slab construction: (a) deep slab and large column; (b) slab with flared column head; (c) slab with drop panel.*

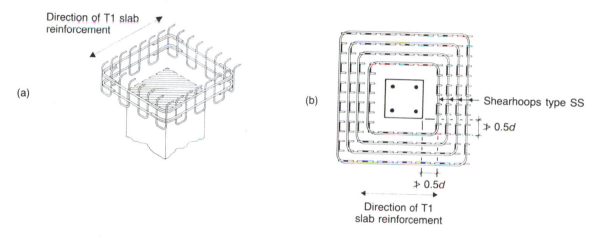

Fig. 3.43 *Shearhead: (a) Shearhoop type SS; (b) typical arrangement for internal square column.*

a 'shearhead' which can resist the shear stresses in the slab. Basically, it consists of a series of prefabricated cages of shear reinforcement known as shearhoops which can simply be attached to the main steel (*Fig. 3.43*). Shearheads are available in a range of diameters and are suitable for use with internal and edge columns. The main advantage of shearheads is that they avoid the use of individual shear links, which are especially difficult to fix in slabs. The early reactions from designers and contractors appear to be favourable and it likely that future revisions of BS 8110 will include design recommendations for the use of shearheads in flat slab construction.

With the emphasis on reducing production costs, the use of composite construction has become very popular over recent years. This form of construction involves casting a concrete slab on profiled steel sheets (*Fig. 3.44*). The steel sheets serve two purposes. Firstly, they provide permanent shuttering to the wet concrete and can support the construction loads. Secondly, the sheets combine structurally with the hardened concrete and replace the tensile reinforcement in the slab. Mesh reinforcement may be included in the top face of the slab to control cracking of the concrete. Composite construction is primarily used for floors but occasionally for roofs. This flooring system is suitable for most building works but is especially suitable for multi-storey steel framework commercial developments. Guidance on the design of floors with profiled steel sheeting can be found in Part 4 of BS 5950.

Slabs that are supported along two opposite edges are said to be one-way spanning (*Fig. 3.45*). A slab is said to be two-way spanning when it is supported on all four sides and the ratio of the length of the longer side to the length of the shorter side is equal to 2 or less (*Fig. 3.46*).

This book considers only the design of one-way and two-way spanning simply supported solid slabs supporting uniformly distributed loads. The reader is referred to more specialized books on this subject for guidance on the design of the other slab types discussed above.

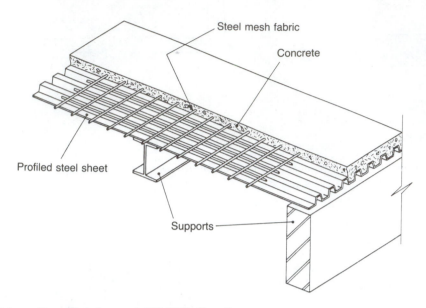

Fig. 3.44 *Typical floor with profiled sheet steel (BS 5950: Part 4).*

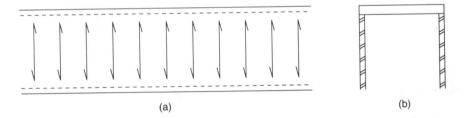

Fig. 3.45 *One-way spanning solid slab: (a) plan; (b) section.*

3.10.2 DESIGN OF ONE-WAY SPANNING SOLID SLABS

The general procedure to be adopted for slab design is as follows:

1. Determine a suitable depth of slab.
2. Calculate main and secondary reinforcement areas.
3. Check critical shear stresses.
4. Check detailing requirements.

3.10.2.1 Depth of slab (clause 3.5.7, BS 8110)

Solid slabs are designed as if they consist of a series of beams of 1 m width. The effective span of the slab is taken as the smaller of (a) the distance between centres of bearings, A, or (b) the clear distance between supports, D, plus the effective depth d of the slab (*Fig. 3.47*).

The deflection requirements for slabs, which are the same as those for beams, will often control the depth of slab needed. The minimum effective depth of slab, d_{min}, can be calculated using

$$d_{min} = \frac{\text{span}}{\text{basic ratio} \times \text{modification factor}} \quad (3.20)$$

The basic (span/depth) ratios are given in *Table 3.11*. The modification factor is a function of the amount of reinforcement in the slab which is itself a function of the depth of the slab. Therefore, in order to make a first estimate of the effective depth, d_{min}, of the slab, a value of (say) 1.4 is assumed for the modification factor. The main steel areas can then be calculated (*section 3.10.2.2*), and used to determine the actual value of the modification factor. If the assumed value is slightly greater than the actual value, the depth of the slab will satisfy the deflection requirements in BS 8110. Otherwise, the calculation

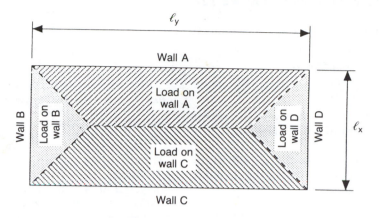

Fig. 3.46 *Plan of two-way spanning slab. l_x = length of shorter side, l_y = length of longer side. Provided $l_y/l_x \leqslant 2$ slab will span in two directions as indicated.*

must be repeated using a revised value of the modification factor.

3.10.2.2 Steel areas (clause 3.5.4, BS 8110)

The overall depth of slab, h, is determined by adding allowances for cover (*Table 3.5*) and half the (assumed) main steel bar diameter to the effective depth. The self-weight of the slab together with the dead and live loads are used to calculate the design moment, M.

The ultimate moment of resistance of the slab, M_u, is calculated using equation 3.11, developed in *section 3.9.1.1*, namely

$$M_u = 0.156 f_{cu} b d^2$$

If $M_u \geqslant M$, which is the usual condition for slabs, compression reinforcement will not be required and the area of tensile reinforcement, A_s, is determined using equation 3.12 developed in *section 3.9.1.1*, namely

$$A_s = \frac{M}{0.87 f_y z}$$

where $z = d [0.5 + \sqrt{(0.25 - K/0.9)}]$ and $K = \dfrac{M}{f_{cu} b d^2}$

Secondary or distribution steel is required in the transverse direction and this is usually based on the

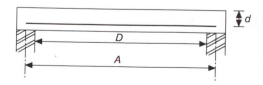

Fig. 3.47 *Effective span of simply supported slab.*

minimum percentages of reinforcement ($A_{s\,min}$) given in Table 3.27 of BS 8110:

$A_{s\,min} = 0.24\% A_c$ when $f_y = 250$ N mm^{-2}

$A_{s\,min} = 0.13\% A_c$ when $f_y = 460$ N mm^{-2}

where A_c is the total area of concrete.

3.10.2.3 Shear (clause 3.5.5, BS 8110)

Shear resistance is generally not a problem in solid slabs subject to uniformly distributed loads and, in any case, shear reinforcement should not be provided in slabs less than 200 mm deep. As discussed for beams in *section 3.9.1.3*, the design shear stress, v, is calculated from $v = \dfrac{V}{bd}$. The ultimate shear resistance, v_c, is determined using *Table 3.8*. If $v < v_c$, no shear reinforcement is required. Where $v > v_c$, the form and area of shear reinforcement in solid slabs should be provided in accordance with the requirements contained in Table 3.16.

Table 3.16 Form and area of shear reinforcement in solid slabs (Table 3.17, BS 8110)

Values of v (N mm^{-2})	Area of shear reinforcement to be provided
$v < v_c$	None
$v_c < v < (v_c + 0.4)$	Nominal links in areas where $v > v_c$ $A_{sv} \geqslant 0.4 b s_v/0.87 f_{yv}$
$(v_c + 0.4) < v < 0.8\sqrt{f_{cu}}$ or 5 N mm^{-2}	Design links $A_{sv} \geqslant b s_v (v - v_c)/0.87 f_{yv}$

3.10.2.4 Reinforcement details (clause 3.12, BS 8110)

For reasons of durability the code specifies limits in respect of:

1. minimum reinforcement areas
2. spacing of reinforcement
3. maximum crack widths.

These are outlined below together with the simplified rules for curtailment of reinforcement.

Reinforcement areas (clause 3.12.5, BS 8110). The area of tension reinforcement, A_s, should not be less than the following limits:

$$A_s \not< 0.24\% \; A_c \quad \text{when} f_y = 250 \text{ N mm}^{-2}$$
$$A_s \not< 0.13\% \; A_c \quad \text{when} f_y = 460 \text{ N mm}^{-2}$$

where A_c is the total area of concrete.

Spacing of reinforcement (clause 3.12.11.2.7, BS 8110). The clear distance between tension bars, s_b, should lie within the following limits: $h_{agg} + 5$ mm or bar diameter $\leq s_b \leq 3d$ or 750 mm whichever is the lesser where h_{agg} is the maximum aggregate size. (See also below, section on crack widths.)

Crack width (clause 3.12.11.2.7, BS 8110). Unless the actual crack widths have been checked by direct calculation, the following rules will ensure that crack widths will not generally exceed 0.3 mm. This

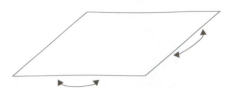

Fig. 3.49 *Bending of two-way spanning slabs.*

limiting crack width is based on considerations of appearance and durability:

1. No further check is required on bar spacing if either:
   (a) $f_y = 250$ N mm^{-2} and slab depth ≤ 250 mm, or
   (b) $f_y = 460$ N mm^{-2} and slab depth ≤ 200 mm, or
   (c) the reinforcement percentage ($100A_s / bd$) < 0.3%.
2. Where none of conditions (a), (b) or (c) apply and the percentage of reinforcement in the slab exceeds 1%, then the maximum clear distance between bars (s_{max}) given in Table 3.30 of BS 8110 should be used, namely

$$s_{max} \leq 300 \text{ mm} \quad \text{when} f_y = 250 \text{ N mm}^{-2}$$
$$s_{max} \leq 160 \text{ mm} \quad \text{when} f_y = 460 \text{ N mm}^{-2}$$

Curtailment of reinforcement (clause 3.12.10.3, BS 8110). Simplified rules for the curtailment of reinforcement are given in clause 3.12.10.3 of BS 8110. These are shown diagrammatically in *Fig. 3.48* for simply supported and cantilever solid slabs.

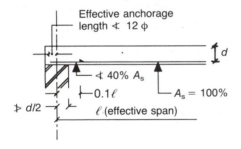
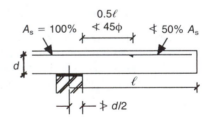

Fig. 3.48 *Simplified rules for curtailment of bars in slabs (Fig. 3.25(b), BS 8110): (a) simply supported ends; (b) cantilever slab.*

Example 3.7 Design of one-way spanning concrete floor

A reinforced concrete floor subject to an imposed load of 3.5 kN m^{-2} spans between brick walls as shown below. Design the floor for mild exposure conditions assuming the following material strengths:

$$f_{cu} = 35 \text{ N mm}^{-2} \qquad f_y = 460 \text{ N mm}^{-2}$$

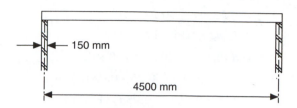

150 mm

4500 mm

DEPTH OF SLAB AND MAIN STEEL AREA

Overall depth of slab, h

Minimum effective depth, d_{min}, is

$$d_{min} = \frac{span}{basic\ ratio \times modification\ factor}$$
$$= \frac{4500}{20 \times (say)\ 1.4} = 161mm$$

Hence, assume effective depth of slab (d) = 165 mm. Assume diameter of main steel (Φ) = 10 mm. From *Table 3.5*, cover to all steel for mild conditions of exposure (c) = 20 mm.

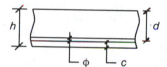

Overall depth of slab (h) = $d + \Phi/2 + c$
$\qquad\qquad = 165 + 10/2 + 20 = 190$ mm

Loading

Dead Self-weight of slab (g_k) = $0.19 \times$ 24 kN m^{-3} = 4.56 kN m^{-2}

Imposed Total imposed load (q_k) = 3.5 kN m^{-2}

Ultimate load For 1 m width of slab total ultimate load is

$(1.4g_k + 1.6q_k)$ width of slab $\times$ span = $(1.4 \times 4.56 + 1.6 \times 3.5)\ 1 \times 4.5 = 53.93$ kN

Design moment

$$M = \frac{Wl}{8} = \frac{53.93 \times 4.5}{8} = 30.34 \text{ kN m}$$

Ultimate moment

$$M_u = 0.156 f_{cu} bd^2 = 0.156 \times 35 \times 10^3 \times 165^2$$
$$= 148.6 \times 10^6 \text{ N mm} = 148.6 \text{ kN m}$$

Since $M_u > M$, no compression reinforcement is required.

Main steel

$$K = \frac{M}{f_{cu}bd^2} = \frac{30.34 \times 10^6}{35 \times 10^3 \times 165^2} = 0.0318$$

$$z = d\,[0.5 + \sqrt{(0.25 - K/0.9)}]$$

$$= 165[0.5 + \sqrt{(0.25 - 0.0318/0.9)}]$$

$$= 165 \times 0.963 \not> 0.95d = 157 \text{ mm}$$

Hence $z = 157$ mm.

$$A_s = \frac{M}{0.87f_{y}z} = \frac{30.34 \times 10^6}{0.87 \times 460 \times 157} = 483 \text{ mm}^2 \text{ m}^{-1} \text{ width of slab}$$

For detailing purposes this area of steel has to be transposed into bars of a given diameter and spacing using steel area tables. Thus from *Table 3.17*, provide 10 mm diameter bars spaced at 150 mm, i.e. T10 at 150 mm centres ($A_s = 523 \text{ mm}^2 \text{ m}^{-1}$).

Table 3.17 Cross-sectional area per metre width for various bar spacings (mm^2)

Bar size (mm)	Spacing of bars								
	50	75	100	125	150	175	200	250	300
6	566	377	283	226	189	162	142	113	94.3
8	1010	671	503	402	335	287	252	201	168
10	1570	1050	785	628	523	449	393	314	262
12	2260	1510	1130	905	754	646	566	452	377
16	4020	2680	2010	1610	1340	1150	1010	804	670
20	6280	4190	3140	2510	2090	1800	1570	1260	1050
25	9820	6550	4910	3930	3270	2810	2450	1960	1640
32	16100	10700	8040	6430	5360	4600	4020	3220	2680
40	25100	16800	12600	10100	8380	7180	6280	5030	4190

Actual modification factor

The actual value of the modification can now be calculated using equations 7 and 8 given in *Table 3.13* (*section 3.9.1.4*):

Design service stress $(f_s) = \dfrac{5f_{y}A_{s,req}}{8A_{s,prov}}$ (equation 8, *Table 3.13*)

$$= \frac{5 \times 460 \times 483}{8 \times 523} = 266 \text{ N mm}^{-2}$$

Modification factor $= 0.55 + \dfrac{(477 - f_s)}{120\,(0.9 + M/bd^2)}$ (equation 7, *Table 3.13*)

$$= 0.55 + \frac{(477 - 266)}{120\,(0.9 + 30.34 \times 10^6/10^3 \times 165^2)}$$

$$= 1.42$$

Hence,

New $d_{min} = \dfrac{4500}{20 \times 1.42} = 159$ mm $<$ assumed $d = 165$ mm

Minimum area of reinforcement, $A_{s\,min}$, is equal to

$$A_{s\,min} = 0.13\%\ bh = 0.13\%\ 10^3 \times 190 = 247\ \text{mm}^2\ \text{m}^{-1} < A_s$$

Therefore take $d = 165$ mm and provide T10 at 150 mm centres as main steel.

SECONDARY STEEL
Based on minimum steel area $= 247\ \text{mm}^2\ \text{m}^{-1}$. Hence from *Table 3.17*, provide T8 at 200 mm centres ($A_s = 252\ \text{mm}^2\ \text{m}^{-1}$).

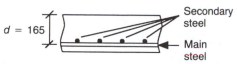

SHEAR REINFORCEMENT

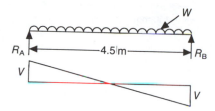

Design shear stress, v
Since slab is symmetrically loaded

$$R_A = R_B = W/2 = 27\ \text{kN}$$

Ultimate shear force (V) = 27 kN and design shear stress, v, is $v = \dfrac{V}{bd} = \dfrac{27 \times 10^3}{10^3 \times 165} = 0.16\ \text{N mm}^{-2}$

Design concrete shear stress, v_c
Assuming that 50% of main steel is curtailed at the supports, $A_s = \dfrac{523}{2} = 262\ \text{mm}^2\ \text{m}^{-1}$.

$$\frac{100 A_s}{bd} = \frac{100 \times 262}{10^3 \times 165} = 0.16$$

From *Table 3.8*, designing concrete shear stress for grade 25 concrete is 0.42 N mm^{-2}.

Hence

$$v_c = \left(\frac{35}{25}\right)^{1/3} \times 0.42 = 0.47\ \text{N mm}^{-2}$$

From *Table 3.16*, since $v < v_c$, no shear reinforcement is required.

REINFORCEMENT DETAILS
The sketch below shows the main reinforcement requirements for the slab. For reasons of buildability the actual reinforcement details may well be slightly different.

Check spacing between bars
Maximum spacing between bars should not exceed the lesser of $3d$ (= 495 mm) or 750 mm. Actual spacing = 150 mm main steel and 200 mm secondary steel. OK.

Maximum crack width
Since the slab depth does not exceed 200 mm, the above spacing between bars will automatically ensure that the maximum permissible crack width of 0.3 mm will not be exceeded.

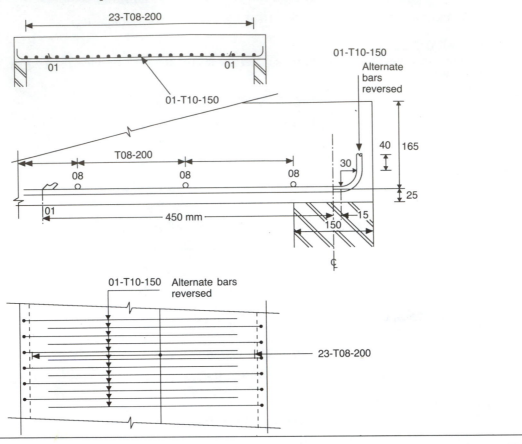

Example 3.8 Analysis of a one-way spanning concrete floor

A concrete floor reinforced with 10 mm diameter mild steel bars ($f_y = 250$ N mm^{-2}) at 125 mm centres spans between brick walls as shown in *Fig. 3.50*. Calculate the maximum uniformly distributed imposed load the floor can carry.

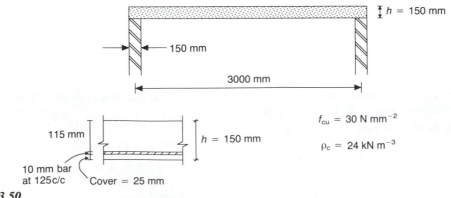

$f_{cu} = 30$ N mm^{-2}

$\rho_c = 24$ kN m^{-3}

Fig. 3.50

EFFECTIVE SPAN

Effective depth of slab, d, is $d = h$ - cover - $\Phi/2$ = 150 - 25 - 10/2 = 120 mm

Effective span is the lesser of (a) centre-to-centre (c/c) distance between bearings = 3000 mm and (b) clear distance between supports plus effective depth = 2850 + 120 = 2970 mm. Hence effective span = 2970 mm.

MOMENT CAPACITY, M

Assume $z = 0.95\ d = 0.95 \times 120 = 114$ mm.

$$A_s = \frac{M}{0.87 f_y z}$$

Hence

$$M = A_s.0.87 f_y z = 628 \times 0.87 \times 250 \times 114$$
$$= 15.5 \times 10^6 \text{ N mm} = 15.5 \text{ kN m/width of slab}$$

MAXIMUM UNIFORMLY DISTRIBUTED IMPOSED LOAD (q_k)

Loading

Dead load Self-weight of slab (g_k) = 0.15×24 kN m^{-3} = 3.6 kN m^{-2}

Ultimate load Total ultimate load (W) = $(1.4 g_k + 1.6 q_k)$ span
$$= (1.4 \times 3.6 + 1.6 q_k)\ 2.970$$

Imposed load Design moment (M) = $\dfrac{Wl}{8}$

From above, $M = 15.5$ kN m = $(5.04 + 1.6 q_k)\dfrac{2.970^2}{8}$

Rearranging gives

$$q_k = \frac{15.5 \times 8/2.970^2 - 5.04}{1.6} = 5.6 \text{ kN m}^{-2}$$

Lever arm (z)

Check that assumed value of z is correct, i.e $z = 0.95d$.

$$K = \frac{M}{f_{cu} b d^2} = \frac{15.5 \times 10^6}{30 \times 10^3 \times 120^2} = 0.0359$$

$$z = d\,[0.5 + \sqrt{(0.25 - K/0.9)}] \ngtr 0.95d$$
$$= d[0.5 + \sqrt{(0.25 - 0.0359/0.9)}] = 0.958d$$

Hence, assumed value of z is correct and the maximum uniformly distributed load that the floor can carry is 5.6 kN m^{-2}.

3.10.3 DESIGN OF TWO-WAY SPANNING SOLID SLABS

The design of two-way spanning slabs supporting uniformly distributed loads and simply supported on all sides is generally similar to that outlined above for one-way spanning slabs. The extra complication arises from the fact that it is rather difficult to analyse the bending moments in these plate-like structures (*Fig. 3.49*). However, Table 3.14 of BS 8110 (not reproduced here) gives coefficients which can be used for this purpose.

3.11 Foundations

Foundations are required primarily to carry the dead and imposed loads due to the structure's floors, beams, walls, columns, etc. and transmit and distribute the loads safely to the ground (*Fig. 3.51*). The purpose of distributing the load is to avoid the safe bearing capacity of the soil being exceeded otherwise excessive settlement of the structure may occur.

Foundation failure can produce catastrophic effects on the overall stability of a structure so that it may

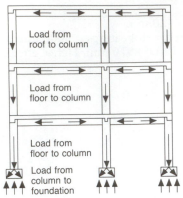

Fig. 3.51 *Loading on foundations.*

slide or even overturn (*Fig. 3.52*). Such failures are likely to have tremendous financial and safety implications. It is essential, therefore, that much attention is paid to the design of this element of a structure.

3.11.1 FOUNDATION TYPES

There are many types of foundations which are commonly used, namely strip, pad and raft. The foundations may bear directly on the ground or be supported on piles. The choice of foundation type will largely depend upon (1) ground conditions (i.e. strength and type of soil) and (2) type of structure (i.e. layout and level of loading).

Pad footings are usually square or rectangular slabs and used to support a single column (*Fig. 3.53*). The pad may be constructed using mass concrete or reinforced concrete depending on the relative size of the loading. Detailed design of pad footings is discussed in *section 3.11.2.1*.

Continuous strip footings are used to support loadbearing walls or under a line of closely spaced columns (*Fig. 3.54*). Strip footings are designed as pad footings in the transverse direction and in the longitudinal direction as an inverted continuous beam subject to the ground bearing pressure.

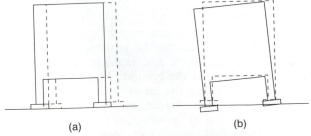

Fig. 3.52 *Foundation failures: (a) sliding failure; (b) overturning failure.*

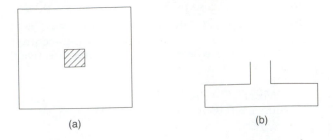

Fig. 3.53 *Pad footing: (a) plan; (b) elevation.*

Where the ground conditions are relatively poor, a raft foundation may be necessary in order to distribute the loads from the walls and columns over a large area. In its simplest form this may consist of a flat slab, possibly strengthened by upstand or downstand beams for the more heavily loaded structures (*Fig. 3.55*).

Where the ground conditions are so poor that it is not practical to use strip or pad footings but better-quality soil is present at lower depths, the use of pile foundations should be considered (*Fig. 3.56*). The piles may be made of precast reinforced concrete, prestressed concrete or *in situ* reinforced concrete. Loads are transmitted from the piles to the surrounding strata by end bearing and/or friction. End-bearing piles derive most of their carrying capacity from the penetration resistance of the soil at

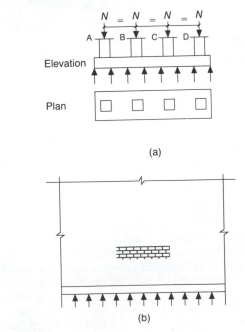

Fig. 3.54 *Strip footings: (a) footing supporting columns; (b) footing supporting wall.*

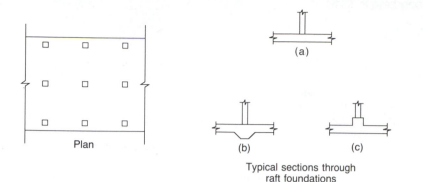

Plan

(a)

(b)   (c)

Typical sections through
raft foundations

Fig. 3.55 *Raft foundations. Typical sections through raft foundations: (a) flat slab; (b) flat slab and downstand; (c) flat slab and upstand.*

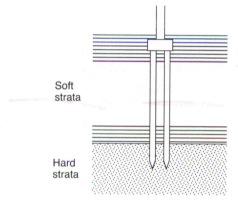

Soft
strata

Hard
strata

Fig. 3.56 *Piled foundation.*

the toe of the pile, while friction piles rely on the adhesion or friction between the sides of the pile and the soil.

3.11.2 FOUNDATION DESIGN
Foundation failure may arise as a result of (a) allowable bearing capacity of the soil being exceeded, or (b) bending and/or shear failure of the base. The first condition allows the plan area of the base to be calculated, being equal to the design load divided by the bearing capacity of the soil, i.e.

$$\text{Ground pressure} = \frac{\text{design load}}{\text{plan area}} < \frac{\text{bearing capacity}}{\text{of soil}} \quad (3.21)$$

Since the settlement of the structure occurs during its working life, the design loadings to be considered when calculating the size of the base should be taken as those for the serviceability limit state (i.e. $1.0G_k + 1.0Q_k$). The calculations to determine the thickness of the base and the bending and shear reinforcement should, however, be based on ultimate loads (i.e. $1.4G_k + 1.6Q_k$). The design of a pad footing only will be considered here. The reader

is referred to more specialized books on this subject for the design of the other foundation types discussed above. However, it should be borne in mind that in most cases the design process will be similar to that for beams and slabs.

3.11.2.1 Pad footing
The general procedure to be adopted for the design of pad footings is as follows:

1. Calculate the plan area of the footing using serviceability loads.
2. Determine the reinforcement areas required for bending using ultimate loads (*Fig. 3.57*).
3. Check for punching, face and transverse shear failures (*Fig. 3.58*).

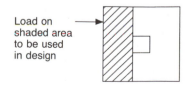

Load on
shaded area
to be used
in design

Fig. 3.57 *Critical section for bending.*

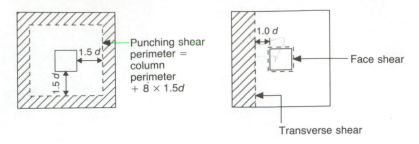

Fig. 3.58 *Critical sections for shear. (Load on shaded areas to be used in design.)*

Example 3.9 Design of a pad footing

A 400 mm square column carries a dead load (G_k) of 900 kN and imposed load (Q_k) of 300 kN (*Fig. 3.59*). The safe bearing capacity of the soil is 150 kN m^{-2}. Design a square pad footing to resist the loads assuming the following material strengths:

$$f_{cu} = 35 \text{ N mm}^{-2} \qquad f_y = 460 \text{ N mm}^{-2}$$

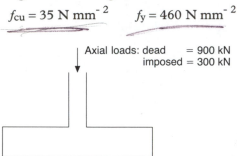

Fig. 3.59

PLAN AREA OF BASE

Loading

Dead load Assume a footing weight of 130 kN.

$$\text{Total dead load } (G_k) = 900 + 130 = 1030 \text{ kN}$$

Serviceability load Design axial load (N) = $1.0G_k + 1.0Q_k$
$$= 1.0 \times 1030 + 1.0 \times 300 = 1330 \text{ kN}$$

Plan area

$$\text{Plan area of base} = \frac{N}{\text{bearing capacity of soil}}$$

$$= \frac{1330}{150} = 8.87 \text{m}^2$$

Hence provide a 3 m square base (plan area = 9 m^2).

Self-weight of footing
Assume the overall depth of footing (h) = 600 mm.

$$\text{Self-weight of footing} = \text{area} \times h \times \text{density of concrete}$$
$$= 9 \times 0.6 \times 24 = 129.6 \text{ kN} < \text{assumed (130 kN)}$$

BENDING REINFORCEMENT

Design moment, M

Total ultimate load $(W) = 1.4G_k + 1.6Q_k$
$$= 1.4 \times 900 + 1.6 \times 300 = 1740 \text{ kN}$$

$$\text{Earth pressure } (p_s) = \frac{W}{\text{plan area of base}}$$

$$= \frac{1740}{9} = 193 \text{ kN m}^{-2}$$

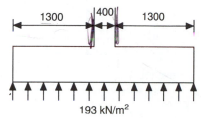

Maximum design moment occurring at face of column (M) is $M = \dfrac{p_s l^2}{2} = \dfrac{193 \times 1.300^2}{2}$

$$= 163 \text{ kN m m}^{-1} \text{ width of slab}$$

Ultimate moment

Effective depth Base to be cast against blinding, hence cover (c) to reinforcement = 40 mm (see clause 3.3.1.4, BS 8110). Assume 20 mm diameter (Φ) bars will be needed as bending reinforcement in both directions.

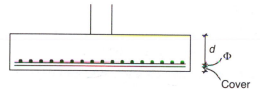

Hence, effective depth, d, is

$$d = h - c - \Phi = 600 - 40 - 20 = 540 \text{ mm}$$

Ultimate moment $M_u = 0.156 f_{cu} b d^2 = 0.156 \times 35 \times 10^3 \times 540^2$
$$= 1592 \times 10^6 \text{ N mm} = 1592 \text{ kN m}$$
Since $M_u > M$ no compression reinforcement is required.

Main steel

$$K = \frac{M}{f_{cu} b d^2} = \frac{163 \times 10^6}{35 \times 10^3 \times 540^2} = 0.016$$

$$z = d[0.5 + \sqrt{(0.25 - K/0.9)}] \not> 0.95d$$
$$= d[0.5 + \sqrt{(0.25 - 0.016/0.9)}]$$
$$= 0.981d \not> 0.95d = 0.95 \times 540 = 513 \text{ mm}$$

$$A_s = \frac{M}{0.87 f_y z} = \frac{163 \times 10^6}{0.87 \times 460 \times 513} = 794 \text{ mm}^2 \text{m}^{-1}$$

Minimum steel area is

$$0.13bh/100 = 780 \text{ mm}^2 \text{ m}^{-1} < A_s \quad \text{OK}$$

Hence from *Table 3.17* provide T20 at 300 mm centres ($A_s = 1050 \text{ mm}^2 \text{ m}^{-1}$) distributed uniformly across the full width of the footing parallel to the x–x and y–y axes (see clause 3.11.3.2, BS 8110).

CRITICAL SHEAR STRESSES
Punching shear

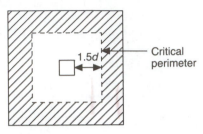

Critical perimeter, p_{crit}, is

$$= \text{column perimeter} + 8 \times 1.5d$$
$$= 4 \times 400 + 8 \times 1.5 \times 540 = 8080 \text{ mm}$$

Area within perimeter is

$$(400 + 3d)^2 = (400 + 3 \times 540)^2 = 4.08 \times 10^6 \text{ mm}^2$$

Ultimate punching force, V, is

$$V = \text{load on shaded area} = 193 \times (9 - 4.08) = 950 \text{ kN}$$

Design punching shear stress, v, is

$$v = \frac{V}{p_{crit}d} = \frac{950 \times 10^3}{8080 \times 540} = 0.22 \text{ N mm}^{-2}$$

$$\frac{100A_s}{bd} = \frac{100 \times 1050}{10^3 \times 540} = 0.19$$

Hence from *Table 3.8*, design concrete shear stress, v_c, is

$$v_c = 0.36 \times (35/25)^{1/3} = 0.40 \text{ N mm}^{-2}$$

Since $v_c > v$, punching failure is unlikely and a 600 mm depth of slab is acceptable.

Maximum shear
Maximum shear stress (v_{max}) occurs at face of column. Hence

$$v_{max} = \frac{W}{\text{column perimeter} \times d} = \frac{1740 \times 10^3}{(4 \times 400) \times 540}$$

$$= 2.0 \text{ N mm}^{-2} < \text{permissible} (= 0.8 \sqrt{35} = 4.73 \text{ N mm}^{-2})$$

Face shear

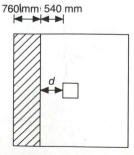

760 mm 540 mm

Ultimate shear force (V) = load on shaded area
$$= p_s \times \text{area} = 193 \,(3 \times 0.760) = 440 \text{ kN}$$
Design shear stress, v, is

$$v = \frac{V}{bd} = \frac{440 \times 10^3}{3 \times 10^3 \times 540} = 0.27 \text{ N mm}^{-2} < v_c$$

Hence no shear reinforcement is required.

REINFORCEMENT DETAILS
The sketch below shows the main reinforcement requirements for the pad footing.

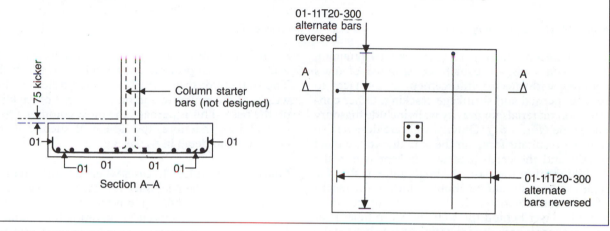

3.12   Retaining walls

Sometimes it is necessary to maintain a difference in ground levels between adjacent areas of land. Typical examples of this include road and railway embankments, reservoirs and ramps. A common solution to this problem is to build a natural slope between the two levels. However, this is not always possible because slopes are very demanding of space. An alternative solution which allows an immediate change in ground levels to be effected is to build a vertical wall which is capable of resisting the pressure of the retained material. These structures are commonly referred to as retaining walls (*Fig. 3.60*). Retaining walls are important elements in many building and civil engineering projects and the purpose of the following sections is to briefly describe the various types of retaining walls available and outline the design procedure associated with one common type, namely cantilever retaining walls.

3.12.1 TYPES OF RETAINING WALLS
Retaining walls are designed on the basis that they are capable of withstanding all horizontal pressures and forces without undue movement arising from deflection, sliding or overturning. There are two main categories of concrete retaining walls: (a) gravity walls and (b) flexible walls.

3.12.1.1 Gravity walls
Where walls up to 2 m in height are required, it is

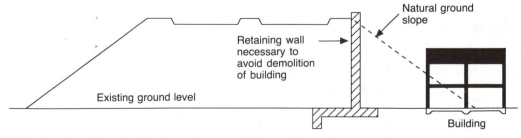

Fig. 3.60 *Section through road embankment incorporating a retaining wall.*

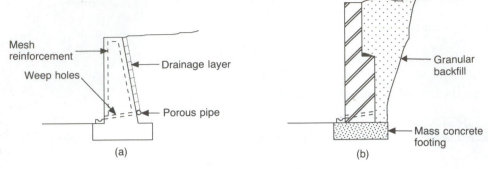

Fig. 3.61 *Gravity retaining walls: (a) mass concrete wall; (b) masonry wall.*

generally economical to choose a gravity retaining wall. Such walls are usually constructed of mass concrete with mesh reinforcement in the faces to reduce thermal and shrinkage cracking. Other construction materials for gravity walls include masonry and stone (*Fig. 3.61*). Gravity walls are designed so that the resultant force on the wall due to the dead weight and the earth pressures is kept within the middle third of the base. A rough guide is that the width of base should be about a third of the height of the retained material. It is usual to include a granular layer behind the wall and weep holes near the base to minimize hydrostatic pressure behind the wall. Gravity walls rely on their dead weight for strength and stability. The main advantages with this type of wall are simplicity of construction and ease of maintenance.

3.12.1.2 Flexible walls
These retaining walls may be of two basic types, namely (i) cantilever and (ii) counterfort.

Cantilever walls. Cantilevered reinforced concrete retaining walls are suitable for heights up to about 7 m. They generally consist of a uniform vertical stem monolithic with a base slab (*Fig. 3.62*).

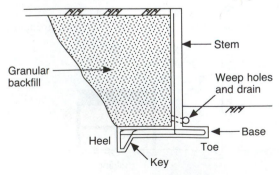

Fig. 3.62 *Cantilever wall.*

A key is sometimes incorporated at the base of the wall in order to prevent sliding failure of the wall. The stability of these structures often relies on the weight of the structure and on the weight of backfill on the base. This is perhaps the most common type of wall and, therefore, the design of such walls is considered in detail in *section 3.12.2*.

Counterfort walls. In cases where a higher stem is needed, it may be necessary to design the wall as a counterfort (*Fig. 3.63*). Counterfort walls can be designed as continuous slabs spanning horizontally between vertical supports known as counterforts. The counterforts are designed as cantilevers and will normally have a triangular or trapezoidal shape. As with cantilever walls, stability is provided by the weight of the structure and earth on the base.

3.12.2 DESIGN OF CANTILEVER WALLS
Generally, the design process involves ensuring that the wall will not fail either due to foundation failure or structural failure of the stem or base. Specifically, the design procedure involves the following steps:

1. Calculate the soil pressures on the wall.
2. Check the stability of the wall.
3. Design the bending reinforcement.

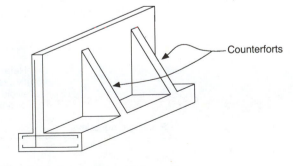

Fig. 3.63 *Counterfort retaining wall.*

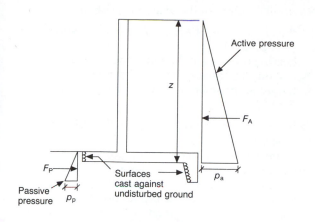

Fig. 3.64 *Active and passive pressure acting on a wall.*

As in the case of slabs, the design of retaining walls is usually based on a 1 m width of section.

3.12.2.1 Soil pressures

The method most commonly used for determining the soil pressures is based on Rankin's formula, which may be considered to be conservative but is straightforward to apply. The pressure on the wall resulting from the retained fill has a destabilizing effect on the wall and is normally termed active pressure (*Fig. 3.64*). The earth in front of the wall resists the destabilizing forces and is termed passive pressure.

The active pressure (p_a) is given by

$$p_a = \rho \, k_a z \qquad (3.22)$$

where ρ is the unit weight of soil (kN m^{-3}), k_a the coefficient of active pressure, and z the height of retained fill. Here k_a is calculated using

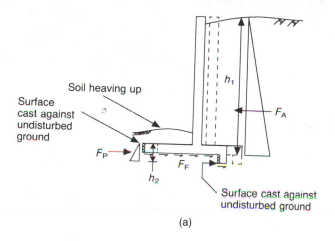

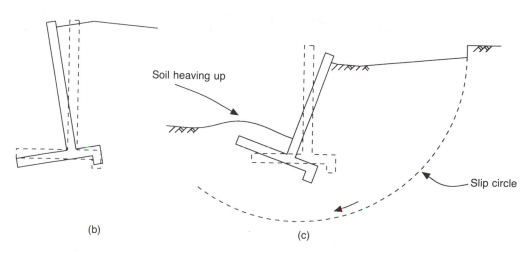

Fig. 3.65 *Modes of failure: (a) sliding; (b) overturning; (c) slip circle.*

Table 3.18 Values of ρ and ϕ

Material	ρ $(kN\ m^{-3})$	ϕ
Sandy gravel	17–22	35–40°
Loose sand	15–16	30–35°
Crushed rock	12–22	35–40°
Ashes	9–10	35–40°
Broken brick	15–16	35–40°

$$k_a = \frac{1 - \sin\phi}{1 + \sin\phi} \qquad (3.23)$$

where ϕ is the internal angle of friction of retained soil. Typical values of ρ and ϕ for various soil types are shown in Table 3.18.

The passive pressure (p_p) is given by

$$p_p = \rho\,k_p z \qquad (3.24)$$

where ρ is the unit weight of soil, z the height of retained fill and k_p the coefficient of passive pressure and is calculated using

$$k_p = \frac{1 + \sin\phi}{1 - \sin\phi} = \frac{1}{k_a} \qquad (3.25)$$

3.12.2.2 Stability

Foundation failure of the wall may arise due to (a) sliding or (b) rotation. Sliding failure will occur if the active pressure force (F_A) exceeds the passive pressure force (F_P) plus the friction force (F_F) arising at the base/ground interface (*Fig. 3.65(a)*) where

$$F_A = 0.5 p_a h_1 \qquad (3.26)$$

$$F_P = 0.5 p_p h_2 \qquad (3.27)$$

$$F_F = \mu\,W_t \qquad (3.28)$$

The factor of safety against this type of failure occurring is normally taken to be at least 1.5:

$$\frac{F_F + F_P}{F_A} \not< 1.5 \qquad (3.29)$$

Rotational failure of the wall may arise due to:

1. the overturning effect of the active pressure force (*Fig. 3.65(b)*);
2. bearing pressure of the soil being exceeded which will display similar characteristics to (1); or
3. failure of the soil mass surrounding the wall (*Fig. 3.65(c)*).

Failure of the soil mass (type (3)) will not be considered here but the reader is referred to any standard book on soil mechanics for an explanation of the procedure to be followed to avoid such failures.

Failure type (1) can be checked by taking moments about the toe of the foundation (A) as shown in *Figure 3.66* and ensuring that the ratio of sum of restoring moments (ΣM_{res}) and sum of overturning moments (ΣM_{over}) exceeds 2.0, i.e.

$$\frac{\Sigma M_{res}}{\Sigma M_{over}} \not< 2.0 \qquad (3.30)$$

Failure type (2) can be avoided by ensuring that the ground pressure does not exceed the allowable bearing pressure for the soil. The ground pressure under the toe (p_{toe}) and the heel (p_{heel}) of the base can be calculated using

$$p_{toe} = \frac{N}{D} + \frac{6M}{D^2} \qquad (3.31)$$

$$p_{heel} = \frac{N}{D} - \frac{6M}{D^2} \qquad (3.32)$$

provided that the load eccentricity lies within the middle third of the base, that is

$$M/N \leqslant D/6 \qquad (3.33)$$

where M is the moment about centre line of base, N the total vertical load (W_t) and D the width of base.

3.12.2.3 Reinforcement areas

Structural failure of the wall may arise if the base and

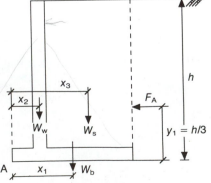

$$\frac{W_b x_1 + W_w x_2 + W_s x_3}{F_A y_1} > 2.0$$

Fig. 3.66

stem are unable to resist the vertical and horizontal forces due to the retained soil. The areas of steel reinforcement needed in the wall can be calculated by considering the ultimate limit states of bending and shear. As was pointed out at the beginning of this chapter, cantilever retaining walls can be regarded for design purposes as three cantilever beams (*Fig. 3.2*) and thus the equations developed in *section 3.9* for rectangular beams can be used here.

The areas of main reinforcement (A_s) can be calculated using

$$A_s = \frac{M}{0.87f_y z}$$

where M is the design moment, f_y the reinforcement

grade, z the lever arm which is $= d[0.5 + \sqrt{(0.25 - K/0.9)}]$ and K the coefficient which is equal to

$$\frac{M}{f_{cu}bd^2}$$

The area of distribution steel is based on the minimum steel area (A_s) given in *Table 3.27* of BS 8110, i.e.

$$A_s = \frac{0.24A_c}{100} \text{ when } f_y = 250 \text{ N mm}^{-2}$$

$$A_s = \frac{0.13A_c}{100} \text{ when } f_y = 460 \text{ N mm}^{-2}$$

where A_c is the gross cross-sectional area of concrete.

Example 3.10 Design of a cantilever retaining wall

The cantilever retaining wall shown below is backfilled with granular material having a unit weight, ρ, of 19 kN m^{-3} and an internal angle of friction, ϕ, of 30°. Assuming that the allowable bearing pressure of the soil is 120 kN m^{-2}, the coefficient of friction is 0.4 and the unit weight of reinforced concrete is 24 kN m^{-3}

1. Determine the factors of safety against sliding and overturning.
2. Calculate ground bearing pressures.
3. Design the wall and base reinforcement assuming $f_{cu} = 35$ N mm^{-2} and $f_y = 460$ N mm^{-2} and the cover is 35 mm.

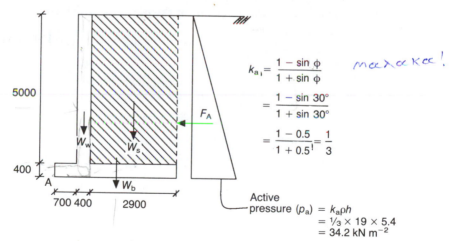

$$k_{a_1} = \frac{1 - \sin\phi}{1 + \sin\phi}$$

$$= \frac{1 - \sin 30°}{1 + \sin 30°}$$

$$= \frac{1 - 0.5}{1 + 0.5} = \frac{1}{3}$$

Active pressure (p_a) $= k_a \rho h$
$= \frac{1}{3} \times 19 \times 5.4$
$= 34.2$ kN m^{-2}

SLIDING

Consider the forces acting on a 1 m length of wall. Horizontal force on wall due to backfill, F_A, is

$$F_A = 0.5 p_a h = 0.5 \times 34.2 \times 5.4 = 92.34 \text{ kN}$$

and

Weight of wall (W_w) $= 0.4 \times 5 \times 24 = 48.0$ kN
Weight of base (W_b) $= 0.4 \times 4 \times 24 = 38.4$ kN
Weight of soil (W_s) $= 2.9 \times 5 \times 19 = 275.5$ kN
Total vertical force (W_t) $= 361.9$ kN

Friction force, F_F is $F_F = \mu W_t = 0.4 \times 361.9 = 144.76$ kN

Assume passive pressure force (F_p) = 0. Hence factor of safety against sliding is

$$\frac{144.76}{92.34} = 1.56 > 1.5 \quad \text{OK}$$

OVERTURNING

Taking moments about point A (see above), sum of overturning moments (M_{over}) is

$$\frac{F_A \times 5.4}{3} = \frac{92.34 \times 5.4}{3} = 166.2 \text{ kN m}$$

Sum of restoring moments (M_{res}) is

$$M_{res} = W_w \times 0.9 + W_b \times 2 + W_s \times 2.55$$
$$= 48 \times 0.9 + 38.4 \times 2 + 275.5 \times 2.55 = 822.5 \text{ kN m}$$

Factor of safety against overturning is

$$\frac{822.5}{166.2} = 4.9 > 2.0 \quad \text{OK}$$

GROUND BEARING PRESSURE

Moment about centre line of base (M) is

$$M = \frac{F_A \times 5.4}{3} + W_w \times 1.1 - W_s \times 0.55$$
$$= \frac{92.34 \times 5.4}{3} + 48 \times 1.1 - 275.5 \times 0.55 = 67.5 \text{ kN m}$$
$$N = 361.9 \text{ kN}$$
$$\frac{M}{N} = \frac{67.5}{361.9} = 0.187 \text{ m} < \frac{D}{6} = \frac{4}{6} = 0.666 \text{ m}$$

Therefore, the maximum ground pressure occurs at the toe, p_{toe}, which is given by

$$p_{toe} = \frac{361.9}{4} + \frac{6 \times 67.5}{4^2} = 116 \text{ kN m}^{-2} < \text{allowable (120 kN m}^{-2})$$

Ground pressure at the heel, p_{heel}, is

$$p_{heel} = \frac{361.9}{4} - \frac{6 \times 67.5}{4^2} = 65 \text{kN m}^{-2}$$

BENDING REINFORCEMENT

Wall

Height of stem of wall, h_s = 5 m. Horizontal force on stem due to backfill, F_s, is

$$F_s = 0.5 k_a \rho \, h_s^2 = 0.5 \times \frac{1}{3} \times 19 \times 5^2 = 79.17 \text{ kN m}^{-1} \text{ width}$$

Design moment at base of stem, M is

$$M = \frac{\gamma_f F_s h_s}{3} = 1.4 \times 79.17 \times \frac{5}{3} = 184.7 \text{ kN m}$$

Effective depth

Assume diameter of main steel (ϕ) = 20 mm.
Hence effective depth, d, is d = 400 - cover - ϕ/2 = 400 - 35 - 20/2 = 355 mm

Ultimate moment of resistance

$M_u = 0.156 f_{cu} b d^2 = 0.156 \times 35 \times 10^3 \times 355^2 \times 10^{-6} = 688$ kN m

Since $M_u > M$, no compression reinforcement is required.

Steel area

$$K = \frac{M}{f_{cu}bd^2} = \frac{184.7 \times 10^6}{35 \times 10^3 \times 355^2} = 0.0419$$

$$z = d[0.5 + \sqrt{(0.25 - K/0.9)}]$$

$$= 355[0.5 + \sqrt{(0.25 - 0.0419/0.9)}] = 337 \text{ mm}$$

$$A_{st} = \frac{M}{0.8f_{yz}} = \frac{184.7 \times 10^6}{0.87 \times 460 \times 337} = 1369 \text{ mm}^2 \text{ m}^{-1}$$

Hence from *Table 3.17*, provide T20 at 200 mm centres ($A_s = 1570$ mm^2 m^{-1}) in near face (NF) of wall. Steel is also required in the front face (FF) of wall in order to prevent excessive cracking. This is based on the minimum steel area, i.e.

$$\frac{0.13bh}{100} = \frac{0.13 \times 10^3 \times 400}{100} = 520 \text{ mm}^2 \text{ m}^{-1}$$

Hence provide T12 at 200 centres ($A_s = 566$ mm^2).

Base

Heel

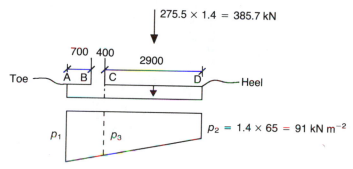

$$p_3 = 91 + \frac{2.9 (162.4 - 91)}{4} = 142.8 \text{ kN m}^{-2}$$

Design moment at point C, M_c, is

$$\frac{385.7 \times 2.9}{2} + \frac{2.9 \times 38.4 \times 1.4 \times 1.45}{4} - \frac{91 \times 2.9^2}{2}$$

$$- \frac{51.8 \times 2.9 \times 2.9}{2 \times 3} = 160.5 \text{ kN m}$$

$$K = \frac{160.5 \times 10^6}{35 \times 10^3 \times 355^2} = 0.036$$

$$z = 355[0.5 + \sqrt{(0.25 - 0.036/0.9)}] \ngtr 0.95d = 337 \text{ mm}$$

$$A_s = \frac{160.5 \times 10^6}{0.87 \times 460 \times 337} = 1190 \text{ mm}^2 \text{ m}^{-1}$$

Hence from *Table 3.17*, provide T20 at 200 mm centres ($A_s = 1570$ mm^2 m^{-1}) in top face (T) of base.

Toe. Design moment at point B, M_B, is given by

$$M_B \approx \frac{162.4 \times 0.7^2}{2} - \frac{0.7 \times 38.4 \times 1.4 \times 0.7}{4 \times 2} = 36.5 \text{ kN m}$$

$$A_s = \frac{36.5 \times 1190}{160.5} = 270 \text{ mm}^2 \text{ m}^{-1} \ll \text{minimum steel area} = 520 \text{ mm}^2 \text{ m}^{-1}$$

Hence provide T12 at 200 mm centres ($A_s = 566$ mm^2 m^{-1}), in bottom face (B) of base and as distribution steel in base and stem of wall.

REINFORCEMENT DETAILS

The sketch below shows the main reinforcement requirements for the retaining wall. For reasons of buildability the actual reinforcement details may well be slightly different.

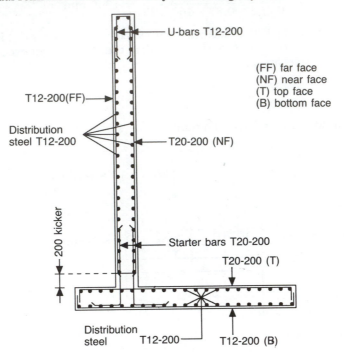

3.13 Design of short-braced columns

The function of columns in a structure is to act as vertical supports to suspended members such as beams and roofs and to transmit the loads from these members down to the foundations (*Fig. 3.67*). Columns are primarily compression members although they may also have to resist bending moments transmitted by beams.

Columns may be classified as short or slender, braced or unbraced, depending on various dimensional and structural factors which will be discussed below. However, due to limitations of space, the study will be restricted to the design of the most common type of column found in building structures, namely short-braced columns.

3.13.1 COLUMN SECTIONS

Some common column cross-sections are shown in *Fig. 3.68*. Any section can be used, however, provided

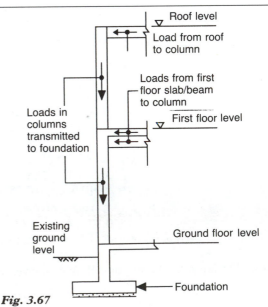

Fig. 3.67

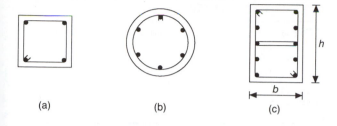

(a)   (b)   (c)

Fig. 3.68 *Column cross-sections.*

that the greatest overall cross-sectional dimension does not exceed four times its smaller dimension (i.e. $h \leqslant 4b$, *Fig. 3.68(c)*). With sections where $h > 4b$ the member should be regarded as a wall for design purposes (clause 1.2.4.1, BS 8110).

3.13.2. SHORT AND SLENDER COLUMNS (CLAUSE 3.8.1.3, BS 8110)

Columns may fail due to one of three mechanisms:

1. compression failure of the concrete/steel reinforcement (*Fig. 3.69*);
2. buckling (*Fig. 3.70*);
3. combination of buckling and compression failure.

For any given cross-section, failure mode (1) is most likely to occur with columns which are short and stocky, while failure mode (2) is probable with columns which are long and slender. It is important, therefore, to be able to distinguish between columns which are short and those which are slender since the failure mode and hence the design procedures for the two column types are likely to be different. Clause 3.8.1.3 of BS 8110 classifies a column as being short if

$$\frac{l_{ex}}{h} < 15 \quad \text{and} \quad \frac{l_{ey}}{b} < 15$$

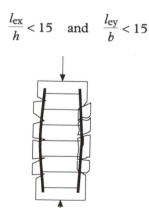

Fig. 3.69

Fig. 3.70

where l_{ex} is the effective height of the column in respect of the major axis (i.e. x–x axis), l_{ey} the effective height of the column in respect of the minor axis, (i.e. y–y axis) b the width of the column cross-section and h the depth of the column cross-section. It should be noted that the above definition applies only to columns which are braced, rather than unbraced. This distinction is discussed more fully in *section 3.13.3*. Effective heights of columns is covered in *section 3.13.4*.

3.13.3 BRACED AND UNBRACED COLUMNS (CLAUSE 3.8.1.5, BS 8110)

A column may be considered braced if the lateral loads, due to wind for example, are resisted by shear walls or some other form of bracing rather than by the column. For example, all the columns in the reinforced concrete frame shown in *Fig. 3.71* are braced in the y direction. A column may be considered to be unbraced if the lateral loads are resisted by the sway action of the column. For example, all the columns shown in *Fig. 3.71* are unbraced in the x direction. Depending upon the layout of the structure, it is possible for the columns to be braced or unbraced in both directions as shown in *Figs 3.72* and *3.73* respectively.

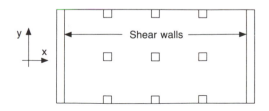

Fig. 3.71 *Columns braced in y direction and unbraced in the x direction*

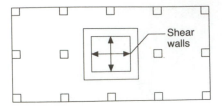

Fig. 3.72 *Columns braced in both directions.*

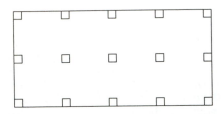

Fig. 3.73 *Columns unbraced in both directions.*

Table 3.19 Values of β for braced columns (Table 3.21, BS 8110)

End condition at top	End condition at bottom		
	1	2	3
1	0.75	0.80	0.90
2	0.80	0.85	0.95
3	0.90	0.95	1.00

3.13.4 EFFECTIVE HEIGHT (CLAUSE 3.8.1.6, BS 8110)

The effective height (l_e) of a column in a given plane is obtained by multiplying the clear height between lateral restraints (l_0) by a coefficient (β) which is a function of the fixity at the column ends and is obtained from *Table 3.19*.

$$l_e = \beta\, l_0 \tag{3.34}$$

End condition 1 signifies that the column end is fully restrained. End condition 2 signifies that the column end is partially restrained and end condition 3 signifies that the column end is nominally restrained. In practice it is possible to infer the degree of restraint at the column ends simply by reference to the diagrams shown in *Fig. 3.74*.

End condition 1

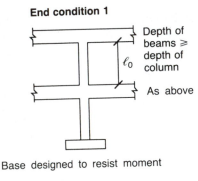

Depth of beams ⩾ depth of column

l_0

As above

Base designed to resist moment

End condition 2

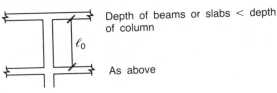

Depth of beams or slabs < depth of column

l_0

As above

End condition 3

l_0

Nominal restraint between beams and column, e.g. beam designed and detailed as if simply supported

Base not designed to resist moment

Fig. 3.74 *Column end restraint conditions.*

Example 3.11 Classification of a concrete column

Determine if the column shown in *Fig. 3.75* is short.

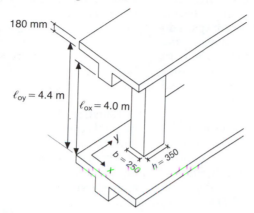

Fig. 3.75

For bending in the y direction: end condition at top of column = 1, end condition at bottom of column = 1. Hence from *Table 3.19*, $\beta_x = 0.75$.

$$\frac{l_{ex}}{h} = \frac{\beta_x l_{ox}}{h} = \frac{0.75 \times 4000}{350} = 8.57$$

For bending in the x direction: end condition at top of column = 2, end condition at bottom of column = 2. Hence from *Table 3.19*, $\beta_y = 0.85$.

$$\frac{l_{ey}}{b} = \frac{\beta_y l_{oy}}{b} = \frac{0.85 \times 4400}{250} = 14.96$$

Since both l_{ex}/h and l_{ey}/b are both less than 15, the column is short.

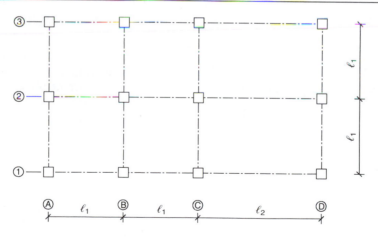

Fig. 3.76 *Floor plan.*

3.13.5 SHORT-BRACED COLUMN DESIGN

For design purposes, BS 8110 divides short-braced columns into three categories. These are:

1. columns resisting axial loads only;
2. columns supporting an approximately symmetrical arrangement of beams;
3. columns resisting axial loads and uniaxial or biaxial bending.

Referring to the floor plan shown in *Fig. 3.76*, it can be seen that column B2 supports beams which are equal in length and symmetrically arranged. Provided the floor is uniformly loaded, column B2 will resist an axial load only and is an example of category 1. Column C2 supports a symmetrical arrangement of beams but which are unequal in length. Column C2 will, therfore, resist an axial load and moment. However, provided that (a) the loadings on the beams are uniformly distributed, and (b) the beam spans do not differ by more than 15% of the longer, the moment will be small. As such, column C2 belongs to category 2 and it can safely be designed by considering the axial load only but using slightly reduced values of the design stresses in the concrete and steel reinforcement (*section 3.13.5.2*).

Columns belong to category 3 if conditions (a) and (b) are not satisfied. The moment here becomes significant and the column may be required to resist an axial load and uni-axial bending, e.g. columns A2, B1, B3, C1, C3 and D2, or an axial load and bi-axial bending, e.g. columns A1, A3, D1 and D3.

The design procedures associated with each of these categories are discussed in the subsections below.

3.13.5.1 Axially loaded columns (clause 3.8.4.3, BS 8110)

Consider a column having a net cross-sectional area of concrete A_c and a total area of longitudinal reinforcement A_{sc} (*Fig. 3.77*). As discussed in *section 3.7*, the design stresses for concrete and steel in compression are $0.67f_{cu}/1.5$ and $f_y/1.15$ respectively, i.e.

$$\text{Concrete design stress} = \frac{0.67f_{cu}}{1.5} = 0.45f_{cu}$$

$$\text{Reinforcement design stress} = \frac{f_y}{1.15} = 0.87f_y$$

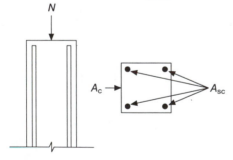

Fig. 3.77

Both the concrete and reinforcement assist in carrying the load. Thus, the ultimate load N which can be supported by the column is the sum of the loads carried by the concrete (F_c) and the reinforcement (F_s), i.e.

$$N = F_c + F_s$$
$$F_c = \text{stress} \times \text{area} = 0.45f_{cu}A_c$$
$$F_s = \text{stress} \times \text{area} = 0.87f_yA_{sc}$$

Hence

$$N = 0.45f_{cu}A_c + 0.87f_yA_{sc} \qquad (3.35)$$

Equation 3.35 assumes that the load is applied perfectly axially to the column. However, in practice, perfect conditions never exist. To allow for a small eccentricity BS 8110 reduces the design stresses in equation 3.35 by about 10%, giving the following expression:

$$N = 0.4f_{cu}A_c + 0.75f_yA_{sc} \qquad (3.36)$$

This is equation 38 in BS 8110 which can be used to design short-braced axially loaded columns.

3.13.5.2 Columns supporting an approximately symmetrical arrangement of beams (clause 3.8.4.4, BS 8110)

Where the column is subject to an axial load and 'small' moment (*section 3.13.5*), the latter is taken into account simply by decreasing the design stresses for concrete and steel in equation 3.36 by around 10%, giving the following expression for the load-carrying capacity of the column:

$$N = 0.35f_{cu}A_c + 0.67f_yA_{sc} \qquad (3.37)$$

This is equation 39 in BS 8110 and can be used to design columns supporting an approximately symmetrical arrangement of beams provided (a) the loadings on the beams are uniformly distributed, and (b) the beam spans do not differ by more than 15% of the longer. Equations 3.36 and 3.37 are not only used to determine the load-carrying capacities of short-braced columns predominantly supporting axial loads but can also be used for initial sizing of these elements as illustrated in Example 3.12.

Example 3.12 Sizing a concrete column

A short-braced column in which $f_{cu} = 30$ N mm^{-2} and $f_y = 460$ N mm^{-2} is required to support an ultimate axial load of 2000 kN. Determine a suitable section for the column assuming that the area of longitudinal steel, A_{sc} is of the order of 3% of the gross cross-sectional area of column, A_{col}.

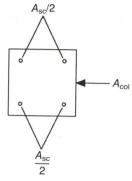

Since the column is axially loaded use equation 3.36:

$$N = 0.4f_{cu}A_c + 0.75f_yA_{sc}$$

$$2000 \times 10^3 = 0.4 \times 30 \left(A_{col} - \frac{3A_{col}}{100}\right) + 0.75 \times 460 \times \frac{3A_{col}}{100}$$

$$A_{col} = 90\ 950\ \text{mm}^2$$

Assuming that the column is square,

$$b = h \approx \sqrt{90\ 950} = 302\ \text{mm}$$

Hence a 300 mm square column constructed of grade 30 concrete would be suitable.

3.13.5.3 Columns resisting axial forces and moments

The area of longitudinal steel for columns resisting axial forces and uniaxial or biaxial bending is normally calculated using the design charts given in Part 3 of BS 8110. These charts are only available for columns having a rectangular cross-section and a symmetrical arrangement of reinforcement. A typical column design chart is shown in *Fig. 3.78*. Each chart is particular for a selected

1. characteristic strength of concrete f_{cu}
2. characteristic strength of reinforcement f_y
3. d/h ratio.

Design charts have been prepared for concrete grades 25, 30, 35, 40, 45 and 50 and reinforcement grade 460. For specified concrete and steel strengths there are a series of charts for different d/h ratios in the range 0.75–0.95 in 0.05 increments.

Uniaxial bending. With columns which are subject to an axial load (N) plus uniaxial bending moment (M) the procedure simply involves plotting the N/bh and M/bd^2 ratios on the appropriate chart and reading off the corresponding area of reinforcement as a percentage of the gross-sectional area of concrete ($100A_{sc}/bh$) (Example 3.15). Where the actual d/h ratio for the section being designed lies between two charts, both charts may be read and the

longitudinal steel area found by linear interpolation.

Biaxial bending (clause 3.8.4.5, BS 8110). Where the column is subject to biaxial bending, the problem is reduced to one of uniaxial bending simply by increasing the moment about one of the axes using the procedure outlined below. Referring to *Fig. 3.79*, if $M_x/M_y \geqslant h'/b'$ the enhanced design moment, about the x – x axis, M_x, is

$$M_x' = M_x + \frac{\beta h'}{b'} M_y$$

$M_x/M_y < h'/b'$; the enhanced design moment about the y – y axis, M_y', is

$$M_y' = M_y + \frac{\beta b'}{h'} M_x$$

where b' and h' are the effective depths (*Fig. 3.79*) and β is the enhancement coefficient for biaxial bending obtained from *Table 3.20*. The area of longitudinal steel can then be determined using the ultimate axial load (N) and enhanced moment (M_x' or M_y') in the same way as that described for uniaxial bending.

3.13.6 REINFORCEMENT DETAILS

In order to ensure structural stability, durability and practicability of construction BS 8110 lays down

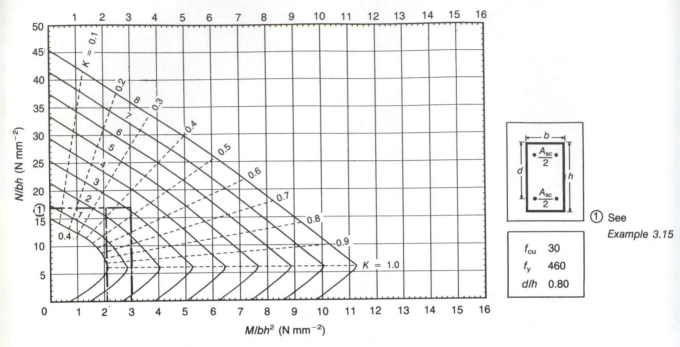

Fig. 3.78 *Column design chart (chart no.27, BS 8110).*

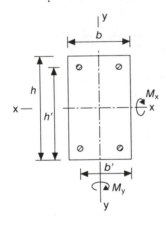

Fig. 3.79

various rules governing the minimum size, amount and spacing of (i) longitudinal reinforcement and (ii) links. These are discussed in the subsections below.

3.13.6.1 Longitudinal reinforcement

Size and minimum number of bars. Columns with rectangular cross-sections should be reinforced with a minimum of four longitudinal bars; columns with circular cross-sections should be reinforced with a minimum of six longitudinal bars. Each of the bars should not be less than 12mm in diameter.

Reinforcement areas (clause 3.12.5, BS 8110). The code recommends that for columns with a gross cross-sectional area A_{col}, the area of longitudinal reinforcement (A_{sc}) should lie within the following limits:

$$0.4\% \ A_{col} < A_{sc} < 6\% \ A_{col}$$

in a vertically cast column and

$$0.4\% \ A_{col} \leqslant A_{sc} \leqslant 8\% \ A_{col}$$

Table 3.20 Values of the coefficient β (Table 3.24, BS 8110)

$\dfrac{N}{bhf_{cu}}$	0	0.1	0.2	0.3	0.4	0.5	⩾0.6
β	1.00	0.88	0.77	0.65	0.53	0.42	0.30

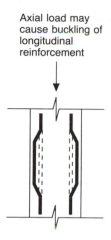

Fig. 3.80

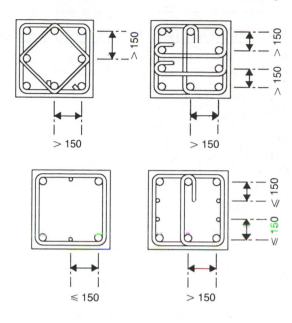

Fig. 3.82 *Arrangement of links in columns.*

in a horizontally cast column. At laps the maximum area of longitudinal reinforcement may be increased to 10% of the gross cross-sectional area of the column for both types of columns.

Spacing of reinforcement. The minimum distance between adjacent bars should not be less then the diameter of the bars or $h_{agg} + 5$ mm, where h_{agg} is the maximum size of the coarse aggregate. The code does not specify any limitations with regards to the maximum spacing of bars, but for practical reasons it should not normally exceed 250 mm.

3.13.6.2 Links (clause 3.12.7, BS 8110)

The axial loading on the column may cause buckling of the longitudinal reinforcement and subsequent

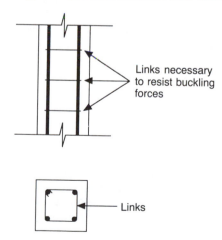

Links necessary to resist buckling forces

Links

Fig. 3.81

cracking and spalling of the adjacent concrete cover (*Fig. 3.80*). In order to prevent such a situation from occurring, the longitudinal steel is normally laterally restrained at regular intervals by links passing round the bars (*Fig. 3.81*).

Size and spacing of links. Links should be at least one-quarter of the size of the largest longitudinal bar or 6 mm, whichever is the greater. However, in practice 6 mm bars may not be freely available and a minimum bar size of 8 mm is preferable. Links should be provided at a maximum spacing of 12 times the size of the smallest longitudinal bar or the smallest cross-sectional dimension of the column. The latter condition is not mentioned in BS 8110 but was referred to in CP 114 and is still widely observed in order to reduce the risk of diagonal shear failure of columns.

Arrangement of links. The code further requires that links should be so arranged that every corner and alternate bar in an outer layer of reinforcement is supported by a link passing around the bar and having an included angle of not more than 135°. All other bars should be within 150 mm of a restrained bar (*Fig. 3.82*).

Example 3.13 Axially loaded column

Design the longitudinal steel and links for a 350 mm square, short braced column which supports the following axial loads:

$$G_k = 1000 \text{ kN} \quad Q_k = 1000 \text{ kN}$$

Assume $f_{cu} = 40 \text{ N mm}^{-2}$, $f_{yv} = 250 \text{ N mm}^{-2}$ and $f_y = 460 \text{ N mm}^{-2}$.

LONGITUDINAL STEEL

Since column is axially loaded, use equation 3.36, i.e.

$$N = 0.4f_{cu}A_c + 0.75f_yA_{sc}.$$

Total ultimate load, N, is

$$N = 1.4G_k + 1.6Q_k = 1.4 \times 1000 + 1.6 \times 1000 = 3000 \text{ kN}$$

Substituting this into the above equation for N gives

$$3000 \times 10^3 = 0.4 \times 40 \times (350^2 - A_{sc}) + 0.75 \times 460A_{sc}$$

$$A_{sc} = 3161 \text{ mm}^2$$

Hence from *Table 3.7*, provide 4T32 ($A_{sc} = 3220 \text{ mm}^2$).

LINKS

The diameter of the links is one-quarter times the diameter of the largest longitudinal bar, that is, $\frac{1}{4} \times 32 = 8$ mm, but not less than 8 mm diameter. The spacing of the links is the lesser of (a) 12 times the diameter of the smallest longitudinal bar, that is, $12 \times 32 = 384$ mm, or (b) the smallest cross-sectional dimension of the column (= 350 mm). Hence provide R8 links at 350 mm centres.

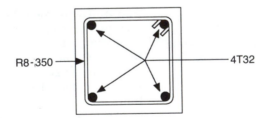

R8-350   4T32

Example 3.14 Column supporting an approximately symmetrical arrangement of beam

An internal column in a braced two-storey building supporting an approximately symmetrical arrangement of beams (350 mm wide × 600 mm deep) results in characteristic dead and imposed loads each of 1000 kN being applied to the column. The column is 350 mm square and has a clear height of 4.5 m as shown in Figure 3.83. Design the longitudinal reinforcement and links assuming

$$f_{cu} = 40 \text{ N mm}^{-2} \qquad f_y = 460 \text{ N mm}^{-2} \qquad f_{yv} = 250 \text{ N mm}^{-2}$$

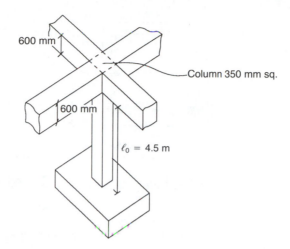

Fig. 3.83

CHECK IF COLUMN IS SHORT

Effective height

Depth of beams (600 mm) > depth of column (350 mm), therefore end condition at top of column = 1. Assuming that the pad footing is not designed to resist any moment, end condition at bottom of column = 3. Therefore from *Table 3.19*, $\beta = 0.9$.

$$l_{ex} = l_{ey} = \beta l_0 = 0.9 \times 4500 = 4050 \text{ mm}$$

Short or slender

$$\frac{l_{ex}}{h} = \frac{l_{ey}}{b} = \frac{4050}{350} = 11.6$$

Since both ratios are less than 15, the column is short.

LONGITUDINAL STEEL

Since column supports an approximately symmetrical arrangement of beams use equation 3.37, i.e.

$$N = 0.35 f_{cu} A_c + 0.67 f_y A_{sc}$$

Total axial load, N, is

$$N = 1.4 G_k + 1.6 Q_k = 1.4 \times 1000 + 1.6 \times 1000 = 3000 \text{ kN}$$

Substituting this into the above equation for N

$$3000 \times 10^3 = 0.35 \times 40(350^2 - A_{sc}) + 0.67 \times 460 A_{sc} = 4368 \text{ mm}^2$$

Hence from *Table 3.7*, provide 4T32 and 4T25 ($A_{sc} = 3220 + 1960 = 5180 \text{ mm}^2$).

LINKS

The diameter of the links is one-quarter times the diameter of the largest longitudinal bar, that is, $\frac{1}{4} \times 32 = 8$ mm, but not less than 8 mm diameter. The spacing of the links is the lesser of (a) 12 times the diameter of the smallest longitudinal bar, that is, $12 \times 25 = 300$ mm, or (b) the smallest cross-sectional dimension of the column (= 350 mm). Provide R8 links at 300 mm centres.

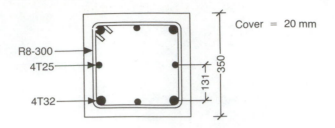

Cover = 20 mm

R8-300
4T25

350
131

4T32

Example 3.15 Column resisting an axial load and bending

Design the longitudinal and shear reinforcement for a 275 mm square, short-braced column which supports either (a) an ultimate axial load of 1280 kN and a moment of 62.5 kN m about the x–x axis or (b) an ultimate axial load of 1280 kN and bending moments of 35 kN m about the x–x axis and 25 kN m about the y–y axis.

Assume $f_{cu} = 30$ N mm^{-2}, $f_y = 460$ N mm^{-2} and cover to all reinforcement is 35 mm.

LOAD CASE (A)

Longitudinal steel

$$\frac{M}{bh^2} = \frac{62.5 \times 10^6}{275 \times 275^2} = 3$$

$$\frac{N}{bh} = \frac{1280 \times 10^3}{275 \times 275} = 17$$

Assume diameter of longitudinal bars (ϕ) = 20 mm, diameter of links (ϕ') = 8 mm.

$$d = h - \text{cover} - \phi' - \phi/2$$
$$= 275 - 35 - 8 - 20/2 = 222 \text{ mm}$$
$$d/h = 222/275 = 0.8$$

From *Fig. 3.78* (chart no. 27, BS 8110: Part 3) $100A_{sc}/bh = 3.2$, $A_{sc} = 3.2 \times 275 \times 275/100 = 2420$ mm^2

Provide 8T20 ($A_{sc} = 2510$ mm^2, *Table 3.7*).

Links

The diameter of the links is one-quarter times the diameter of the largest longitudinal bar, that is, $\frac{1}{4} \times 20 = 5$ mm, but not less than 8 mm diameter. The spacing of the links is the lesser of (a) 12 times the diameter of the smallest longitudinal bar, that is, $12 \times 20 = 240$ mm, or (b) the smallest cross-sectional dimension of the column (= 275 mm). Provide R8 links at 240 mm centres.

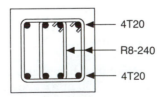

4T20
R8-240
4T20

LOAD CASE (B)

Longitudinal steel

Assume diameter of longitudinal bars (ϕ) = 25 mm, diameter of link (ϕ') = 8 mm.

$$b' = h' = h - \phi/2 - \phi' - \text{cover}$$

$$= 275 - 25/2 - 8 - 35 = 220 \text{ mm}$$

$$M_x/M_y = 35/25 = 1.4 > h'/b' = 1$$

$$\frac{N}{bhf_{cu}} = \frac{1280 \times 10^3}{275 \times 275 \times 30} = 0.56$$

Hence, $\beta = 0.35$ (*Table 3.20*).
Enhanced design moment about the x–x axis, (M_x'), is

$$M_x' = M_x + \frac{\beta h' M_y}{b'}$$

$$= 35 + \frac{0.35 \times 220 \times 25}{220} = 43.8 \text{ kN m}$$

$$\frac{M_x'}{bh^2} = \frac{43.8 \times 10^6}{275 \times 275^2} = 2.1$$

$$\frac{N}{bh} = \frac{1280 \times 10^3}{275 \times 275} = 17.0$$

$$d/h = 220/275 = 0.8$$

From *Fig. 3.78* (chart no. 27, BS 8110: Part 3) $100A_{sc}/bh = 2.4$,

$$A_{sc} = 2.4 \times 275 \times 275/100 = 1815 \text{ mm}^2$$

Provide 4T25 ($A_{sc} = 1960 \text{ mm}^2$, *Table 3.7*).

Links

The diameter of the links is one-quarter times the diameter of the largest longitudinal bar, that is, ¼ × 25 ≈ 6 mm but not less than 8 mm. The spacing of the links is the lesser of (a) 12 times the diameter of the smallest longitudinal bar, that is, 12 × 25 = 300 mm, or (b) the smallest cross-sectional dimension of the column (= 275 mm). Provide R8 links at 275 mm centres.

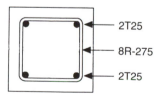

3.14 Summary

This chapter has considered the design of a number of reinforced concrete elements to BS 8110: *Structural Use of Concrete*. The elements considered primarily either resist bending or support compressive loading. The latter category includes columns subject to axial loading and axial loading and small moments. Elements which fall into the first category include beams, slabs, retaining walls and foundations. They are generally designed for the ultimate limit states of bending and shear and checked for the serviceability limit states of deflection and cracking. The notable exception to this is slabs where the deflection requirements will usually be used to determine the depth of the member. Furthermore, where slabs resist point loads, e.g. pad foundations or in flat slab construction, checks on punching shear will also be required.

Questions

1. (a) Derive from first principles the following equation for the ultimate moment of resistance (M_u) of a singly reinforced section assuming a rectangular stress-block distribution

$$M_u = 0.156 f_{cu} b d^2$$

   (b) Draw a fully annotated sketch of the assumed stress block used in the derivation and list any assumptions/limitations of the formula.

2. (a) Explain the difference between M and M_u.

   (b) Design the bending and shear reinforcement for the beam shown below using the following information:
   $f_{cu} = 25$ N mm^{-2} $f_y = 460$ N mm^{-2}
   $f_{yv} = 250$ N mm^{-2} $b = 300$ mm
   span/depth ratio = 12.

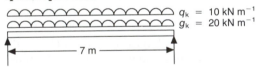

$q_k = 10$ kN m^{-1}
$g_k = 20$ kN m^{-1}

7 m

3. (a) Discuss how shear failure can arise in reinforced concrete members and how such failures can be avoided.

   (b) Describe the measures proposed in BS 8110 regarding design for durability of structures.

4. Redesign the slab in *Example 3.7* assuming that the characteristic strength of the reinforcement is 250 N mm^{-2}. Comment on your results.

5. (a) Explain the difference between columns which are short and slender and those which are braced and unbraced.

   (b) Calculate the ultimate axial load capacity of a short-braced column supporting an approximately symmetrical arrangement of beams assuming that it is 500 mm square and is reinforced with 8T 20 mm diameter bars. Assume that $f_{cu} = 40$ N mm^{-2}, $f_y = 460$ N mm^{-2} and the concrete cover is 20 mm. Design the shear reinforcement for the column.

6. (a) A braced column which is 300 mm square is restrained such that it has an effective height of 4.5 m. Classify the column as short or slender.

   (b) The column supports: (1) an ultimate axial load of 500 kN and a bending moment of 200 kN m, or (2) an ultimate axial load of 800 kN and bending moments of 75 kN m about the x axis and 50 kN m about the y axis. The characteristic strengths are $f_{cu} = 40$ N mm^{-2} for the concrete and $f_y = 460$ N mm^{-2} for the reinforcement. The cover is 35 mm. Using design charts, calculate the longitudinal steel required for both loading cases.

7. An internal column in a multi-storey building supporting an approximately symmetrical arrangement of beams carries an ultimate load of 2000 kN. The storey height is 5.2 m and the effective height factor is 0.85, $f_{cu} = 30$ N mm^{-2} and $f_y = 460$ N mm^{-2}. Assuming that the column is square, short and braced, calculate:
   1. a suitable cross-section for the column;
   2. the area of the longitudinal reinforcement;
   3. the size and spacing of the links.
   Sketch the reinforcement detail in cross-section.

Chapter 4

Design in structural steelwork to BS 5950

This chapter is concerned with the detailed design of structural steelwork elements to BS 5950: Part 1. The chapter describes with the aid of a number of fully worked examples the design of the following elements: beams and joists, struts and columns and bolted and welded connections. The design of beams and joists covers those members which are fully laterally restrained and those which may be subject to lateral torsional buckling. The section on columns includes the design of cased columns and column baseplates.

4.1 Introduction

Several codes of practice are currently in use in the UK for design in structural steelwork. For buildings the older, but still quite popular, permissible stress code BS 449 has now been largely superseded by BS 5950, which is a limit state code introduced in 1985. For steel bridges the limit state code BS 5400: Part 3 is used. Since the primary aim of this chapter is to give guidance on the design of structural steelwork elements, this is best illustrated by considering the contents of BS 5950.

BS 5950 is divided into the following nine parts:

Part 1: *Code of Practice for Design in Simple and Continuous Construction: Hot Rolled Sections.*
Part 2: *Specification for Materials, Fabrication and Erection: Hot Rolled Sections.*
Part 3: *Design in Composite Construction.*
Part 4: *Code of Practice for Design of Floors with Profiled Steel Sheeting.*
Part 5: *Code of Practice for Design of Cold Formed Sections.*
Part 6: *Code of Practice for Design in Light Gauge Sheeting, Decking and Cladding.*
Part 7: *Specification for Materials and Workmanship: Cold Formed Sections.*
Part 8: *Code of Practice for Fire-resistant Design.*
Part 9: *Code of Practice for Stressed Skin Design.*

Part 1 covers most of the material required for everyday design. Since most of the chapter is concerned with the contents of Part 1, it should be assumed that all references to BS 5950 refer to Part 1 exclusively. Part 4 deals with the design of floors made from profiled steel sheet and concrete. It was briefly referred to in *Chapter 3* in the context of slab design (*section 3.10*). The remaining parts of the code generally either relate to specification or to specialist types of construction, which are not relevant to the present discussion and will not be mentioned further.

4.2 Iron and steel: manufacture and properties

Iron has been produced for several thousands of years but it was not until the eighteenth century that it began to be used as a structural material. The first cast-iron bridge by Darby was built in 1779 at Coalbrookdale in Shropshire. Fifty years later wrought iron chains were used in Thomas Telford's Menai Straits suspension bridge. However, it was not until 1898 that the first steel-framed building was constructed. Nowadays most construction is carried out using steel which combines the best properties of cast and wrought iron.

Cast iron is basically remelted pig iron which has been cast into definite useful shapes. Charcoal was used for smelting iron but in 1740 Abraham Darby found a method of converting coal into coke which revolutionized the iron-making process. A latter development to this process was the use of limestone to combine with the impurities in the ore and coke and form a slag which could be run off independently of the iron. Nevertheless, the pig iron contained many impurities which made the material brittle and weak in tension.

In 1784, a method known as 'puddling' was

developed which could be used to convert pig iron into a tough and ductile metal known as wrought iron. Essentially this process removed many of the impurities present in pig iron, e.g. carbon, manganese, silicon and phosphorus, by oxidation. However, it was difficult to remove the wrought iron from the furnace without contaminating it with the slag. Thus although wrought iron was tough and ductile, it was rather soft and was eventually superseded by steel.

Bessemer discovered a way of making steel from iron. This involved filling a large vessel, lined with calcium silicate bricks, with molten pig iron which was then blown from the bottom to remove the impurities. The vessel is termed a converter and the process is known as acid Bessemer. However, this converter was unable to remove the phosphorus from the molten iron which resulted in the metal being weak and brittle and completely unmalleable. Later in 1878 Gilchrist Thomas proposed an alternative lining for the converter in which dolomite was used instead of silica and which overcame the problems associated with the acid Bessemer. His process became known as basic Bessemer. The resulting metal, namely steel, was found to be superior to cast iron and wrought iron as it had the same high strength in tension and compression and was ductile.

If a rod of steel is subjected to a tensile test (*Fig. 4.1(a)*), and the stress in the rod (load/cross-sectional area in $N\,mm^{-2}$) is plotted against the strain (change in length/original length), as the load is applied, a graph similar to that shown in *Fig. 4.1(b)* would be obtained. Note that the stress–strain curve is linear up to a certain value, known as the yield point. This is the elastic range. Beyond this point the steel yields without an increase in load, although there is significant 'strain hardening' as the bar continues to strain towards failure. This is the plastic range.

In the elastic range the bar will return to its original length if unloaded. However, once past the yield point, in the plastic range, the bar will be permanently strained after unloading. *Fig. 4.1(c)* shows the idealized stress–strain curve for structural steel which is used in the design of steel members.

4.3 Structural steel and steel sections

Structural steel is manufactured in three basic grades, 43, 50 and 55. Grade 55 is the strongest, but the lower strength grade 43 is the most commonly used in structural applications, for reasons that will later become apparent.

Figure 4.2 shows in end view a selection of sections commonly used in steel design. Some sections may be rolled into shape directly at a steel rolling mill, while others may be fabricated in a welding shop or on site using (usually) electric arc welding.

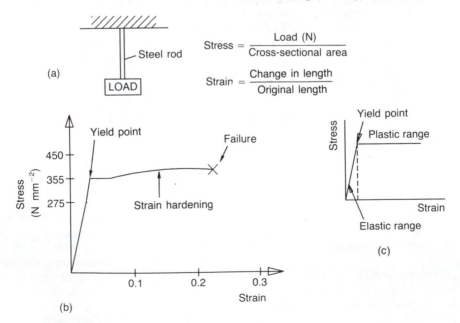

Fig. 4.1 *Stress–strain curves for structural steel: (a) schematic arrangement of test; (b) actual stress–strain curve from experiment; (c) idealized stress–strain relationship.*

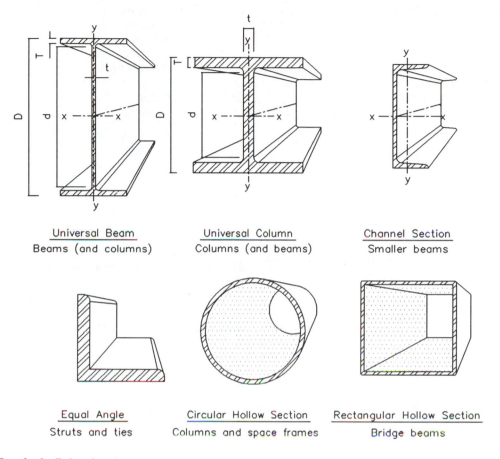

Universal Beam
Beams (and columns)

Universal Column
Columns (and beams)

Channel Section
Smaller beams

Equal Angle
Struts and ties

Circular Hollow Section
Columns and space frames

Rectangular Hollow Section
Bridge beams

Fig. 4.2 *Standard rolled steel sections.*

These sections are designed to achieve economy of material while maximizing strength, particularly in bending. Bending strength can be maximized by concentrating the metal at the extremities of the section, where it can sustain the tensile and compressive stresses associated with bending. The most commonly used sections are still universal beams (UBs) and universal columns (UCs). While boxes and tubes have some popularity in specialist applications, they tend to be expensive to make and are difficult to maintain, particularly in small sizes.

The geometric properties of these steel sections, including the principal dimensions, area, second moment of area, radius of gyration and elastic and plastic section moduli, have been tabulated in a booklet entitled *Structural Sections to BS 4: Part 1 and BS 4848: Part 4* which is published by British Steel PLC. *Appendix B* contains extracts from these tables for UBs and UCs, and will be frequently referred to. The axes x–x and y–y shown in *Fig. 4.2* and referred to in the steel tables in *Appendix B* denote the strong and weak bending axes respectively.

This chapter will concentrate on the design of UBs and UCs and their connections to suit various applications. Irrespective of the element being designed, the designer will need an understanding of the following aspects which are discussed next:

1. symbols
2. general principles and design methods
3. loadings
4. design strengths.

4.4 Symbols

For the purpose of this chapter, the following symbols have been used. These have largely been taken from BS 5950.

GEOMETRIC PROPERTIES

A area of section

A_g gross sectional area of steel section

B	breadth of section
b	outstand of flange
D	depth of section
d	depth of web
I_x, I_y	second moment of area about the major and minor axes
$\mathcal{J}$	torsion constant of section
L	length of span
r	root radius
r_x, r_y	radius of gyration of a member about its major and minor axes
S_x, S_y	plastic modulus about the major and minor axes
T	thickness of flange
t	thickness of web
u	buckling parameter of the section
x	torsional index of section
Z_x, Z_y	elastic modulus about major and minor axes

BENDING

A_v	shear area
b_1	stiff bearing length
E	modulus of elasticity
F_t	tensile force
F_v	shear force
L	actual length
l_e, L_E	effective length
M	large end moment
M_A	maximum moment
M_c	moment capacity
M_b	buckling resistance moment
M_E	elastic critical moment
M_o	mid-span moment on a simply supported span equal to the unrestrained length
$\overline{M_x}, \overline{M_y}$	equivalent uniform moment about the major and minor axes
P_{crip}	bearing resistance
P_w	buckling resistance of an unstiffened web
P_x	buckling resistance of stiffener
m	equivalent uniform moment factor
n	slenderness correction factor
n_1	$= D/2$
n_2	$= 2.5(T + r)$
P_{cs}	contact stress
P_v	shear capacity of a section
p_c	compressive strength of steel
p_b	bending strength of steel
p_y	design strength of steel
v	slenderness factor for beam
β	ratio of smaller to larger end moment
γ_f	overall load factor
γ_l	load variation factor
γ_m	material strength factor

γ	ratio M/M_o
δ	deflection
ε	constant $= (275/p_y)^{1/2}$
λ	slenderness ratio
ν	Poisson's ratio

COMPRESSION

A_g	gross sectional area of steel section
F_c	ultimate applied axial load
L	actual length
L_E	effective length
M_b	buckling resistance moment
M_{cx}, M_{cy}	moment capacity of section about the major and minor axes in the absence of axial load
M_{ex}, M_{ey}	eccentricity moment about the major and minor axes
M_x, M_y	applied moment about major and minor axes
P_s	compression resistance of column
P_c	compression resistance of cased column
P_{cs}	compression resistance of short strut
p_c	compressive strength
P_E	Euler load
t	thickness of baseplate
λ	slenderness ratio
λ_{LT}	equivalent slenderness

CONNECTIONS

a	effective throat size of weld
A_e	effective area
A_n	net area of steel section
A_s	effective area of bolt
A_t	tensile stress area of a bolt
d_b	diameter of bolt
D_h	diameter of bolt hole
e_1	edge distance
e_2	end distance
p	pitch
t	thickness of ply
F_s	applied shear force
F_t	applied tension force
f_v	shear stress
P_{bb}	bearing capacity of a bolt
P_{bg}	bearing capacity of parts connected by friction grip fasteners
P_{bs}	bearing capacity of parts connected by ordinary bolts
P_s	shear capacity of a bolt
P_{sL}	slip resistance provided by a friction grip fastener
P_t	tension capacity of a member or fastener
P_o	minimum shank tension
p_{bb}	bearing strength of a bolt

p_{bg}	bearing strength of parts connected by friction grip fasteners
p_{bs}	bearing strength of parts connected by ordinary bolts
p_s	shear strength of a bolt
p_t	tension strength of a bolt
p_w	design strength of a fillet weld
s	leg length of a fillet weld
K_e	coefficient = 1.2 for grade 43 steel
K_s	coefficient = 1.0 for clearance holes
μ	slip factor

4.5 General principles and design methods

As stated at the outset, BS 5950 is based on limit state philosophy. Table 1 of BS 5950, reproduced as *Table 4.1*, outlines typical limit states appropriate to steel structures. As this book is principally about the design of structural elements we will be concentrating on the ultimate limit state of strength (1), and the serviceability limit state of deflection (5). Stability (2) is an aspect of complete structures or substructures that will not be examined at this point, except to say that structures must be robust enough not to overturn or sway excessively under wind or other sideways loading. Fatigue (3 and 7) is generally taken account of by the provision of adequate safety factors to prevent occurrence of the high stresses associated with fatigue. Brittle fracture (4) can be avoided by selecting the correct grade of steel for the expected ambient conditions. Avoidance of excessive vibration (6) is an aspect of structural dynamics and is beyond the scope of this book. Corrosion (8) can be a serious problem for exposed steelwork, but correct preparation and painting of the steel will ensure maximum durability and minimum maintenance during the design life of the structure.

Although BS 5950 does not specifically mention fire resistance, this is an important aspect that fundamentally affects steel's economic viability compared to its chief rival, concrete. Exposed structural steelwork does not perform well in a fire. The high conductivity of steel together with the thin sections used causes high temperatures to be quickly reached in steel members, resulting in premature failure due to softening at around 600 °C. Structural steelwork has to be insulated to provide adequate fire resistance in multi-storey structures. Insulation may consist of sprayed treatment, concrete encasement or boxing using a proprietary boarding system. All insulation treatments are expensive. However, where the steel member has been encased in concrete it may be possible to take structural advantage of the concrete, thereby mitigating some of the additional expenditure incurred (*section 4.9.5*).

For steel structures three principal methods of design are identified in clause 2.1.2 of BS 5950:

1. *Simple design*. The structure is regarded as having pinned joints, and significant moments are not developed at connections (*Fig. 4.3 (a)*). The structure is prevented from becoming a mechanism by appropriate bracing using shear walls. This apparently conservative assumption is a very popular method of design.
2. *Rigid design*. The joints in the structure are assumed to be able to fully transfer the forces and moments in the members which they attach (*Fig. 4.3(b)*). Analysis of the structure may be by elastic or plastic methods, and will be more complex than simple design. However, the increasing use of microcomputers has made this method more viable. In theory a more economic design can be achieved by this method, but unless the joints are truly rigid the analysis will give an upper bound (unsafe) solution.
3. *Semi-rigid design*. The joints in the structure are assumed to transmit some restraint moment. However, without experimental verification this restraint moment is limited to 10% of the free moment applied to the beam, which is usually regarded as hardly worth the additional work involved.

4.6 Loading

As for structural design in other media, the designer needs to estimate the loading to which the structure may be subject during its design life. The characteristic loads can be obtained from BS 6399: Part 1 and CP 3: Chapter 5: Part 2 (which is to be superseded in due course by BS 6399: Part 2). In general a characteristic load is expected to be exceeded in only

Table 4.1 Limit states (Table 1, BS 5950)

Ultimate	Serviceability
1. Strength (including general yielding, rupture, buckling and transformation into a mechanism)	5. Deflection
	6. Vibration (e.g. wind-induced oscillation)
2. Stability against overturning and sway	7. Repairable damage due to fatigue
3. Fracture due to fatigue	8. Corrosion and durability
4. Brittle fracture	

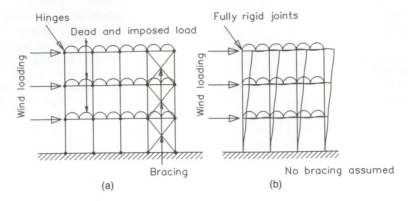

Fig. 4.3 *(a) Simple and (b) rigid design methods.*

5% of instances, or for 5% of the time, but in the case of wind loads it represents a gust expected only once every 50 years.

To obtain design loading at **ultimate limit state** for strength and stability calculations the characteristic load is multiplied by a load factor obtained from Table 2 of BS 5950, reproduced as *Table 4.2*. Generally, dead load is multiplied by 1.4 and imposed vertical (or live) load by 1.6, except when the load case considers wind load also, in which case, dead, imposed and wind loads are all multiplied by 1.2. Several loading cases may be specified to give a 'worst case' envelope of forces and moments around the structure.

Table 4.2 Load factors and combinations (Table 2, BS 5950)

Loading	Factor, γ_f
Dead load	1.4
Dead load restraining uplift or overturning	1.0
Dead load acting with wind and imposed loads combined	1.2
Imposed load	1.6
Imposed load acting with wind load	1.2
Wind load	1.4
Wind load acting with imposed load or crane load	1.2
Forces due to temperature effects	1.2
Crane loading effects	
Vertical load	1.6
Vertical load acting with horizontal load (crabbing or surge)	1.4
Horizontal load	1.6
Horizontal load acting with vertical load	1.4
Crane load acting with wind load	1.2

To obtain design loading at **serviceability limit state** for calculation of deflections the most adverse realistic combination of unfactored characteristic imposed loads is usually used. In the case of wind loads acting together with dead and imposed loads, only 80% of imposed and wind loads need to be considered.

4.7 Design strengths

In BS 5950 no distinction is made between characteristic and design strength. In effect the material safety factor $\gamma_m = 1.0$. Structural steel used in the UK is specified by BS 5950: Part 2, and strengths of the more commonly used steels are given in Table 6 of BS 5950, reproduced here as *Table 4.3*. As a result of the residual stresses locked into the metal during the rolling process, the thicker the material, the lower the design strength.

Having discussed these more general aspects relating to structural steelwork design, the following will consider the detailed design of beams and joists (*section 4.8*), struts and columns (*section 4.9*) and some simple bolted and welded connections (*section 4.10*).

4.8 Design of steel beams and joists

Structural design of steel beams and joists primarily involves predicting the strength of the member. This requires the designer to imagine all the ways in which the member may fail during its design life. It would be useful at this point, therefore, to discuss some of the more common modes of failure associated with beams and joists.

Table 4.3   Design strengths p_y (Table 6, BS 5950)

Design grade	Thickness (mm), less than or equal to	Sections, plates and hollow sections, p_y (N mm^{-2})
43	16	275
	40	265
	63	255
	80	245
50	100	235
	16	355
	40	345
	63	335
	80	325
55	100	315
	16	450
	25	430
	40	
	63	400

4.8.1 MODES OF FAILURE

4.8.1.1 Bending

The vertical loading gives rise to bending of the beam. This results in longitudinal stresses being set up in the beam. These stresses are tensile in one half of the beam and compressive in the other. As the bending moment increases, more and more of the steel reaches its yield stress. Eventually, all the steel yields in tension and/or compression across the entire cross-section of the beam. At this point the beam cross-section has become plastic and it fails by

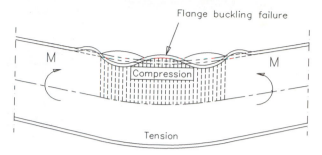

Fig. 4.5 *Local flange buckling failure.*

formation of a plastic hinge at the point of maximum moment induced by the loading. *Figure 4.4* reviews this process. *Chapter 2* summarizes how classical beam theory is derived from these considerations.

4.8.1.2 Local buckling

During the bending process outlined above, if the compression flange or the part of the web subject to compression is too thin, the plate may actually fail by buckling or rippling, as shown in *Fig. 4.5*, before the full plastic moment is reached.

4.8.1.3 Shear

Due to excessive shear forces, usually adjacent to supports, the beam may fail in shear. The beam web, which resists shear forces, may fail as shown in *Fig. 4.6(a)*, as steel yields in tension and compression in the shaded zones. The formation of plastic hinges in the flanges accompanies this process.

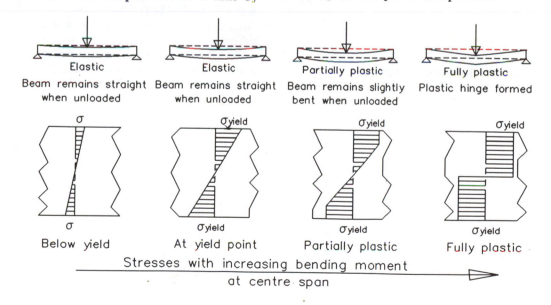

Fig. 4.4 *Bending failure of a beam.*

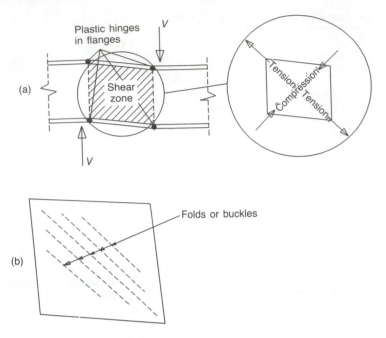

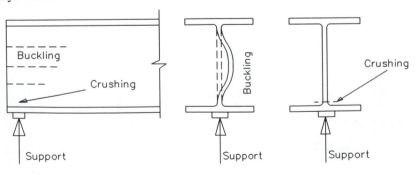

Fig. 4.6 *Shear and shear buckling failures: (a) shear failure; (b) shear buckling.*

4.8.1.4 Shear buckling
During the shearing process described above, if the web is too thin it will fail by buckling or rippling in the shear zone, as shown in *Fig. 4.6(b)*.

4.8.1.5 Web bearing and buckling
Due to high vertical stresses directly over a support or under a concentrated load, the beam web may actually crush, or buckle as a result of these stresses, as illustrated in *Fig. 4.7*.

4.8.1.6 Lateral torsional buckling
When the beam has a higher bending stiffness in the vertical plane compared to the horizontal plane, the beam can twist sideways under the load. This is perhaps best visualized by loading a scale rule on its edge, as it is held as a cantilever – it will tend to twist and deflect sideways. This is illustrated in *Fig. 4.8*. Where a beam is not prevented from moving sideways, by a floor for instance, it is necessary to check that it is laterally stable under load.

4.8.1.7 Deflection
Although a beam cannot fail as a result of excessive deflection alone, it is necessary to ensure that deflections are not excessive under unfactored imposed loading. Excessive deflections are those resulting in severe cracking in finishes which would render the building unserviceable.

Fig. 4.7 *Web buckling and web bearing failures.*

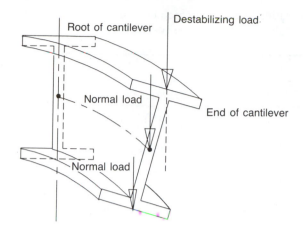

Fig. 4.8 *Lateral torsional buckling of cantilever.*

4.8.2 SUMMARY OF DESIGN PROCESS

The design process for a beam can be summarized as follows:

1. determination of design shear forces, F_v, and bending moments, M, at critical points on the element (*Chapter 2*);
2. selection of UB or UC;
3. classification of section;
4. check shear strength; if unsatisfactory return to (2);
5. check bending capacity; if unsatisfactory return to (2);
6. check deflection; if unsatisfactory return to (2);
7. check web bearing and buckling at supports or concentrated load; if unsatisfactory provide web stiffener or return to (2);
8. check lateral torsional buckling; if unsatisfactory return to (2) or provide lateral restraints;
9. summarize results.

4.8.3 INITIAL SECTION SELECTION

It is perhaps most often the case in the design of skeletal building structures, that bending is the critical mode of failure, and so beam bending theory can be used to make an initial selection of section. Readers should refer to *Chapter 2* for more clarification on bending theory if necessary.

To avoid bending failure, it is necessary to ensure that the design moment, M, does not exceed the moment capacity of the section, M_c, i.e.

$$M < M_c \qquad (4.1)$$

Generally, the moment capacity for a rolled steel section is given by

$$M_c = p_y S \qquad (4.2)$$

where p_y is the assumed design strength of the steel and S the plastic modulus of the section. Combining the above equations gives an expression for S:

$$S > M/p_y \qquad (4.3)$$

This can be used to select UB sections from steel tables (*Appendix B*) with the plastic modulus of section S greater than the calculated value.

4.8.4 CLASSIFICATION OF SECTION

Having selected a suitable section, or proposed a suitable section fabricated by welding, it must be classified.

4.8.4.1 Strength classification

In making the initial choice of section, a steel strength will have been assumed. If grade 43 steel is to be used, for example, it will have been assumed that the strength is 275 N mm^{-2}. By referring to the flange thickness T from the steel tables, the design strength can be obtained from Table 6 of BS 5950, reproduced as *Table 4.3*.

If the section is fabricated from welded plate, the strength of the web and flange may be taken separately from Table 6 of BS 5950 as that for the web thickness t and flange thickness T respectively.

4.8.4.2 Section classification

As previously noted, the bending strength of the section depends on how the section performs in bending. If the section is stocky, i.e. has thick flanges and web, it can sustain the formation of a plastic hinge. On the other hand, a slender section, i.e. one with thin flanges and web, will fail by local buckling before the yield stress can be reached. Four classes of section are identified in clause 3.5.2 of BS 5950:

Class 1 plastic cross-sections are those in which a plastic hinge can be developed with significant rotation capacity. If the plastic design method is used in the structural analysis, all members must be of this type.

Class 2 compact cross-sections are those in which the full plastic moment capacity can be developed, but local buckling may prevent production of a plastic hinge with sufficient rotation capacity to permit plastic design.

Class 3 semi-compact cross-sections can develop their elastic moment capacity, but local buckling may prevent the production of the full plastic moment.

Class 4 slender cross-sections contain slender elements subject to compression due to moment

Table 4.4   Limiting width to thickness ratios (elements which exceed these limits are to be taken as class 4, slender cross-sections) (based on Table 7, BS 5950)

Type of element (all rolled sections)	Class of section		
	(1) Plastic	(2) Compact	(3) Semi-compact
Outstand element of compression flange	$\dfrac{b}{T} \leq 8.5\,\varepsilon$	$\dfrac{b}{T} \leq 9.5\,\varepsilon$	$\dfrac{b}{T} \leq 15\,\varepsilon$
Web with neutral axis at mid-depth	$\dfrac{d}{t} \leq 79\,\varepsilon$	$\dfrac{d}{t} \leq 98\,\varepsilon$	$\dfrac{d}{t} \leq 120\,\varepsilon$
Web subject to compression throughout	$\dfrac{d}{t} \leq 39\,\varepsilon$	$\dfrac{d}{t} \leq 39\,\varepsilon$	$\dfrac{d}{t} \leq 39\,\varepsilon$

Note. $\varepsilon = (275/p_y)^{1/2}$ (4.4)

or axial load. Local buckling may prevent the full elastic moment capacity from being developed.

Limiting width to thickness ratios for elements for the above classes are given in Table 7 of BS 5950, part of which is reproduced as *Table 4.4*. (Refer to *Fig. 4.2* for details of UB and UC dimensions.) Once the section has been classified, the various strength checks can be carried out to assess its suitability as discussed below.

4.8.5 SHEAR
According to clause 4.2.3 of BS 8110, the shear force, F_v, should not exceed the shear capacity of the section, P_v, i.e.

$$F_v \leq P_v \qquad (4.5)$$

where
$$P_v = 0.6 p_y A_v \qquad (4.6)$$

in which A_v is the shear area ($= tD$ for rolled I-, H- and channel sections). Clause 4.2.3 also states that when the buckling ratio (d/t) of the web exceeds 63ε (equation 4.4), then the web should be additionally checked for shear buckling. However, only a few UB sections in grade 55 steel are affected, just one in grade 50 steel, and none in grade 43.

4.8.6 LOW SHEAR AND MOMENT CAPACITY
As stated in clause 4.2.1.3 of BS 5950, at critical

points the combination of (i) maximum moment and coexistent shear and (ii) maximum shear and coexistent moment, should be checked.

If the shear force F_v is less than $0.6P_v$, then this is a **low shear load**. Otherwise, if $0.6P_v < F_v < P_v$, then it is a **high shear load.** When the shear load is low, the moment capacity of the section is calculated according to clause 4.2.5 of BS 5950 as follows. For class 1 plastic or class 2 compact sections, the moment capacity

$$M_c = p_y S \leq 1.2\,p_y Z \qquad (4.7)$$

The additional check ($M_c \leq 1.2\,p_y Z$) is to guard against plastic stress reversals under working dead, imposed and wind loads. For class 3 semi-compact sections

$$M_c = p_y Z \qquad (4.8)$$

and for class 4 slender sections

$$M_c = p_y' Z \qquad (4.9)$$

where p_y is the design strength of the steel, p_y' the design strength reduced for slender sections, S the plastic modulus of the section and Z the elastic modulus of section. It should be noted that nearly all sections in grade 43 steel are plastic and that only a few sections in higher strength steel are semi-compact. No British rolled UB sections in pure bending, no matter what the design strength, are slender.

Example 4.1 Selection of a beam section in grade 43 steel

The simply supported beam shown in *Fig. 4.9* supports uniformly distributed characteristic dead and imposed loads of 5 kN m^{-1} each, as well as a characteristic imposed point load of 30 kN at mid-span. Assuming the beam is fully laterally restrained, select a suitable UB section in grade 43 steel to satisfy bending and shear considerations.

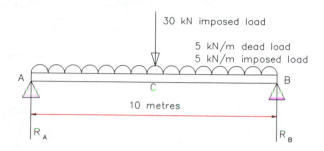

Fig. 4.9 *Loading for example 4.1.*

DESIGN BENDING MOMENT AND SHEAR FORCE

Total loading $= (30 \times 1.6) + (5 \times 1.4 + 5 \times 1.6)\,10$

$\qquad\qquad = 48 + 15 \times 10 = 198$ kN

Because the structure is symmetrical $R_A = R_B = 198/2 = 99$ kN. The central bending moment, M, is

$$M = \frac{Wl}{4} + \frac{wl^2}{8}$$

$$= \frac{(48 \times 10)}{4} + \frac{(15 \times 10^2)}{8}$$

$$= 120 + 187.5 = 307.5 \text{ kN m}$$

Shear force and bending moment diagrams are shown in *Fig. 4.10.*

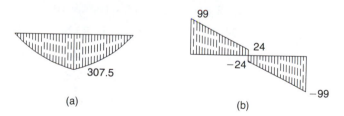

Fig. 4.10 *(a) Bending moment and (b) Shear force diagrams.*

INITIAL SECTION SELECTION

Assuming $p_y = 275$ N mm^{-2}

$$S_x > \frac{M}{p_y} = \frac{307.5 \times 10^6}{275} = 1.118 \times 10^6 \text{ mm}^3 = 1118 \text{ cm}^3$$

From steel tables (*Appendix B*), suitable sections are:

1. $356 \times 171 \times 67$ UB: $S_x = 1210$ cm^3;
2. $406 \times 178 \times 60$ UB: $S_x = 1190$ cm^3;
3. $457 \times 152 \times 60$ UB: $S_x = 1280$ cm^3.

The above illustrates how steel beam sections are specified. For section 1, for instance, 356×171 represents the serial size in the steel tables; 67 represents the mass per metre in kilograms; and UB stands for universal beam.

All the above sections give a value of plastic modulus about axis x – x, S_x, just greater than that required. Whichever one is selected will depend on economic and engineering considerations. For instance, if lightness were the primary consideration, perhaps section 3 would be selected, which is also the strongest (largest S_x). However, if minimizing the depth of the member were the main consideration then section 1 would be chosen. Let us choose the compromise candidate, section 2.

CLASSIFICATION

Strength classification
Because the flange thickness $T = 12.8$ mm (< 16 mm), then $p_y = 275$ N mm^{-2} (as assumed) from *Table 4.3* and $\varepsilon = (275/p_y) = 1$ (*Table 4.4*).

Section classification
$b/T = 6.95$ which is less than $8.5\varepsilon = 8.5$. Hence from *Table 4.4*, section is plastic. $d/t = 46.2$ which is less than $79\varepsilon = 79$. Hence from *Table 4.4*, section is plastic. Therefore $406 \times 178 \times 60$ UB section is class 1 plastic.

SHEAR STRENGTH
As $d/t = 46.2 < 63\varepsilon$, shear buckling need not be considered. Shear capacity of section, P_v, is

$$P_v = 0.6 p_y A = 0.6 p_y tD = 0.6 \times 275 \times 7.8 \times 406.4$$
$$= 523 \times 10^3 \text{ N} = 523 \text{ kN}$$

Now, as F_v (99 kN) $< 0.6 P_v = 0.6 \times 523 = 314$ kN (low shear load).

BENDING MOMENT
Moment capacity of section, M_c, is

$$M_c = p_y S = 275 \times 1190 \times 10^3$$
$$= 327 \times 10^6 \text{ N mm} = 327 \text{ kN m}$$
$$\leqslant 1.2\, p_y Z = 1.2 \times 275 \times 1060 \times 10^3$$
$$= 349.8 \times 10^6 \text{ N mm} = 349.8 \text{ kN m} \quad \text{OK}$$

Moment M due to imposed loading = 307.5 kN m. Extra moment due to self-weight, M_{sw}, is

$$M_{sw} = 1.4\, (60 \times 9.81/10^3)\, 10^2/8 = 10.3 \text{ kN m}$$

Total imposed moment $M_t = 307.5 + 10.3 = 317.8$ kN m < 327 kN m. Hence the proposed section is suitable.

Example 4.2 Selection of beam section in grade 55 steel

Repeat the above design in grade 55 steel.

INITIAL SECTION SELECTION
Since section is of grade 55 steel, assume $p_y = 450$ N mm^{-2}

$$S_x > \frac{M}{p_y} = \frac{307.5 \times 10^6}{450} = 683 \times 10^3 \text{ mm}^3 = 683 \text{ cm}^3$$

Suitable sections (with classifications) are:

1. $305 \times 127 \times 48$ UB: $S = 706$ cm^3, $p_y = 450$, plastic;
2. $305 \times 165 \times 46$ UB: $S = 723$ cm^3, $p_y = 450$, compact;
3. $356 \times 171 \times 45$ UB: $S = 774$ cm^3, $Z = 687$ cm^3, $p_y = 450$, semi-compact;
4. $406 \times 140 \times 39$ UB: $S = 721$ cm^3, $Z = 627$ cm^3, $p_y = 450$, semi-compact.

Section 4 must be discounted. As it is semi-compact and Z is less than 683 cm^3, it fails in bending. Section 3 is also semi-compact, and Z is not sufficiently larger than 683 to take care of its own self-weight. Section 2 looks the best choice, being the lightest of the two remaining, and with the greater strength.

SHEAR STRENGTH

As $d/t < 63\varepsilon$, no shear buckling check is required. Shear capacity of section, P_v, is

$$P_v = 0.6 p_y t D = 0.6 \times 450 \times 6.7 \times 307.1$$
$$= 556 \times 10^3 \text{ N} = 556 \text{ kN}$$

As F_v (99 kN) $< 0.6 P_v = 333$ kN (low shear load).

BENDING MOMENT

Moment capacity of section subject to low shear load, M_c, is given by

$$M_c = p_y S = 450 \times 723 \times 10^3$$
$$= 325 \times 10^6 \text{ N mm} = 325 \text{ kN m}$$
$$\leqslant 1.2\, p_y Z = 1.2 \times 450 \times 648 \times 10^3$$
$$= 349.9 \times 10^6 \text{ N mm} = 349.9 \text{ kN m} \quad \text{OK}$$

Moment M due to dead and imposed loading = 307.5 kN m. Extra moment M_{sw} due to self-weight is

$$M_{sw} = 1.4(46 \times 9.81/10^3)10^2/8 = 7.9 \text{ kN m}$$

Total imposed moment $M_t = 307.5 + 7.9 = 315.4$ kN m, which is less than the section's moment of resistance $M_c = 325$ kN m. This section is satisfactory.

4.8.7 HIGH SHEAR AND MOMENT CAPACITY

When the shear load is high, i.e. $F_v > 0.6 P_v$, the moment-carrying capacity of the section is reduced. This is because the web cannot take the full tensile or compressive stress associated with the bending moment as well as sustaining a substantial shear stress due to the shear force. This is not a problem with semi-compact and slender sections, where $M_c = p_y Z$ even when there is a high shear load. However, for plastic and compact sections clause 4.2.6 of BS 5950 specifies how M_c is modified:

Example 4.3 Selection of a cantilever beam section

A proposed cantilever beam 1 m long is to be built into a concrete wall as shown in *Fig. 4.11*. It supports characteristic dead and imposed loading of 450 and 270 kN m^{-1} respectively. Select a suitable UB section in grade 43 steel to satisfy bending and shear criteria only.

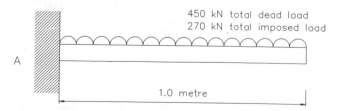

450 kN total dead load
270 kN total imposed load

A

1.0 metre

Fig. 4.11

DESIGN BENDING MOMENT AND SHEAR FORCE

Shear force F_v at A is

$$(450 \times 1.4) + (270 \times 1.6) = 1062 \text{ kN}$$

Bending moment M at A is

$$\frac{Wl}{2} = \frac{1062 \times 1}{2} = 531 \text{ kN m}$$

INITIAL SECTION SELECTION

Assuming $p_y = 275 \text{ N mm}^{-2}$

$$S_x > \frac{M}{p_y} = \frac{531 \times 10^6}{275} = 1931 \times 10^3 \text{ mm}^3 = 1931 \text{ cm}^3$$

Suitable sections (with classifications) are:

1. $457 \times 191 \times 89$ UB: $S_x = 2010 \text{ cm}^3$, $p_y = 275$, plastic;
2. $533 \times 210 \times 82$ UB: $S_x = 2060 \text{ cm}^3$, $p_y = 275$, plastic.

SHEAR STRENGTH

Shear capacity of $533 \times 210 \times 82$ UB section, P_v, is

$$P_v = 0.6 p_y t D = 0.6 \times 275 \times 9.6 \times 528.3$$
$$= 837 \times 10^3 \text{ N} = 837 \text{ kN} < 1062 \text{ kN} \quad \text{Not OK}$$

In this case, because the length of the cantilever is so short, the selection of section will be determined from the shear strength, which is more critical than bending. Try a new section:

$$610 \times 229 \times 113 \text{ UB: } S_x = 3290 \text{ cm}^3 \qquad p_y = 265, \text{ plastic.}$$

$$P_v = 0.6 \times 265 \times 607.3 \times 11.2 = 1081 \times 10^3 \text{ N}$$
$$= 1081 \text{ kN} \quad \text{OK but high shear load}$$

BENDING MOMENT

From above

$$\rho_1 = \frac{2.5 F_v}{P_v} - 1.5 = \frac{2.5 \times 1062}{1081} - 1.5$$
$$= 2.46 - 1.5 = 0.96$$
$$M_c = p_y(S - S_v \rho_1)$$
$$= 265[3290 \times 10^3 - (11.2 \times 607.3^2/4)0.96]$$
$$= 265(2299 \times 10^3) = 609 \times 10^6 \text{ N mm} = 609 \text{ kN m}$$
$$\leqslant 1.2 \, p_y Z = 1.2 \times 265 \times 2880 \times 10^3$$
$$= 915.8 \times 10^6 \text{ N mm} = 915.8 \text{ kN m} \quad \text{OK}$$
$$M_{sw} = \frac{1.4 (113 \times 9.81/10^3) \, 1^2}{2} = 0.8 \text{ kN m}$$
$$M_t = M + M_{sw} = 531 + 0.8 = 532 \text{ kN m} < M_c$$

This section is satisfactory.

Table 4.5 Deflection limits on beams due to unfactored imposed load (based on Table 5, BS 5950)

Cantilevers	Length/180
Beams carrying plaster or other brittle finish	Span/360
All other beams	Span/200

$$M_c = p_y(S - S_v\rho_1) \leqslant 1.2p_yZ \qquad (4.10)$$

where

$$\rho_1 = \frac{2.5F_v}{P_v} - 1.5 \qquad (4.11)$$

and S_v for rolled sections with equal flanges is taken as the plastic modulus of the shear area of section equal to $tD^2/4$. The effect of this modification is to proportionately reduce the moment-carrying capacity of the web as the shear load rises from 60 to 100% of the web's shear capacity.

4.8.8 DEFLECTION

A check should be carried out on the maximum deflection of the beam due to the most adverse realistic combination of unfactored imposed serviceability loading. In BS 5950 this is covered by clause 2.5.1 and Table 5, part of which is reproduced as *Table 4.5*. *Table 4.5* outlines recommended limits to these deflections, compliance with which should avoid significant damage to the structure and finishes.

Calculation of deflections from first principles has to be done using the area–moment method, Macaulay's method, or some other similar approach, a subject which is beyond the scope of this book. The reader is referred to a suitable structural analysis text for more detail on this subject. However, many calculations of deflection are carried out using formulae for standard cases, which can be combined as necessary to give the answer for more complicated situations. Figure 4.12 summarizes some of the more useful formulae.

Example 4.4 Deflection checks on steel beams

Carry out a deflection check for *Examples 4.1 – 4.3* above.

FOR EXAMPLE 4.1

$$\delta_c = \frac{5wl^4}{384EI} + \frac{Wl^3}{48EI}$$

$$= \frac{5 \times 5 \times 10^4}{384 \times 205 \times 10^6 \times 21\,500 \times 10^{-8}} + \frac{30 \times 10^3}{48 \times 205 \times 10^6 \times 21\,500 \times 10^{-8}}$$

$$= 0.0148 + 0.0142 = 0.029 \text{ m} = 29 \text{ mm}$$

From *Table 4.5*, the recommended maximum deflection for beams carrying plaster is span/360 which equals 10 000/360 = 27.8 mm, and for other beams span/200 = 10 000/200 = 50 mm. Therefore if the beam was carrying a plaster finish one might consider choosing a larger section.

FOR EXAMPLE 4.2

$$\delta_c = \frac{5wl^4}{384EI} + \frac{Wl^3}{48EI}$$

$$= \frac{5 \times 5 \times 10^4}{384 \times 205 \times 10^6 \times 9950 \times 10^{-8}} + \frac{30 \times 10^3}{48 \times 205 \times 10^6 \times 9950 \times 10^{-8}}$$

$$= 0.0319 + 0.0306 = 0.0625 \text{ m} = 62.5 \text{ mm}$$

The same recommended limits from *Table 4.5* apply as above. This grade 55 beam therefore fails the deflection test. This partly explains why the higher strength steel beams are not particularly popular.

FOR EXAMPLE 4.3

$$\delta_B = \frac{wl^4}{8EI} = \frac{270 \times 1^4}{8 \times 205 \times 10^6 \times 87\,400 \times 10^{-8}}$$

$$= 0.00019 \text{ m} = 0.19 \text{ mm}$$

The recommended limit from *Table 4.5* is

$$\text{Length}/180 = 5.6 \text{ mm}$$

So deflection is no problem for this particular beam.

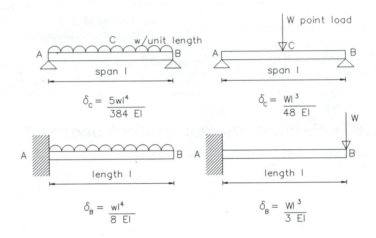

Fig. 4.12 *Deflections for standard cases. E = elastic modulus of steel (205 kN mm^{-2}) and I = second moment of area (x–x) of section.*

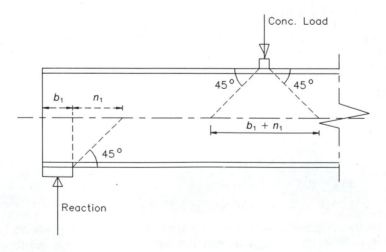

Fig. 4.13 *Web buckling.*

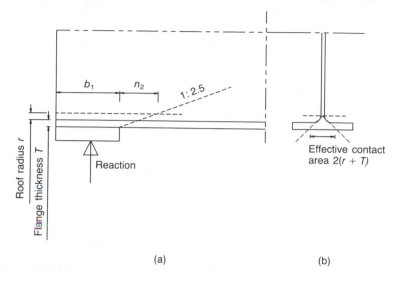

Fig. 4.14 *Web bearing.*

4.8.9 WEB BEARING AND BUCKLING

Clause 4.5 of BS 5950 covers all aspects of web bearing, web buckling and stiffener design. Usually most critical at the position of a support or concentrated load is the problem of web buckling.

4.8.9.1 Web buckling

Figure 4.13 illustrates how the load is assumed to be transmitted through the web at a support or concentrated load. The critical point will be at the centre of the web, where the maximum buckling moment will occur. According to clause 4.5.2.1 of BS 5950, the buckling resistance of an unstiffened web, P_w, which should be more than the reaction or concentrated load at that point, is given by

$$P_w = (b_1 + n_1)tp_c \qquad (4.12)$$

where b_1 is the stiff bearing length, $n_1 = D/2$ (*Fig. 4.13*), t is the web thickness and p_c the buckling compressive strength from clause 4.7.5 of BS 5950 using Table 27(c) of BS 5950 (*Table 4.19*) and

$$\lambda = 2.5d/t \qquad (4.13)$$

Essentially, this is a strut buckling problem and will be discussed more fully in *section 4.9* of this book on the design of compression members.

4.8.9.2 Web bearing

Figure 4.14 shows how the loading is assumed to be transmitted through the flange/web connection in this case. Clause 4.5.3 of BS 5950 shows how the bearing resistance P_{crip} is calculated at the flange/web connection:

$$P_{crip} = (b_1 + n_2) tp_{yw} \qquad (4.14)$$

where b_1 is the stiff bearing length, $n_2 = 2.5(T + r)$ (Fig. 4.14(a)), t is the thickness of the web and p_{yw} the design strength of the web. This is essentially a much simpler check which ensures that the stress at the critical point on the flange/web connection does not exceed the strength of the steel. It should also be checked that the contact stresses between load or support and flange do not exceed p_y (Fig. 4.14(b)).

Example 4.5 Checks on web buckling and bearing for steel beams

Check web buckling and bearing for *Example 4.1*, assuming the beam sits on 100 mm bearings at each end.

WEB BUCKLING AT SUPPORTS
$\lambda = 2.5d/t = 2.5 \times 46.2 = 115.5$
Then from Table 27(c) (reproduced here as *Table 4.19*) with $p_y = 275$, $p_c = 102$ N mm^{-2}

$$P_w = (b_1 + n_1)tp_c$$
$$= (100 + 203.2) \times 7.8 \times 102$$
$$= 241 \times 10^3 \text{ N} = 241 \text{ kN} > 99 \text{ kN} \quad \text{OK}$$

where $n_1 = D/2 = 406.4/2 = 203.2$ mm.

WEB BEARING AT SUPPORTS

$$P_{crip} = (b_1 + n_2)tp_{yw}$$
$$= (100 + 57.5) \times 7.8 \times 275$$
$$= 338 \times 10^3 \text{ N} = 338 \text{ kN} > 99 \text{ kN} \quad \text{OK}$$

where $n_2 = 2.5(T + r) = 2.5(12.8 + 10.2) = 57.5$ mm.
So no web stiffeners are required at supports.

CONTACT STRESS AT SUPPORTS

$$P_{cs} = (b_1 \times 2(r + T))p_y = (100 \times 46) \times 275$$
$$= 1265 \times 10^3 \text{ N} = 1265 \text{ kN} > 99 \text{ kN} \quad \text{OK}$$

A web check at the concentrated load should also be carried out, but readers can confirm that this aspect is not critical in this case.

4.8.10 STIFFENER DESIGN

If it is found that the web fails in buckling or bearing, it is not always necessary to select another section; larger supports can be designed, or load-carrying stiffeners can be locally welded between the flanges and the web. Clause 4.5 of BS 5950 gives guidance on the design of such stiffeners and the following example illustrates this.

Example 4.6 Design of a steel beam with web stiffeners

A simply supported beam is to span 5 m and support uniformly distributed characteristic dead and imposed loads of 200 and 100 kN m^{-1} respectively. The beam sits on 150 mm long bearings at supports, and both flanges are laterally and torsionally restrained (*Fig. 4.15*). Select a suitable UB section to satisfy bending, shear and deflection criteria. Also check web bearing and buckling at supports, and design stiffeners if they are required.

200 kN/m dead load
100 kN/m imposed load

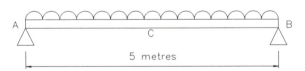

5 metres

Fig. 4.15

DESIGN SHEAR FORCE AND BENDING MOMENT

Factored loading $= (200 \times 1.4) + (100 \times 1.6) = 440$ kN m^{-1}
Reactions $R_A = R_B = 440 \times 5 \times 0.5 = 1100$ kN

Bending moment $M = \dfrac{wl^2}{8} = \dfrac{440 \times 5^2}{8} = 1375$ kN m

INITIAL SECTION SELECTION

Assuming $p_y = 275$ N mm^{-2}

$$S_x > M/p_y = \frac{1375 \times 10^6}{275} = 5000 \times 10^3 \text{ mm}^3 = 5000 \text{ cm}^3$$

From *Appendix B*, suitable sections (with classifications) are:

1. $610 \times 305 \times 179$ UB: $S = 5520$ cm^3, $p_y = 265$, plastic;
2. $686 \times 254 \times 170$ UB: $S = 5620$ cm^3, $p_y = 265$, plastic;
3. $762 \times 267 \times 173$ UB: $S = 6200$ cm^3, $p_y = 265$, plastic;
4. $838 \times 292 \times 176$ UB: $S = 6810$ cm^3, $p_y = 265$, plastic.

Section 2 looks like a good compromise; it is the lightest section, and its depth is not excessive.

SHEAR STRENGTH

$P_v = 0.6p_ytD = 0.6 \times 265 \times 14.5 \times 692.9$
$\quad = 1597 \times 10^3$ N $= 1597$ kN > 1100 kN
However, as F_v (= 1100 kN) $> 0.6P_v$ (= 958 kN, high shear load).

BENDING MOMENT

As the shear force is zero at the centre, the point of maximum bending moment, M_c is calculated from

$$M_c = p_yS = \frac{265 \times 5620 \times 10^3}{1 \times 10^6} = 1489.3 \text{ kN m}$$

$$\leqslant 1.2p_yZ = \frac{1.2 \times 265 \times 4910 \times 10^3}{1 \times 10^6}$$

$$= 1561.3 \text{ kN m} \quad \text{OK}$$

$$M_{sw} = \frac{1.4 \times 170 \times 9.81}{1000} \times \frac{5^2}{8} = 7.3 \text{ kN m}$$

Total $M_t = 1375 + 7.3 = 1382.3$ kN m < 1489.3 kN m $\quad$ OK

DEFLECTION

Calculated deflection, δ_c, is

$$\delta_c = \frac{5wl^4}{384EI} = \frac{5 \times 100 \times 5^4}{384 \times 205 \times 10^6 \times 170\,000 \times 10^{-8}}$$

$$= 2.3 \times 10^{-3} \text{ m} = 2.3\text{mm}$$

Maximum recommended deflection limit for a beam carrying plaster from *Table 4.5* = span/360 = 13.8 mm $\quad$ OK.

WEB BUCKLING

$$\lambda = 2.5d/t = 2.5 \times 42.4 = 106$$

Then from Table 27(c) of BS 5950 (*Table 4.19*) $p_c = 114$ N mm^{-2}

$$P_w = (b_1 + n_1)tp_c = (150 + 346.5)14.5 \times 114$$

$$= 820.7 \times 10^3 \text{ N} = 820.7 \text{ kN}$$

The web's buckling resistance P_w (= 820.7 kN) $< R_A$ (= 1100 kN), and so load-carrying stiffeners are required. Clause 4.5.4 of BS 5950 stipulates that load-carrying stiffeners should be checked for (a) buckling and (b) bearing.

BEARING CHECK

Let us propose 20 mm thick stiffeners each side of the web and welded continuously to it. According to clause 4.5.1.2 of BS 5950 the outstand of the stiffener is limited to $13t_s = 260$ mm, where t_s is the thickness of the stiffener. However, as the width of section $B = 255.8$ mm, the stiffener outstand is effectively limited to 120 mm.

The plan of the web and stiffener at the support position is shown in *Fig. 4.16*. Note that a length of web on each side of the centre line of the stiffener not exceeding 20 times the web thickness should be included in calculating the buckling resistance (clause 4.5.1.5 of BS 5950). Hence, second moment of area (I) of the effective section (shown cross-hatched in *Fig. 4.16*) (based on $bd^3/12$) for stiffener buckling about the z–z axis is

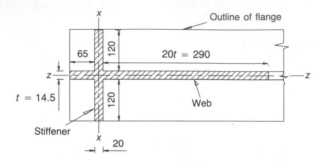

Fig. 4.16 *Plan of web and stiffener.*

$$I = \frac{20 \times (120 + 120 + 14.5)^3}{12} + \frac{(290 + 65) \times 14.5^3}{12}$$

$$= 27.56 \times 10^6 \text{ mm}^4$$

Area of the effective section, A, is

$$A = 20(120 + 120 + 14.5) + (290 + 65)14.5 = 10\ 237.5 \text{ mm}^2$$

Radius of gyration, r, is

$$r = (I/A)^{0.5} = (27.56 \times 10^6/10\ 237.5)^{0.5} = 51.88 \text{ mm}$$

According to clause 4.5.1.5 of BS 5950, the effective length (l_e) of load-carrying stiffeners when the compression flange is laterally restrained is given by

$$l_e = 0.7L \tag{4.15}$$

where L = length of stiffener (= depth between fillet weld, d)

$$l_e = 0.7 \times 615.1 = 430.6 \text{ mm}$$
$$\lambda = l_e/r = 430.6/51.88 = 8.3$$

Then from Table 27(c), BS 5950 (*Table 4.19*) $p_c = p_y = 265$ N mm^{-2}. The buckling resistance of the stiffener, P_x, is given by

$$P_x = p_c A_s \tag{4.16}$$

$$= 265 \times 10\ 237.5 = 2713 \times 10^3 \text{ N}$$
$$= 2713 \text{ kN} > \text{reaction, } F_x \ (= 1100 \text{ kN}) \quad \text{OK}$$

BEARING CHECK

There is an additional stipulation in clause 4.5.4.2 of BS 5950 which, in effect, says that 80% of the reaction should be carried directly through the stiffener. This is the so-called bearing check, and BS 5950 requires that the load-carrying stiffeners should have an area in contact with the flange, A, which exceeds the following expression:

$$A > \frac{0.8F_x}{p_{ys}} \tag{4.17}$$

where F_x is the reaction, A the area of stiffener in contact with the flange and p_{ys} the design strength of the stiffener.

$$A > \frac{0.8 \times 1100 \times 10^3}{265} = 3320 \text{ mm}^2$$

The actual stiffener area in contact with the flange, if a 15 mm fillet is cut out for the root radius, is
$A = (120 - 15)2 \times 20 = 4200 \text{ mm}^2 >$ required $(= 3320 \text{ mm}^2)$ OK

Hence, the stiffener provided satisfies web buckling. However, the web must still be checked for bearing.

WEB BEARING
Bearing resistance of the unstiffened web, P_{crip}, is given by

$$P_{crip} = (b_1 + n_2)tp_{yw} = (150 + 97) \times 14.5 \times 265$$
$$= 949 \times 10^3 \text{ N} = 949 \text{ kN} < R_A = 1100 \text{ kN}$$

where $n_2 = 2.5 (23.7 + 15.2) = 97$ mm. Therefore, a bearing stiffener is required which is capable of resisting a design load of

$$F_v - P_{crip} = 1100 - 949 = 151 \text{ kN}$$

Bearing capacity of stiffener alone is

$$p_{ys}A = 265 \times 4200 = 1113 \times 10^3 \text{ N} > 151 \text{ kN}   \text{OK}$$

Hence the stiffener is also adequate for web bearing.

4.8.11 LATERAL TORSIONAL BUCKLING
If a beam is not effectively laterally restrained, by a concrete floor fixed to the beam, for instance, it is possible for the beam to twist sideways under a load less than that which would cause the beam to fail in bending, shear or deflection. This is called **lateral torsional buckling** which is covered by clause 4.3 of BS 5950. It is illustrated by *Fig. 4.17* for a cantilever, but readers can experiment with this phenomenon very easily using a scale rule or ruler loaded on its edge. They will see that although the

problem occurs more readily with a cantilever, it also applies to beams supported at each end. Keen experimenters will also find that the more restraint they provide at the beam supports, the higher the load required to make the beam twist sideways. This effect is taken into account by using the concept of 'effective length', as discussed below.

4.8.11.1 Effective length
The concept of **effective length** is introduced in clause 4.3.5 of BS 5950. For a beam supported at

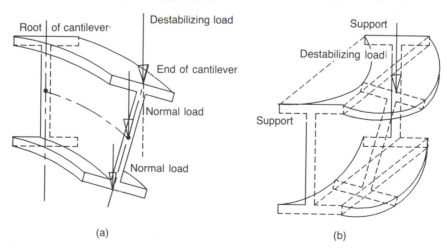

(a)   (b)

Fig. 4.17 *Lateral torsional buckling: (a) cantilever beam; (b) simply support beam.*

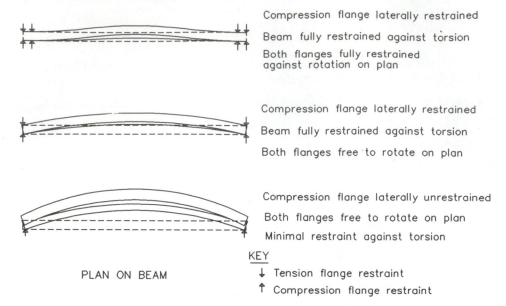

Compression flange laterally restrained

Beam fully restrained against torsion

Both flanges fully restrained
against rotation on plan

Compression flange laterally restrained

Beam fully restrained against torsion

Both flanges free to rotate on plan

Compression flange laterally unrestrained

Both flanges free to rotate on plan

Minimal restraint against torsion

KEY

↓ Tension flange restraint

↑ Compression flange restraint

PLAN ON BEAM

Fig. 4.18 *Restraint conditions.*

its ends only, with no intermediate lateral restraint, and standard restraint conditions at supports, the effective length is equal to the actual length between supports. When a greater degree of lateral and torsional restraint is provided at supports, the effective length is less than the actual length, and vice versa. The effective length appropriate to different end restraint conditions is specified in Table 9 of BS 5950, reproduced as *Table 4.6*, and illustrated in *Fig. 4.18*. The table gives different values of effective length depending on whether the load is

normal or destabilizing. This is perhaps best explained using *Fig. 4.17(a)*, in which it can readily be seen that a load applied to the top of the beam will cause it to twist further, thus worsening the situation. If the load is applied below the centroid of the section, however, it has a slightly restorative effect, and is conservatively assumed to be normal.

Determining the effective length for real beams, when it is difficult to define the actual conditions of restraint, and whether the load is normal or destabilizing, is perhaps one of the greatest problems in the

Table 4.6   Effective length, L_E, for beams (Table 9, BS 5950)

Conditions of restraint at supports	Loading conditions	
	Normal	Destabilizing
Compression flange laterally restrained		
Beam fully restrained against torsion		
Both flanges fully restrained against rotation on plan	0.7L	0.85L
Both flanges partially restrained against rotation on plan	0.85L	1.0L
Both flanges free to rotate on plan	1.0L	1.2L
Compression flange laterally unrestrained		
Both flanges free to rotate on plan		
Restraint against torsion provided only by positive connection of bottom flange to supports	1.0L+2D	1.2L+2D
Restraint against torsion provided only by dead bearing of bottom flange on supports	1.2L + 2D	1.4L + 2D

Table 4.7 Slenderness correction factor n for standard load conditions (Table 20, BS5950)

Load conditions	Bending moment diagrams	n
		0.86
		0.94
		0.94
		0.94

design of steelwork. It is an aspect that has been amended in BS 5950 in order to relate more to practical circumstances, but the fundamental problems still remain.

4.8.11.2 Lateral torsional buckling resistance

Checking of lateral torsional buckling for rolled UB sections can be carried out in two different ways. Firstly, there is a slightly conservative, but quite simple, check in clause 4.3.7.7 of BS 5950, which only applies to equal flanged rolled sections. The approach is similar to that for struts (section 4.9.1), and involves the calculation of an equivalent slenderness

$$\lambda = \frac{L_E n}{r_y} \qquad (4.18)$$

where n is the slenderness correction factor from Tables 15, 16 (*section 4.8.11.3*) or (provided the load is not destabilizing) 20 of BS 5950, which depends on the loading and bending moment diagram, and r_y is the radius of gyration about the y–y axis. Table 20 of BS 5950 is partly reproduced as *Table 4.7*. Then p_b, the buckling strength, is read from Tables 19(a)–(d) of BS 5950, knowing λ and x ($\approx D/T$, a torsional index which can be obtained from steel tables. Tables 19(a) and (b) of BS 5950 are reproduced as *Tables 4.8* and *4.9* respectively. Finally the buckling moment, M_b, is

$$M_b = p_b S_x \qquad (4.19)$$

where S_x is the plastic section modulus about the x–x axis.

Example 4.7 Design of a laterally unrestrained steel beam

Assuming the beam in *Example 4.1* is not laterally restrained, determine whether the selected section is stable, and if not, select one which is. Assume the compression flange is laterally restrained, the beam is fully restrained against torsion but both flanges are free to rotate on plan. The loading is normal.

EFFECTIVE LENGTH
Since beam is pinned at both ends, from *Table 4.6*, $L_E = 1.0 \, L = 10$ m.

BUCKLING RESISTANCE
Using the conservative approach of clause 4.3.7.7

$$\frac{L_E}{r_y} = \frac{10\ 000}{39.7} = 252$$

n from *Table 4.7* = 0.9 (approx.), by interpolation between 0.86 and 0.94

$$\lambda = 0.9 \times 252 = 227$$

Then, with $x = 33.9$, $p_b = 63$ (approx.) from *Table 4.9*

$$M_b = p_b S_x = 63 \times 1190 \times 10^3$$
$$= 75.0 \times 10^6 \text{ N mm} = 75.0 \text{ kN m} << 307.5 \text{ kN m}$$

As the buckling resistance moment is much less than the actual imposed moment, this beam would fail by lateral torsional buckling. Using trial and error, a more suitable section can be found:

Try $305 \times 305 \times 118$ UC: $p_y = 265$ N mm^{-2}, plastic.

$$\frac{L_E}{r_y} = \frac{10\,000}{77.5} = 129 \qquad \lambda = 0.9 \times 129 = 116 \qquad x = 16.2$$

Then from *Table 4.8*, $p_b = 170$ N mm^{-2}. Hence

Table 4.8 Bending strengths, p_b, for rolled sections with equal flanges, $p_y = 265$ N mm^{-2} (Table 19 (a), BS 5950)

λ	x 5	10	15	20	25	30	35	40	45	50
30	265	265	265	265	265	265	265	265	265	265
35	265	265	265	265	265	265	265	265	265	265
40	265	265	265	265	265	264	264	264	263	263
45	265	265	261	258	256	255	254	254	254	254
50	265	261	253	249	247	246	245	244	244	244
55	265	255	246	241	238	236	235	235	234	234
60	265	250	239	233	229	227	226	225	224	224
65	265	245	232	225	221	218	216	215	214	214
70	265	240	225	217	212	209	207	205	204	204
75	263	235	219	210	204	200	198	196	195	194
80	260	230	213	202	196	191	189	187	185	184
85	257	226	207	195	188	183	180	178	176	175
90	254	222	201	188	180	175	171	169	167	166
95	252	217	196	182	173	167	163	160	158	157
100	249	213	190	176	166	160	156	153	150	149
105	247	209	185	170	160	153	148	145	143	141
110	244	206	180	164	154	147	142	138	136	134
115	242	202	176	159	148	140	135	132	129	127
120	240	198	171	154	142	135	129	125	123	121
125	237	195	167	149	137	129	124	120	117	115
130	235	191	163	144	132	124	119	114	111	109
135	233	188	159	140	128	119	114	109	106	104
140	231	185	155	136	124	115	109	105	102	99
145	229	182	152	132	120	111	105	101	97	95
150	227	179	148	129	116	107	101	97	93	91
155	225	176	145	125	112	103	97	93	89	87
160	223	173	142	122	109	100	94	89	86	83
165	221	170	139	119	106	97	91	86	83	80
170	219	167	136	116	103	94	88	83	80	77
175	217	165	133	113	100	91	85	80	77	74
180	215	162	130	110	97	88	82	77	74	71
185	213	160	128	108	95	86	79	75	71	69
190	211	157	125	105	92	83	77	73	69	66
195	209	155	123	103	90	81	75	70	67	64
200	207	153	120	101	88	79	73	68	65	62
210	204	148	116	96	84	75	69	64	61	58
220	200	144	112	93	80	71	65	61	58	55
230	197	140	108	89	77	68	62	58	54	52
240	194	136	104	86	74	65	59	55	52	49
250	190	132	101	83	71	63	57	52	49	47

Table 4.9 Bending strengths, p_b, for rolled sections with equal flanges, $p_y = 275$ N mm^{-2} (Table 19 (b), BS 5950)

λ	x									
	5	10	15	20	25	30	35	40	45	50
30	275	275	275	275	275	275	275	275	275	275
35	275	275	275	275	275	275	275	275	275	275
40	275	275	275	275	274	273	272	272	272	272
45	275	275	269	266	264	263	263	262	262	262
50	275	269	261	257	255	253	253	252	252	251
55	275	263	254	248	246	244	243	242	241	241
60	275	258	246	240	236	234	233	232	231	230
65	275	252	239	232	227	224	223	221	221	220
70	274	247	232	223	218	215	213	211	210	209
75	271	242	225	215	209	206	203	201	200	199
80	268	237	219	208	201	196	193	191	190	189
85	265	233	213	200	193	188	184	182	180	179
90	262	228	207	193	185	179	175	173	171	169
95	260	224	201	186	177	171	167	164	162	160
100	257	219	195	180	170	164	159	156	153	152
105	254	215	190	174	163	156	151	148	146	144
110	252	211	185	168	157	150	144	141	138	136
115	250	207	180	162	151	143	138	134	131	129
120	247	204	175	157	145	137	132	128	125	123
125	245	200	171	152	140	132	126	122	119	116
130	242	196	167	147	135	126	120	116	113	111
135	240	193	162	143	130	121	115	111	108	106
140	238	190	159	139	126	117	111	106	103	101
145	236	186	155	135	122	113	106	102	99	96
150	233	183	151	131	118	109	102	98	95	92
155	231	180	148	127	114	105	99	94	91	88
160	229	177	144	124	111	101	95	90	87	84
165	227	174	141	121	107	98	92	87	84	81
170	225	171	138	118	104	95	89	84	81	78
175	223	169	135	115	101	92	86	81	78	75
180	221	166	133	112	99	89	83	78	75	72
185	219	163	130	109	96	87	80	76	72	70
190	217	161	127	107	93	84	78	73	70	67
195	215	158	125	104	91	82	76	71	68	65
200	213	156	122	102	89	80	74	69	65	63
210	209	151	118	98	85	76	70	65	62	59
220	206	147	114	94	81	72	66	62	58	55
230	202	143	110	90	78	69	63	58	55	52
240	199	139	106	87	74	66	60	56	52	50
250	195	135	103	84	72	63	57	53	50	47

$$M_b = p_b S_x = 170 \times 1950 \times 10^3 = 331.5 \times 10^6 \text{ N mm}$$
$$= 331.5 \text{ kN m}$$

$$M_{sw} = 1.4 \times \frac{118 \times 9.81 \times 10^2}{1000 \times 8} = 20.25 \text{ kN m}$$

Total imposed moment is

$$M_t = 307.5 + 20.25 = 327.75 \text{ kN m} < 331.5 \text{ kN m}$$

So this beam, actually a column section, is suitable. Readers may like to check that there are not any lighter UB sections. Since the column has a greater r_y, it is laterally stiffer than a UB section of the same weight and is more suitable than a beam section in this particular situation.

Table 4.10 Slenderness correction factor, n, for members with applied loading substantially concentrated within the middle fifth of the unrestrained length (based on Table 15, BS 5950)

$\gamma = M/M_o$	β positive					
	1.0	0.8	0.6	0.4	0.2	0.0
+50.00	1.00	0.96	0.92	0.87	0.82	0.77
+10.00	0.99	0.99	0.94	0.90	0.85	0.80
+ 5.00	0.98	0.98	0.97	0.93	0.89	0.84
+ 2.00	0.96	0.95	0.95	0.95	0.94	0.94
+ 1.50	0.95	0.95	0.94	0.94	0.93	0.93
+ 1.00	0.93	0.92	0.92	0.92	0.92	0.91
+ 0.50	0.90	0.90	0.90	0.89	0.89	0.89
0.00	0.86	0.86	0.86	0.86	0.86	0.86

Table 4.11 Slenderness correction factor, n, for members with applied loading other than as for Table 15 of BS 5950 (based on Table 16, BS 5950)

$\gamma = M/M_o$	β positive					
	.0	0.8	0.6	0.4	0.2	0.0
+50.00	1.00	0.96	0.92	0.87	0.83	0.77
+10.00	0.99	0.98	0.95	0.91	0.86	0.81
+ 5.00	0.99	0.98	0.97	0.94	0.90	0.85
+ 2.00	0.98	0.98	0.97	0.96	0.94	0.92
+ 1.50	0.97	0.97	0.97	0.96	0.95	0.93
+ 1.00	0.97	0.97	0.97	0.96	0.96	0.95
+ 0.50	0.96	0.96	0.96	0.96	0.96	0.95
0.00	0.94	0.94	0.94	0.94	0.94	0.94

Note. The above tables are extracts from Tables 15 and 16 of the code showing n for positive values of β and γ only.

4.8.11.3 Equivalent slenderness and uniform moment factors

The more rigorous procedure for calculating values of M_b is covered by clauses 4.3.7.1–4.3.7.6 of BS 5950. There are two alternative approaches depending on the loading conditions and involve the use of the following parameters: (i) equivalent uniform

Table 4.12 Equivalent uniform moment factor, m (Table 18, BS 5950)

	β	m
Beta positive	1.0	1.00
	0.9	0.95
	0.8	0.90
	0.7	0.85
	0.6	0.80
	0.5	0.76
	0.4	0.72
	0.3	0.68
	0.2	0.64
	0.1	0.60
Beta negative	0.0	0.57
	-0.1	0.54
	-0.2	0.51
	-0.3	0.48
	-0.4	0.45
	-0.5	
	to	0.43
	-0.1	

Note 1. The values of m in this table apply only to end moments applied to beams of *uniform* section with *equal* flanges. In all other cases $m = 1.0$.

Note 2. Values of m for intermediate values of β may be interpolated, or determined from the equation: $m = 0.57 + 0.33\beta + 0.10\beta^2$: but not less than 0.43.

Table 4.13 Use of *n* and *m* factors for members of uniform section and with equal flanges (based on Table 13, BS 5950)

Description	Members not subject to destabilizing loads		Members subject to destabilizing loads[a]	
	m	*n*	*m*	*n*
Members loaded between adjacent lateral restraints	1.0	From Table 15 or 16 of BS 5950	1.0	1.0
Members not loaded between adjacent lateral restraints	From Table 18 of BS 5950	1.0	1.0	1.0
Cantilevers without intermediate lateral restraints	1.0	1.0	1.0	1.0

[a]Destabilizing load as defined in 4.3.4 of BS 5950.

Table 4.14

Slenderness factor *v* for beams with equal flanges (based on Table 14, BS 5950)

$\dfrac{\lambda}{x}$	N = 0.5
0.5	1.00
1.0	0.99
1.5	0.97
2.0	0.96
2.5	0.93
3.0	0.91
3.5	0.89
4.0	0.86
4.5	0.84
5.0	0.82
5.5	0.79
6.0	0.77
6.5	0.75
7.0	0.73
7.5	0.72
8.0	0.70
8.5	0.68
9.0	0.67
9.5	0.65
10.0	0.64
11.0	0.61
12.0	0.59
13.0	0.57
14.0	0.55
15.0	0.53
16.0	0.52
17.0	0.50
18.0	0.49
19.0	0.48
20.0	0.47

moment factor, *m*, and (ii) equivalent slenderness correction factor, *n*. The following discusses these design approaches for cantilevers and members with equal flanges.

If the section of beam is loaded between adjacent lateral restraints, the **equivalent slenderness correction factor**, *n*, is used; the factor *m* = 1 and factor *n* is read from Table 15 or 16 of BS 5950, parts of which are reproduced as *Tables 4.10* and *4.11* respectively. Note that *Table 4.10* applies to members with applied loading substantially concentrated within the middle fifth of the unrestrained length (*Fig. 4.19*). Otherwise, values for the slenderness correction factor should be obtained from *Table 4.11*. However, *n* = 1 for destabilizing loads and for cantilevers.

If the section of beam is not loaded between adjacent lateral restraints, the **equivalent uniform moment** approach is used; factor *m* is read from Table 18 of BS 5950 (reproduced as *Table 4.12*) and factor *n* = 1. Then the equivalent uniform moment $M = mM_A$, where M_A is the maximum moment on the section of member under consideration. However, *m* = 1 for destabilizing loads and for cantilevers. This is summarized in Table 13 of BS 5950, part of which is reproduced as *Table 4.13*.

Then, for both methods from clause 4.3.7.5 of BS 5950,

$$\lambda_{LT} = nuv\lambda \qquad (4.20)$$

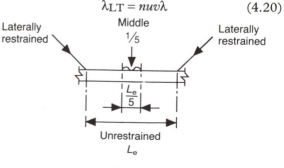

Fig. 4.19 *Member loaded between adjacent points of restraint and concentrated in middle fifth of length.*

where λ_{LT} is the equivalent slenderness, n a slenderness correction factor from *Tables 4.10* and *4.11*, u is the buckling parameter from steel tables and v a slenderness factor from Table 14 of BS 5950, part of which is reproduced as *Table 4.14*. p_b can then be determined from Table 11 of BS 5950, reproduced as *Table 4.15*, knowing λ_{LT} and p_y. Finally, as before $M_b = p_b S_x$ which must be greater than M or $\overline{M}$.

4.9 Design of compression members

4.9.1 STRUTS

Steel compression members, commonly referred to as stanchions, include struts and columns. A strut is a member subject to direct compression only. A column, on the other hand, refers to members subject to a combination of compressive loading and bending. Although most compression members in

---

Example 4.8 Design of a laterally unrestrained beam – rigorous method

Repeat *Example 4.7* using the more rigorous method. Assume selected section is $305 \times 305 \times 118$ UC as before. From *Table 4.10*, with $M/M_0 = 0$, and $\beta = 0$, $n = 0.86$. From *Table 4.11*, with M/M_0 0, and $\beta = 0$, $n = 0.94$. Using the same compromise as before, $n = 0.9$.
From steel tables (*Appendix B*), $u = 0.851$, then

$$\lambda = L_E/r_y = 10\,000/77.5 = 129 \qquad \lambda/x = 129/16.2 = 7.96$$

From *Table 4.14*, $v = 0.7$, then $\lambda_{LT} = nuv\lambda = 0.9 \times 0.851 \times 0.7 \times 129 = 69.2$
From *Table 4.15*, $p_b = 184$ N mm^{-2}, hence

$$M_b = p_b S_x = 184 \times 1950 \times 10^3$$
$$= 358.8 \times 10^6 \text{ N mm} = 358.8 \text{ kN m}$$

This gives, in this case, a buckling moment nearly 10% greater than in *Example 4.7*. This may enable a lighter member to be selected, but for rolled sections it may not be really worth the additional effort. *Note.* An example using the equivalent uniform moment method is included in *section 4.9.2.2*.

4.8.11.4 Cantilever beams

For cantilevers, the effective length is given in clear diagrammatical form in Table 10 of BS 5950, part of which is reproduced as *Table 4.16*.

real structures resist compressive loading and bending, the strut is a convenient starting-point.

Struts (and columns) differ fundamentally in their behaviour under axial load depending on whether they are **slender** or **stocky**. Most real struts (and columns) can neither be regarded as slender nor stocky, but as something in between, but let us look at the behaviour of stocky and slender struts first.

Example 4.9 Checking for lateral instability in a cantilever steel beam

Continue *Example 4.3* to determine whether the cantilever is laterally stable. Assume the load is destabilizing.
From *Table 4.16*, $L_E = 1.4L = 1.4$ m. For $610 \times 229 \times 113$ UB, $L_E/r_y = 1400/48.8 = 29$. From *Table 4.13*, $m = n = 1$ and from steel tables, $u = 0.87$. $\lambda/x = 29/37.9 = 0.76$, then from *Table 4.14*, $v = 0.99$. So,

$$\lambda_{LT} = nuv\lambda = 1 \times 0.87 \times 0.99 \times 29 = 25$$

Then from *Table 4.15*, $p_b = 265$ N mm$^{-2} = p_y$, and so lateral torsional buckling is not a problem with this cantilever.

Table 4.15   Bending strengths, p_b for rolled sections (Table 11, BS 5950)

λ_{LT}	p_y								
	245	265	275	325	340	355	415	430	450
30	245	265	275	325	340	355	408	421	438
35	245	265	273	316	328	341	390	402	418
40	238	254	262	302	313	325	371	382	397
45	227	242	250	287	298	309	350	361	374
50	217	231	238	272	282	292	329	338	350
55	206	219	226	257	266	274	307	315	325
60	195	207	213	241	249	257	285	292	300
65	185	196	201	225	232	239	263	269	276
70	174	184	188	210	216	222	242	247	253
75	164	172	176	195	200	205	223	226	231
80	154	161	165	181	186	190	204	208	212
85	144	151	154	168	172	175	188	190	194
90	135	141	144	156	159	162	173	175	178
95	126	131	134	144	147	150	159	161	163
100	118	123	125	134	137	139	147	148	150
105	111	115	117	125	127	129	136	137	139
110	104	107	109	116	118	120	126	127	128
115	97	101	102	108	110	111	117	118	119
120	91	94	96	101	103	104	108	109	111
125	86	89	90	95	96	97	101	102	103
130	81	83	84	89	90	91	94	95	96
135	76	78	79	83	84	85	88	89	90
140	72	74	75	78	79	80	83	84	84
145	68	70	71	74	75	75	78	79	79
150	64	66	67	70	70	71	73	74	75
155	61	62	63	66	66	67	69	70	70
160	58	59	60	62	63	63	65	66	66
165	55	56	57	59	60	60	62	62	63
170	52	53	54	56	56	57	59	59	59
175	50	51	51	53	54	54	56	56	56
180	47	48	49	51	51	51	53	53	53
185	45	46	46	48	49	49	50	50	51
190	43	44	44	46	46	47	48	48	48
195	41	42	42	44	44	44	46	46	46
200	39	40	40	42	42	42	43	44	44
210	36	37	37	38	39	39	40	40	40
220	33	34	34	35	35	36	36	37	37
230	31	31	31	32	33	33	33	34	34
240	29	29	29	30	30	30	31	31	31
250	27	27	27	28	28	28	29	29	29

Table 4.16 Effective length, L_E, for cantilever of length L (based on Table 10, BS 5950)

Restraint conditions at supports	Restraint conditions at tip	Loading conditions	
		Normal	Destabilizing
Built-in laterally and torsionally	Free	0.8L	1.4L
	Lateral restraint on top flange only	0.7L	1.4L
	Torsionally restrained only	0.6L	0.6L
	Laterally and torsionally restrained	0.5L	0.5L

Stocky struts will fail by crushing or squashing of the material. For stocky struts the 'squash load', P_s, is given by the simple formula

$$P_s = p_y A_g \qquad (4.21)$$

where p_y is the steel design strength and A_g the gross cross-sectional area of the section.

Slender struts will fail by buckling. For elastic slender struts pinned at each end, the 'Euler load', at which a perfect strut buckles elastically, is given by

$$P_E = \frac{\pi^2 EI}{L^2} = \frac{\pi^2 EA_g r_y^2}{L^2} = \frac{\pi^2 EA_g}{\lambda^2} \qquad (4.22)$$

using $r = \sqrt{\dfrac{I}{A}}$ and $\lambda = \dfrac{L}{r}$. If the compressive strength, p_c, which is given by

$$p_c = P_s/A_g \quad \text{(for stocky struts)} \qquad (4.23)$$

and

$$p_c = P_E/A_g \quad \text{(for slender struts)} \qquad (4.24)$$

are plotted against λ (*Fig. 4.20*), the area above the two dotted lines represents an impossible situation in respect of these struts. In this area, the strut has either buckled or squashed. Struts which fall below the dotted lines should theoretically be able to withstand the applied loading without buckling or squashing. In reality, however, this tends not to be the case because of a combination of manufacturing and practical considerations. For example, struts are never completely straight, or are subject to exactly concentric loading. During manufacture, stresses are locked into steel members which effectively reduce their load-carrying capacities. As a result of these factors, failure of a strut will not be completely

due to buckling or squashing, but a combination, with partially plastic stresses appearing across the member section. These non-ideal factors or imperfections are found in practice, and laboratory tests have confirmed that in fact the failure line for real struts lies along a series of lines such as a,b,c and d in *Fig. 4.20*. These are derived from the Perry–Robertson equation which includes allowances for the various imperfections.

Whichever of the lines a–d is used depends on the shape of section and the axis of buckling. Table 25 of BS 5950, reproduced as *Table 4.17*, specifies which of the lines is appropriate for the shape of section, and Tables 27(a)–(d) enable values of p_c to be read off appropriate to the section used. (Tables 27(b) and (c) of BS 5950 have been reproduced as *Tables 4.18* and *4.19* respectively.) Alternatively, Appendix C of BS 5950 gives the actual Perry–Robertson equations which may be used in place of the tables if considered necessary.

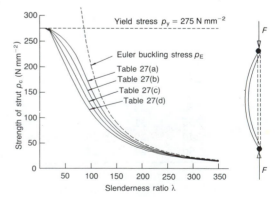

Fig. 4.20 *Buckling curves for ideal and real struts.*

Table 4.17   Strut table selection (based on Table 25, BS 5950)

Type of section	Thickness[a]	Axis of buckling	
		x-x	y-y
Hot-rolled structural hollow section		27(a)	27(a)
Rolled I-section		27(a)	27(b)
Rolled H-section	Up to 40 mm	27(b)	27(c)
	Over 40 mm	27(c)	27(d)

[a] For thicknesses between 40 and 50 mm the value of p_c may be taken as the average of the values for thicknesses up to 40 mm and over 40 mm.

Table 4.18   Compressive strength, p_c (N mm^{-2}) for struts (Table 27(b), BS 5950)

λ	p_y 225	245	255	265	275	λ	p_y 225	245	255	265	275
15	225	245	255	265	275	96	133	140	143	146	148
20	224	243	253	263	272	98	130	137	139	142	145
25	220	239	248	258	267	100	127	133	136	138	141
30	216	234	243	253	262	102	124	130	132	135	137
35	211	229	238	247	256	104	122	127	129	131	133
40	207	224	233	241	250	106	119	124	126	128	130
42	205	222	231	239	248	108	116	121	123	125	126
44	203	220	228	237	245	110	113	118	120	121	123
46	201	218	226	234	242	112	111	115	117	118	120
48	199	215	223	231	239	114	108	112	114	115	117
50	197	213	221	229	237	116	105	109	111	112	114
52	195	210	218	226	234	118	103	106	108	109	111
54	192	208	215	223	230	120	100	104	105	107	108
56	190	205	213	220	227	122	98	101	103	104	105
58	188	202	210	217	224	124	96	99	100	101	102
60	185	200	207	214	221	126	94	96	97	99	100
62	183	197	204	210	217	128	91	94	95	96	97
64	180	194	200	207	213	130	89	92	93	94	95
66	178	191	197	203	210	135	84	86	87	88	89
68	175	188	194	200	206	140	79	81	82	83	84
70	172	185	190	196	202	145	75	77	78	78	79
72	169	181	187	193	198	150	71	72	73	74	74
74	167	178	183	189	194	155	67	69	69	70	70
76	164	175	180	185	190	160	64	65	66	66	66
78	161	171	176	181	186	165	60	61	62	63	63
80	158	168	172	177	181	170	57	58	59	59	60
82	155	164	169	173	177	175	55	56	56	56	57
84	152	161	165	169	173	180	52	53	53	54	54
86	149	157	161	165	169	185	49	50	51	51	51
88	146	154	158	161	165	190	47	48	48	48	49
90	143	150	154	157	161	195	45	46	46	46	47
92	139	147	150	153	156	200	43	44	44	44	44
94	136	143	147	150	152						

Table 4.19 Compressive strength, p_c (N mm^{-2}) for struts (Table 27(c), BS 5950)

λ	p_y 225	245	255	265	275	λ	p_y 225	245	255	265	275
15	225	245	255	265	275	96	118	124	127	129	132
20	224	242	252	261	271	98	115	121	123	126	129
25	217	235	245	254	263	100	112	118	120	123	125
30	211	228	237	246	255	102	110	115	118	120	122
35	204	221	230	238	247	104	107	112	115	117	119
40	198	214	222	230	238	106	105	110	112	114	116
42	195	211	219	227	235	108	102	107	109	111	113
44	193	208	216	224	231	110	100	104	106	108	110
46	190	205	213	220	228	112	98	102	104	106	107
48	187	202	209	217	224	114	96	99	101	103	105
50	184	199	206	213	220	116	93	97	99	101	102
52	181	196	203	210	217	118	91	95	96	98	100
54	179	193	199	206	213	120	89	93	94	96	97
56	176	189	196	202	209	122	87	91	92	93	95
58	173	186	192	199	205	124	85	88	90	91	92
60	170	183	189	195	201	126	83	86	88	89	90
62	167	179	185	191	197	128	82	84	86	87	88
64	164	176	182	188	193	130	80	82	84	85	86
66	161	173	178	184	189	135	75	78	79	80	81
68	158	169	175	180	185	140	71	74	75	76	76
70	155	166	171	176	181	145	68	70	70	71	72
72	152	163	168	172	177	150	64	66	67	68	68
74	149	159	164	169	173	155	61	63	63	64	65
76	146	156	160	165	169	160	58	59	60	61	61
78	143	152	157	161	165	165	55	56	57	58	58
80	140	149	153	157	161	170	52	54	54	55	55
82	137	146	150	154	157	175	50	51	52	52	53
84	134	142	146	150	154	180	48	49	49	50	50
86	132	139	143	146	150	185	46	46	47	47	48
88	129	136	139	143	146	190	43	44	45	45	46
90	126	133	136	139	142	195	42	42	43	43	43
92	123	130	133	136	139	200	40	41	41	41	42
94	120	127	130	133	135						

4.9.2 EFFECTIVE LENGTH

As mentioned in *section 4.9.1*, the compressive strength of struts is primarily related to their slenderness ratio. The slenderness ratio is given by

$$\lambda = \frac{L_E}{r} \qquad (4.25)$$

where L_E is the effective length of the member and r the radius of gyration obtained from steel tables. The concept of effective lengths was discussed earlier in *section 4.8.11.1*, in the context of lateral torsional buckling, and is similarly applicable to the design of struts and columns. It is based on the actual length of the member and the restraint at the member ends.

The formulae in Appendix C of BS 5950 and the graph in *Fig. 4.20* relate to standard restraint conditions in which each end is pinned. In reality each end of the strut may be free, pinned, partially fixed, or fully fixed (rotationally). Also, whether or not the top of the strut is allowed to move laterally with respect to the bottom end is important, i.e. whether the structure is unbraced or braced. *Figure 4.21* summarizes these restraints, and Table 24 of BS 5950, reproduced as *Table 4.20*, stipulates conservative assumptions of effective length L_E from which the slenderness λ can be calculated. Note that the design

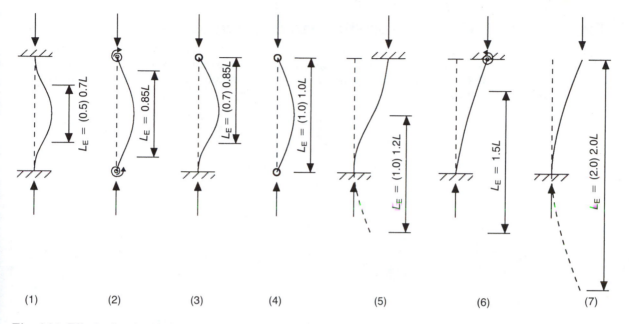

Fig. 4.21 *Effective lengths (the bracketed figures indicate the theoretical effective length for each case).*

effective lengths are greater than the theoretical values where one or both ends of the member are partially or wholly restrained. This is because, in practice, it is difficult if not impossible to guarantee that some rotation of the member will not take place. Furthermore, the effective lengths are always less than the actual length of the compression member except when the structure is unbraced.

Table 4.20 Nominal effective length, L_E, for a strut (Table 24, BS 5950)

Conditions of restraint at ends (in plane under consideration)			Effective length, L_E
Effectively held in position at both ends	Restrained in direction at both ends (1)		0.7L
	Partially restrained in direction at both ends (2)		0.85L
	Restrained in direction at one end (3)		0.85L
	Not restrained in direction at either end (4)		1.0L
One end	*Other end*		
Effectively held in position and restrained in direction	Not held in position	Effectively restrained in direction (5)	1.2L
		Partially restrained in direction (6)	1.5L
		Not restrained in direction (7)	2.0L

Example 4.10 Column resisting an axial load

A proposed 5 m long internal column in a 'rigid' jointed steel structure is to be loaded concentrically with 1000 kN dead and 1000 kN imposed load (*Fig. 4.22*). Assuming that fixity at the top and bottom of the column gives effective rotational restraints, design column sections assuming the structure will be (a) braced and (b) unbraced.

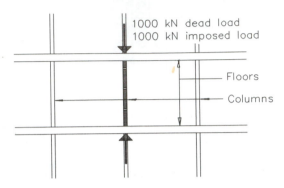

BRACED COLUMN

Design axial loading
Factored loading $= (1.4 \times 1000) + (1.6 \times 1000) = 3000$ kN

Effective length
For the braced case the column is assumed to be effectively held in position at both ends, and restrained in direction at both ends. It will buckle about the weak (y – y) axis. From *Table 4.20* therefore, $L_E = 0.7L = 0.7 \times 5 = 3.5$ m.

Section selection
This column design can only really be done by trial and error.

Initial trial. Try $254 \times 254 \times 107$ UC:

$$p_y = 265 \text{ N mm}^{-2} \qquad r_y = 65.7 \text{ mm} \qquad A_g = 13\ 700 \text{ mm}^2$$
$$\lambda = L_E/r_y = 3500/65.7 = 53$$

From *Table 4.17*, use Table 27(c) of BS 5950 (i.e. *Table 4.19*), from which $p_c = 208$ N mm^{-2}. Then

$$P_c = A_g p_c = 13\ 700 \times 208/10^3$$
$$= 2850 \text{ kN} < 3000 \text{ kN} \quad \text{Not OK}$$

Second trial. Try $305 \times 305 \times 118$ UC:

$$p_y = 265 \text{ N mm}^{-2} \qquad r_y = 77.5 \text{ mm} \qquad A_g = 15\ 000 \text{ mm}^2$$
$$\lambda = L_E/r_y = 3500/77.5 = 45$$

Then from Table 27(c) of BS 5950 (i.e. *Table 4.19*), $p_c = 222$ N mm^{-2}.

$$P_c = A_g p_c = 15\ 000 \times 222/10^3$$
$$= 3330 \text{ kN} > 3000 \text{ kN} \quad \text{OK}$$

UNBRACED COLUMN
For the unbraced case, $L_E = 1.2L = 1.2 \times 5 = 6.0$ m from *Table 4.20*, and the most economic member would appear to be $305 \times 305 \times 158$ UC:

$$p_y = 265 \text{ N mm}^{-2} \qquad r_y = 78.9 \text{ mm} \qquad A_g = 20\,100 \text{ mm}^2$$
$$\lambda = L_E/r_y = 6000/78.9 = 76$$

Then from *Table 4.19*, $p_c = 165 \text{ N mm}^{-2}$.

$$P_c = A_g p_c = 20\,100 \times 167/10^3$$
$$= 3357 \text{ kN} > 3000 \text{ kN} \quad \text{OK}$$

Hence, it can immediately be seen that for a given axial load, a bigger steel section will be required if the column is unbraced.

4.9.3 COLUMNS WITH BENDING MOMENTS

As noted earlier, most columns in steel structures are subject to both axial load and bending. According to BS 5950, such members should be checked for local (yield or buckling) capacity at the points of greatest bending moment and axial load, which usually occur at the member ends. In addition, the member should also be checked for overall buckling.

4.9.3.1 Local capacity check

The purpose of this check is to ensure that nowhere across the section does the steel stress exceed yield. For semi-compact and slender cross-sections and for the simplified approach for compact cross-sections, clause 4.8.3.2 of BS 5950 states that the following relationship should be satisfied:

$$\frac{F}{A_g p_y} + \frac{M_x}{M_{cx}} + \frac{M_y}{M_{cy}} \leqslant 1 \qquad (4.26)$$

where F is the applied axial load, A_g the gross cross-sectional area, p_y the design strength, M_x the applied moment about the major axis, M_{cx} the moment capacity about the major axis in the absence of axial load, M_y the applied moment about the minor axis and M_{cy} the moment capacity about the minor axis in the absence of axial load.

4.9.3.2 Overall buckling check

Buckling due to imposed axial load, lateral torsional buckling due to imposed moment, or a combination of buckling and lateral torsional buckling are additional possible modes of failure in most practical columns in steel structures.

Clause 4.8.3.3.1 of BS 5950 gives a simplified approach for calculating the overall buckling resistance of columns which involves satisfying the following equation:

$$\frac{F}{A_g p_c} + \frac{m M_x}{M_b} + \frac{m M_y}{p_y Z_y} \leqslant 1 \qquad (4.27)$$

where p_c is the buckling compressive strength of the member, m the equivalent uniform moment factor (*Table 4.12*), M_b is the buckling resistance moment capacity (major axis) Z_y is the elastic section modulus about the minor axis and other symbols are as above. Clause 4.8.3.3.2 of BS 5950 gives a more exact approach, but as in practice most designers tend to use the simplified approach, the more exact method is not discussed here.

Example 4.11 Column resisting an axial load and bending

Select a suitable column section in grade 43 steel to support a factored axial concentric load of 2000 kN and factored bending moments of 100 kN m about the major axis, and 20 kN m about the minor axis (*Fig. 4.23*), applied at the top of the column. The column is 10 m long and is fully fixed against rotation at top and bottom, and the floors it supports are braced against sway.

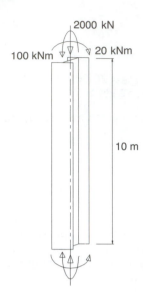

Fig. 4.23

INITIAL SECTION SELECTION
$305 \times 305 \times 118$ UC:

$p_y = 265$ N mm^{-2}, plastic $\quad$ $S_x = 1950$ cm^3
$A_g = 150$ cm^2 $\quad\quad\quad\quad$ $S_y = 892$ cm^3 $\quad$ $Z_y = 587$ cm^3
$r_y = 7.75$ cm $\quad\quad\quad\quad\quad$ $x = 16.2$ $\quad\quad\quad$ $u = 0.851$

Note: In this case the classification procedure is slightly different in respect of web classification. Initially it is conservatively assumed that the web is wholly in compression, the neutral axis lying in the flange. From Table 7 of BS 5950 (*Table 4.4*), for the member to be plastic, compact or semi-compact $d/t \leqslant 39\varepsilon$.

$$M_{cx} = p_y S_x = 265 \times 1950/10^3 = 516.75 \text{ kN m}$$
$$M_{cy} = p_y S_y = 265 \times 892/10^3 = 236.38 \text{ kN m}$$

LOCAL CAPACITY CHECK

$$\frac{F}{A_g p_y} + \frac{M_x}{M_{cx}} + \frac{M_y}{M_{cy}} = \frac{2000 \times 10^3}{150 \times 10^2 \times 265} + \frac{100}{516.75} + \frac{20}{236.38}$$
$$= 0.503 + 0.193 + 0.085$$
$$= 0.781 < 1 \quad \text{OK}$$

OVERALL BUCKLING CHECK

Axial buckling
From *Table 4.20*, effective length $L_E = 0.7L = 0.7 \times 10 = 7$ m and $\lambda = L_E/r_y = 7000/77.5 = 90$
From *Table 4.17*, relevant compression strength values for buckling about the y–y axis are obtained from Table 27(c) of BS 5950 (*Table 4.19*) from which $p_c = 139$ N mm^{-2}. Then

$$\frac{F}{A_g p_c} = \frac{2000 \times 10^3}{15\,000 \times 139} = 0.96$$

Lateral torsional buckling
Use the equivalent uniform moment method. Ratio of end moments, $\beta = 1$. Hence from *Table 4.12*, $m = 1$ and from *Table 4.13*, $n = 1$. $\lambda/x = 90/16.2 = 5.5$. Hence from *Table 4.14*, $v = 0.79$.

$$\lambda_{LT} = nuv\lambda = 1 \times 0.851 \times 0.79 \times 90 = 61$$

From *Table 4.15*, $p_b = 205$ N mm^{-2}.

$$M_b = p_b S_x = 205 \times 1950/10^3 = 400 \text{ kN m}$$

Then

$$\frac{m M_x}{M_b} = \frac{1 \times 100}{400} = 0.25$$

and

$$\frac{m M_y}{p_y Z_y} = \frac{1 \times 20 \times 10^3}{265 \times 587} = 0.128$$

Overall buckling

$$\frac{F}{A_g p_c} + \frac{m M_x}{M_b} + \frac{m M_y}{p_y Z_y} = 0.96 + 0.25 + 0.13 = 1.34 > 1$$

Hence, a bigger section should be selected.

SECOND SECTION SELECTION
Try $356 \times 368 \times 177$ UC:

$p_y = 265 \text{ N mm}^{-2}$ plastic $\qquad S_x = 3460 \text{ cm}^3$

$A_g = 226 \text{ cm}^2 \qquad\qquad\quad S_y = 1670 \text{ cm}^3 \qquad Z_y = 1100 \text{ cm}^3$

$r_y = 9.52 \text{ cm} \qquad\qquad\quad x = 15 \qquad\qquad u = 0.844$

$$M_{cx} = p_y S_x = 265 \times 3460/10^3 = 916.9 \text{ kN m}$$
$$M_{cy} = p_y S_y = 265 \times 1670/10^3 = 442.55 \text{ kN m}$$

LOCAL CAPACITY CHECK

$$\frac{F}{A_g p_y} + \frac{M_x}{M_{cx}} + \frac{M_y}{M_{cy}} = \frac{2000 \times 10^3}{226 \times 10^2 \times 265} + \frac{100}{916.9} + \frac{20}{442.55}$$
$$= 0.334 + 0.109 + 0.045$$
$$= 0.448 < 1 \quad \text{OK}$$

OVERALL BUCKLING CHECK
Axial buckling
$\lambda = L_E/r_y = 7000/95.2 = 74$
As above, use *Table 4.19* from which $p_c = 169 \text{ N mm}^{-2}$. Then

$$\frac{F}{A_g p_c} = \frac{2000 \times 10^3}{22\,600 \times 169} = 0.524$$

Lateral torsional buckling
$\lambda/x = 74/15 = 4.93$. Hence from *Table 4.14*, $v = 0.82$.

$$\lambda_{LT} = nuv\lambda = 1 \times 0.844 \times 0.82 \times 74 = 51$$

From *Table 4.15*, $p_b = 229 \text{ N mm}^{-2}$.

$$M_b = p_b S_x = 229 \times 3460/10^3 = 792 \text{ kN m}$$

Then

$$\frac{m M_x}{M_b} = \frac{1 \times 100}{792} = 0.126$$

and

$$\frac{m M_y}{p_y Z_y} = \frac{1 \times 20 \times 10^3}{265 \times 1100} = 0.069$$

Overall buckling

$$\frac{F}{A_g p_c} + \frac{m M_x}{M_b} + \frac{m M_y}{p_y Z_y} = 0.524 + 0.126 + 0.069 = 0.719 < 1.0$$

Hence a $356 \times 368 \times 177$ UC section is satisfactory.

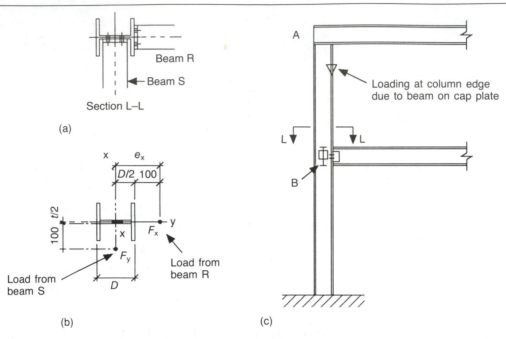

Fig. 4.24 *Load eccentricity for columns in simple construction.*

4.9.4 COLUMN DESIGN IN 'SIMPLE' CONSTRUCTION

At first sight it would appear that columns in so-called 'simple construction' are not subject to moments, as the beams are all joined at connections which allow no moment to develop. In fact, in most cases there is a bending moment due to the eccentricity of the shear load from the beam. This is summarized in clause 4.7.6 of BS 5950 and illustrated in *Fig. 4.24*. Note that where a beam sits on a column cap plate, for example at A (*Fig. 4.24(c)*), it can be assumed that the reaction from the beam acts at the face of the column. However, where the beam is connected to a column by means of a 'simple' connection, e.g. using web cleats, the reaction from the beam can be assumed to act 100 mm from the column (web or flange) face as illustrated in *Fig. 4.24(b)*.

When a roof truss is supported on a column cap plate (*Fig. 4.25*), and the connection is unable to develop significant moments, it can be assumed that the load from the truss is transmitted concentrically to the column. Columns in simple construction will not need to be checked for local capacity but it will

be necessary to carry out the overall buckling check. Since all UC sections are semi-compact, the procedure discussed in *section 4.9.3.2* can be used. The equivalent uniform moment factor, m, should be taken as 1 and the bending strength, p_b, should be based on the following equation for the equivalent slenderness ratio, λ_{LT}:

$$\lambda_{LT} = 0.5 L/r_y \qquad (4.28)$$

where L is the distance between levels at which both axes are restrained and r_y the radius of gyration of the section about its minor axis.

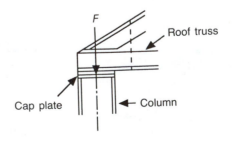

Fig. 4.25 *Column supporting a roof truss.*

Example 4.12 Design of a steel column in 'simple' construction

Select a suitable column section in grade 43 steel to support the ultimate loads from beams A and B shown in *Fig. 4.26*. Assume the column is 7 m long and is effectively held in position at both ends but only restrained in direction at the bottom.

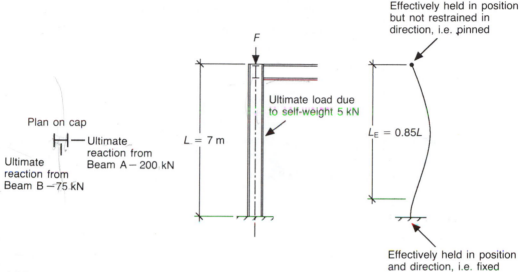

Fig. 4.26

SECTION SELECTION
This can only be done by trial and error. Therefore, try a 203 × 203 × 52 UC section.

DESIGN LOADING AND MOMENTS
Ultimate reaction from beam A, $R_A = 200$ kN; ultimate reaction from beam B, $R_B = 75$ kN; assume self-weight of column = 5 kN. Ultimate axial load, F, is

$$F = R_A + R_B + \text{self-weight of column}$$
$$= 200 + 75 + 5 = 280 \text{ kN}$$

Load eccentricity for beam A,

$$e_x = D/2 + 100 = 206.2/2 + 100 = 203.1 \text{ mm}$$

Load eccentricity for beam B,

$$e_y = t/2 + 100 = 8/2 + 100 = 104 \text{ mm}$$

Moment due to beam A,

$$M_x = R_A e_x = 200 \times 10^3 \times 203.1 = 40.62 \times 10^6 \text{ N mm}$$

Moment due to beam B,

$$M_y = R_B e_y = 75 \times 10^3 \times 104 = 7.8 \times 10^6 \text{ N mm}$$

EFFECTIVE LENGTH
From *Table 4.20*, effective length coefficient = 0.85. Hence, effective length is

$$L_E = 0.85L = 0.85 \times 7000 = 5950 \text{ mm}$$

BENDING STRENGTH

From *Table 4.17*, relevant compressive strength values for buckling about the x – x axis are obtained from Table 27(b) (*Table 4.18*) and from Table 27(c) (*Table 4.19*) for bending about the y – y axis.

$$\lambda_x = L_E/r_x = 5950/89 = 66.8$$

From *Table 4.18*, $p_c = 208$ N mm^{-2}.

$$\lambda_y = L_E/r_y = 5950/51.6 = 115.3$$

From *Table 4.19*, $p_c = 103$ N mm^{-2}. Hence critical compressive strength of column, p_c, is

$$p_c = 103 \text{ N mm}^{-2}$$

OVERALL BUCKLING CHECK

$$\lambda_{LT} = \frac{0.5L}{r_y} = \frac{0.5 \times 7000}{51.6} = 67.8$$

From *Table 4.15*, $p_b = 193$ N mm^{-2}. Buckling resistance moment capacity of column, M_b, is given by

$$M_b = p_b S_x = 193 \times 568 \times 10^3 = 109.6 \times 10^6 \text{ N mm}$$

Hence for stability,

$$\frac{F}{A_g p_c} + \frac{mM_x}{M_b} + \frac{mM_y}{p_y Z_y} \leqslant 1$$

$$= \frac{280 \times 10^3}{66.4 \times 10^2 \times 103} + \frac{1 \times 40.6 \times 10^6}{109.6 \times 10^6}$$

$$+ \frac{1 \times 7.8 \times 10^6}{275 \times 174 \times 10^3}$$

$$= 0.41 + 0.37 + 0.16 = 0.94 < 1$$

Therefore, the 203 × 203 × 52 UC section is suitable.

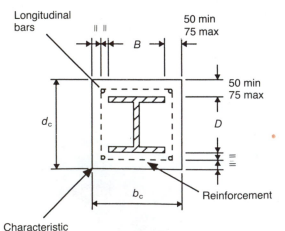

Longitudinal bars

50 min
75 max

B

50 min
75 max

d_c

D

b_c

Reinforcement

Characteristic strength of concrete ≥ 20 N/mm²

Fig. 4.27 *Cased UC section. Reinforcement: steel fabric type D98 (BS 4483) or ≥ 5 mm diameter longitudinal bars and links at a maximum spacing of 200 mm.*

4.9.5 DESIGN OF CASED COLUMNS

As discussed in *section 4.5*, steel columns are sometimes cased in concrete for fire protection. However, the concrete also increases the strength of the section, a fact which can be used to advantage in design provided that the conditions stated in clause 4.14.1 of BS 5950 are met. Some of these conditions are illustrated in *Fig. 4.27*.

BS 5950 gives guidance on the design of UC sections encased in concrete for the following loading conditions which are discussed below: (i) axially loaded columns and (ii) columns subject to axial load and bending.

4.9.5.1 Axially loaded columns

The design procedure for this case is summarized in *Fig. 4.28*.

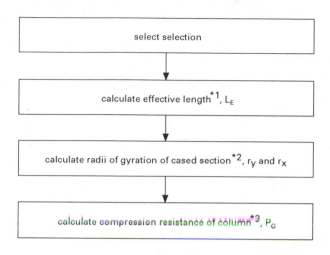

Fig. 4.28 *Design procedure for axially loaded cased columns.*

Notes to Fig. 4.28.

1. The effective length, L_E, should not exceed the least of (i) $40b_c$; (ii) $100b_c^2/d_c$ and (iii) $250r$, where b_c and d_c are as indicated in *Fig. 4.27* and r is the minimum radius of gyration of the uncased UC section, i.e. r_y.

2. The radius of gyration of the cased section about the y–y axis, r_y, is taken as $0.2b_c$ but not more than $0.2(B + 150)$ mm. The radius of gyration of

the cased section about the x–x axis, r_x, is assumed to be equal to the radius of gyration of the uncased steel section about the x–x axis.

3. The compression resistance of the cased section, P_c, is given by

$$P_c = \left(A_g + \frac{0.45f_{cu}A_c}{p_y}\right)p_c \qquad (4.29)$$

However, this should not be greater than the short strut capacity of the section, P_{cs}, which is given by

$$P_{cs} = \left(A_g + \frac{0.25f_{cu}A_c}{p_y}\right)p_y \qquad (4.30)$$

where A_c is the gross sectional area of the concrete but neglecting any casing in excess of 75 mm from the overall dimensions of the UC section or any applied finish, A_g the gross sectional area of the UC section, f_{cu} the characteristic strength of the concrete which should not be greater than 40 N mm^{-2}, p_c the compressive strength of the UC section determined as discussed for uncased columns (*section 4.9.1*) but using r_x and r_y for the cased section (see note 2 above) and p_y the design strength of the UC section which should not exceed 355 N mm^{-2}.

Example 4.13 Encased steel column resisting an axial load

Calculate the compression resistance of a $305 \times 305 \times 118$ kg m^{-1} UC column if it is encased in concrete of compressive strength 20 N mm^{-2} in the manner shown below. Assume that the effective length of the column about both axes is 3.5 m.

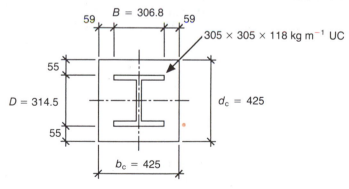

PROPERTIES OF UC SECTION

Area of UC section (A_g)   = 15 000 mm^2 (*Appendix B*)
Radius of gyration (r_x)   = 136 mm
Radius of gyration (r_y)   = 77.5 mm
Design strength (p_y)   = 265 N mm^{-2} (since $T = 18.7$ mm)
Effective length (L_E)   = 3.5 m

EFFECTIVE LENGTH

Check that the effective length of column (= 3500 mm) does not exceed the least of:

$$40b_c = 40 \times 425 = 17\,000 \text{ mm}$$

$$\frac{100b_c^2}{d_c} = \frac{100 \times 425^2}{425} = 42\,500 \text{ mm}$$

$$250r_y = 250 \times 77.5 = 19\,375 \text{ mm} \quad \text{OK}$$

RADII OF GYRATION FOR THE CASED SECTION

For the cased section r_x is same as for the UC section = 136 mm. For the cased section $r_y = 0.2b_c = 0.2 \times 425 = 85$ mm $\not> 0.2(B + 150) = 0.2(306.8 + 150) = 91.36$ mm   OK
Hence $r_y = 85$ mm and $r_x = 136$ mm.

COMPRESSION RESISTANCE

Slenderness ratio

$$\lambda_x = \frac{L_E}{r_x} = \frac{3500}{136} = 25.7 < 180$$

$$\lambda_y = \frac{L_E}{r_y} = \frac{3500}{85} = 41.2 < 180$$

Compressive strength

From *Table 4.17*, relevant compressive strength values for buckling about the x–x axis are obtained from Table 27(b) of BS 5950 (*Table 4.18*) and from Table 27(c) of BS 5950 (*Table 4.19*) for bending about the y–y axis.
For $\lambda_x = 25.7$ and $p_y = 265 \text{ N mm}^{-2}$ compressive strength $(p_c) = 257 \text{ N mm}^{-2}$ (*Table 4.18*). For $\lambda_y = 41.2$ and $p_y = 265 \text{ N mm}^{-2}$ compressive strength $(p_c) = 228 \text{ N mm}^{-2}$ (*Table 4.19*). Hence, p_c is equal to 228 N mm^{-2}.

Compression resistance

$$A_g = 15\,000 \text{ mm}^2$$
$$A_c = d_c b_c = 425 \times 425 = 180\,625 \text{ mm}^2$$
$$p_y = 265 \text{ N mm}^{-2} \quad (\text{since } T = 18.7 \text{ mm})$$
$$p_c = 228 \text{ N mm}^{-2}$$
$$f_{cu} = 20 \text{ N mm}^{-2}$$

Compression resistance of encased column, P_c, is given by

$$P_c = \left[A_g + \frac{0.45 f_{cu} A_c}{p_y} \right] p_c$$

$$= \left[15\,000 + \frac{0.45 \times 20 \times 180\,625}{265} \right] 228 = 4.81 \times 10^6 \text{ N}$$

which should not be greater than the short strut capacity, P_{cs}, given by

$$P_{cs} = \left[A_g + \frac{0.25 f_{cu} A_c}{p_y} \right] p_y$$

$$= \left[15\,000 + \frac{0.25 \times 20 \times 180\,625}{265} \right] 265$$

$$= 4.87 \times 10^6 \text{ N} \quad \text{OK}$$

Hence the compression resistance of the encased column is 4.81×10^6 N. Comparing this with the compression resistance of the uncased column (*Example 4.10*) shows that the load capacity of the column has been increased from 3357 to 4810 kN, which represents an increase of approximately 45%.

4.9.5.2 Columns subject to axial load and bending

The design procedure here is similar to that when the column is axially loaded but also involves checking for local capacity and buckling resistance using the following relationships:

1. *Local capacity.*

$$\frac{F_c}{P_{cs}} + \frac{M_x}{M_{cx}} + \frac{M_y}{M_{cy}} \leqslant 1 \qquad (4.31)$$

2. *Buckling resistance.*

$$\frac{F_c}{P_c} + \frac{m M_x}{M_b} + \frac{m M_y}{M_{cy}} \leqslant 1 \qquad (4.32)$$

where F_c is the compressive force due to axial load, P_c the compression resistance of the cased section (equation 4.29), P_{cs} the short strut capacity (equation 4.30), M_x, M_y the applied moment about the major and minor axes respectively, M_{cx}, M_{cy} the moment capacity of the UC section about the x – x and y – y axes respectively, m is the equivalent uniform moment factor, taken as 1.0 for cased columns (clause 4.7.7) and M_b is the buckling resistance moment of the cased column = $S_x p_b \not> 1.5 M_b$ for the uncased section. To determine p_b, r_y should be taken as the greater of r_y of the uncased section or $0.2(B + 100)$ mm (*Fig. 4.27*).

Example 4.14 Encased steel column resisting a axial load and bending

In *Example 4.11* it was found that a $305 \times 305 \times 118$ kg m^{-1} UC column was incapable of resisting the below design load and moments:

Design axial load	= 2000 kN
Design moment about the x–x axis	= 100 kN m
Design moment about the y–y axis	= 20 kN m

Assuming that the same column is now encased in concrete as shown below, determine its suitability. The effective length of the column about both axes is 7 m.

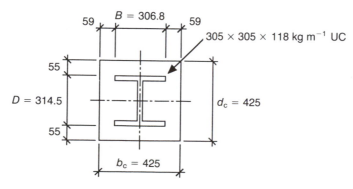

PROPERTIES OF UC SECTION

Area of UC section (A_g) $= 15\,000 \text{ mm}^2$
Radius of gyration: x–x axis (r_x) $= 136 \text{ mm}$
Radius of gyration: y–y axis (r_y) $= 77.5 \text{ mm}$
Elastic modulus: x–x axis (Z_x) $= 1760 \times 10^3 \text{ mm}^3$
Elastic modulus: y–y axis (Z_y) $= 587 \times 10^3 \text{ mm}^3$
Plastic modulus: x–x axis (S_x) $= 1950 \times 10^3 \text{ mm}^3$
Design strength (p_y) $= 265 \text{ N mm}^{-2}$ (since $T = 18.7 \text{ mm}$)
Effective length (L_E) $= 7 \text{ m}$

LOCAL CAPACITY

Axial load (F_c) $= 2000 \text{ kN}$
Applied moment: x–x axis (M_x) $= 100 \text{ kN m}$
Applied moment: y–y axis (M_y) $= 20 \text{ kN m}$

Short strut capacity, P_{cs}, is given by

$$P_{cs} = \left[A_g + \frac{0.25 f_{cu} A_c}{p_y} \right] p_y$$

$$= \left[15\,000 + \frac{0.25 \times 20 \times 425^2}{265} \right] 265$$

$$= 4.878 \times 10^6 \text{ N} = 4878 \text{ kN}$$

Moment capacity of column, M_{cx}, is given by

$$M_{cx} = p_y Z_x = 265 \times 1760 \times 10^3 = 466.4 \times 10^6 \text{ N mm} = 466.4 \text{ kN m}$$

Moment capacity of column, M_{cy}, is given by

$$M_{cy} = p_y Z_y = 265 \times 587 \times 10^3 = 155.6 \times 10^6 \text{ N mm} = 155.6 \text{ kN m}$$

$$\frac{F_c}{P_{cs}} + \frac{M_x}{M_{cx}} + \frac{M_y}{M_{cy}} = \frac{2000}{4878} + \frac{100}{466.4} + \frac{20}{155.6}$$
$$= 0.41 + 0.21 + 0.13 = 0.75 < 1$$

Hence, the local capacity of the section is satisfactory.

BUCKLING RESISTANCE

Radii of gyration for cased section

For the cased section r_x is same as for the UC section $= 136 \text{ mm}$. For the cased section $r_y = 0.2 b_c = 0.2 \times 425 = 85 \text{ mm} \not> 0.2 (B + 150) = 0.2 (306.8 + 150) = 91.36 \text{ mm}$ OK
Hence $r_y = 85 \text{ mm}$ and $r_x = 136 \text{ mm}$.

Slenderness ratio

$$\lambda_x = \frac{L_E}{r_x} = \frac{7000}{136} = 51.5 < 180$$

$$\lambda_y = \frac{L_E}{r_y} = \frac{7000}{85} = 82.4 < 180$$

Compressive strength

For $\lambda_x = 51.5$ and $p_y = 265 \text{ N mm}^{-2}$ compressive strength (p_c) $= 226 \text{ N mm}^{-2}$ (*Table 4.18*). For $\lambda_y = 82.4$ and $p_y = 265 \text{ N mm}^{-2}$ compressive strength (p_c) $= 153 \text{ N mm}^{-2}$ (*Table 4.19*). Hence p_c is equal to 153 N mm^{-2}.

Compression resistance

$$A_g = 15\,000 \text{ mm}^2$$
$$A_c = 425 \times 425 = 180\,625 \text{ mm}^2$$
$$p_y = 265 \text{ N mm}^{-2}$$
$$p_c = 153 \text{ N mm}^{-2}$$
$$f_{cu} = 20 \text{ N mm}^{-2}$$

Compression resistance of encased column, P_c, is given by

$$P_c = \left[A_g + \frac{0.45 f_{cu} A_c}{p_y} \right] p_c$$

$$= \left[15\,000 + \frac{0.45 \times 20 \times 180\,625}{265} \right] 153 = 3.233 \times 10^6 \text{ N}$$

which is **not greater** than short strut capacity, $P_{cs} = 4878$ kN (see above)   OK.

Buckling resistance

For the **uncased** section,

$$\lambda_y = \frac{L_E}{r_y} = \frac{7000}{77.5} = 90$$

$$\lambda/x = 90/16.2 = 5.5$$

From *Table 4.14* $v = 0.79$ and

$$\lambda_{LT} = nuv\lambda_y = 1 \times 0.851 \times 0.79 \times 90 = 61$$

Hence $p_b = 205 \text{ N mm}^{-2}$ from *Table 4.15*.

$$M_b = S_x p_b = 1950 \times 10^3 \times 205 = 400 \times 10^6 \text{ N mm} = 400 \text{ kN m}$$

For the **cased** section,

$$\lambda_y = \frac{L_E}{r_y} = \frac{7000}{85} = 82.4$$

$$\lambda/x = 824/16.2 = 5.08$$

From *Table 4.14* $\upsilon = 0.81$ and

$$\lambda_{LT} = nuv\lambda_y = 1 \times 0.851 \times 0.81 \times 82.4 = 56.8$$

Hence $p_b = 214 \text{ N mm}^{-2}$ from *Table 4.15*.

$$M_b = S_x p_b = 1950 \times 10^3 \times 214 = 417 \times 10^6 \text{ N mm} = 417 \text{ kN m}$$

Hence M_b (for cased UC section = 417 kN m) $\not> 1.5 M_b$ (for uncased UC section = $1.5 \times 400 = 600$ kN m).

Check buckling resistance

From clause 4.7.7 of BS 5950, $m = 1.0$ for cased columns:

$$\frac{F_c}{P_c} + \frac{mM_x}{M_b} + \frac{mM_y}{M_{cy}} = \frac{2000}{3233} + \frac{1 \times 100}{417} + \frac{1 \times 20}{155.6}$$
$$= 0.62 + 0.24 + 0.13 = 0.99 < 1$$

Hence, the section is now just adequate to resist the design axial load of 2000 kN and design moments about the x–x and y–y axes of 100 and 20 kN m respectively.

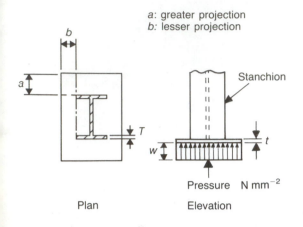

a: greater projection
b: lesser projection

Stanchion

Pressure   N mm^{-2}

Plan

Elevation

Fig. 4.29

4.9.6 DESIGN OF COLUMN BASEPLATES

Clause 4.13.2 of BS 5950 gives guidance on the design of concentrically loaded slab baseplates, which covers most practical design situations (*Fig. 4.29*). The plan area of the baseplate is given by

$$\text{Plan area of baseplate} = \frac{\text{axial load}}{\text{bearing strength}} \quad (4.33)$$

where bearing strength $= 0.4 f_{cu}$ for concrete foundations. The baseplate thickness, t, is given by

$$t = \left[\frac{2.5w}{p_{yp}} (a^2 - 0.3b^2) \right]^{1/2} \nless \text{flange thickness } (T) \quad (4.34)$$

where a is the greater projection of the plate beyond the column (*Fig. 4.29*), b the smaller projection of the plate beyond the column (*Fig. 4.29*), w the pressure on the underside of the plate assuming a uniform distribution (N mm^{-2}) and p_{yp} the design strength of the plate which may be taken from Table 6 of BS 5950 (*Table 4.3*), but not greater than 270 N mm^{-2}.

Example 4.15 Design of a steel column baseplate

Design a baseplate for the axially loaded column shown below, assuming that the foundations are of concrete of compressive characteristic strength 30 N mm^{-2} and the design strength of the plate is 265 N mm^{-2}.

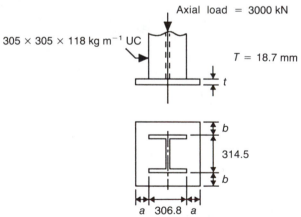

Axial load = 3000 kN

305 × 305 × 118 kg m^{-1} UC

T = 18.7 mm

t

b
314.5
b

a   306.8   a

$$\text{Plan area of baseplate} = \frac{\text{axial load}}{\text{bearing strength}}$$

$$= \frac{3 \times 10^6}{0.4 \times 30} = 2.5 \times 10^5 \text{ mm}^2$$

Assuming that the baseplate is square, the minimum length of the sides of the baseplate is $\sqrt{(2.5 \times 10^5)} = 500$ mm. Pressure on the underside of baseplate, w, is given by

$$w = \frac{\text{axial load}}{\text{plan area}} = \frac{3 \times 10^6}{500^2} = 12 \text{ N mm}^{-2}$$

Baseplate thickness, t, is given by

$$t = \left[\frac{2.5w}{p_{yp}} (a^2 - 0.3b^2) \right]^{1/2} \not< \text{flange thickness}$$

$$= \left[\frac{2.5 \times 12}{265} (96.6^2 - 0.3 \times 92.75^2) \right]^{1/2}$$

$$= 27.6\text{mm} > \text{flange thickness} = 18.7 \text{ mm}$$

where $a = \frac{1}{2} (500 - 306.8) = 96.6$ mm and $b = \frac{1}{2} (500 - 314.5) = 92.75$ mm.
Hence provide baseplate $500 \times 500 \times 30$ mm.

4.10 Design of connections

There are two principal methods for connecting together steel elements of structure, and the various cleats, end plates, etc. also required:

1. Bolting, using ordinary or high strength friction grip (HSFG) bolts, is the principal method of connecting together elements on site.
2. Welding, principally electric arc welding, is an alternative way of connecting elements on site, but most welding usually takes place in factory conditions. End plates and fixing cleats are welded to the elements in the fabrication yard. The elements are then delivered to site where they are bolted together in position.

Figure 4.30 shows some typical connections used in steel structures.

The aim of this section is to describe the design of some commonly used types of bolted and welded connections in steel structures. However, at the outset, it is worth while reiterating some general points relating to connection design given in clause 6 of BS 5950.

The first sentence is vitally important – 'Connections should be designed on the basis of a realistic assumption of the distribution of internal forces, having regard to relative stiffness.' Before any detailed design is embarked upon therefore, a consideration of how forces will be transmitted through the joint is essential.

'Joints should be designed to act in accordance with the assumptions already made in the design.' If, for instance, the structure is designed in 'simple construction', the beam–column joint should be designed accordingly to accept rotations rather than moments. A rigid joint would be completely wrong in this situation, as it would tend to generate a moment in the column for which it has not been designed.

'The ductility of steel assists in the distribution of forces generated within a joint.' This means that residual forces due to initial lack of fit, or due to bolt tightening, do not normally have to be considered.

4.10.1 BOLTED CONNECTIONS
As mentioned above, two types of bolts commonly

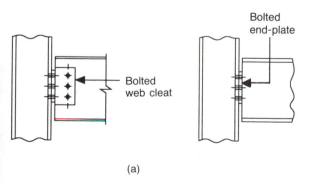

Bolted
web cleat

Bolted
end-plate

(a)

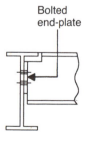

Bolted
end-plate

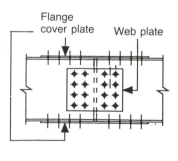

Flange
cover plate   Web plate

(b)

Fig. 4.30 *Typical connections: (a) beam to column; (b) beam to beam.*

used in steel structures are ordinary (or black) bolts and HSFG bolts. Ordinary bolts sustain a shear load by the shear strength of the bolt shank itself, whereas HSFG bolts rely on a high tensile strength to grip the joined parts together so tightly that they cannot slide.

There are two grades of ordinary bolts, namely 4.6 and 8.8. HSFG bolts commonly used in structural connections conform to the general grade. The preferred size of steel bolts are 12, 16, 20, 22, 24 and 30 mm in diameter. Generally, in structural connections, grade 8.8 bolts having a diameter not less than 12 mm are recommended. In any case, as far as possible, only one size and grade of bolt should be used on a project.

The nominal diameter of holes for ordinary bolts, D_h, is equal to the bolt diameter, d_b, plus 2 mm for bolts up to 24 mm in diameter and 3 mm for bolts 27 mm or greater in diameter:

$$D_h = d_b + 2 \text{ mm} \quad \text{for } d_b \leqslant 24 \text{ mm}$$

$$D_h = d_b + 3 \text{ mm} \quad \text{for } d_b \geqslant 27 \text{ mm}$$

4.10.2 FASTENER SPACING AND EDGE/END DISTANCES

Clause 6.2 of BS 5950 contains various recommendations regarding the distance between fasteners and edge/end distances to fasteners, some of which are illustrated in *Fig. 4.31* and summarized below:

1. Spacing between centres of bolts, i.e. pitch (p), in the direction of stress and not exposed to corrosive influences should lie within the following limits:

$$2.5d_b \leqslant p \leqslant 14t$$

where d_b is the diameter of bolts and t the thickness of thinner ply.
2. Minimum edge distance, e_1, and end distance, e_2, to fasteners should conform with the following limits:
Rolled edge $\geq 1.25D_h$

Sheared edge $\geqslant 1.4D_h$
All end distances $\geqslant 1.4D_h$

where D_h is the diameter of the bolt hole. Note that the edge distance, e_1, is the distance from the centre line of the hole to the outside edge of the plate at right angles to the direction of the stress, whereas the end distance, e_2, is the distance from the centre line to the edge of the plate in the direction of stress.
3. Maximum edge distance, e_1, should generally not exceed the following:

$$e_1 \leqslant 11t\,\varepsilon$$

where t is the thickness of the thinner part and $\varepsilon = (275/p_y)^{1/2}$.

4.10.3 STRENGTH CHECKS

Bolted connections may fail due to various mechanisms including shear, bearing, tension and combined shear and tension. The following sections describe these failure modes and outlines the associated design procedures for connections involving (a) ordinary bolts and (b) HSFG bolts.

4.10.3.1 Ordinary bolts

Shear and bearing. Referring to the connection detail shown in *Fig. 4.32*, it can be seen that the loading on bolts A between the web cleat and the column will be in shear, and that there are three principal ways in which the joint may fail. Firstly, the bolts can fail in shear, for example along surface x_1 y_1 (*Fig. 4.32(a)*). Secondly, the bolts can fail in bearing as the web cleat cuts into the bolts (*Fig. 4.32(b)*). This can only happen when the bolts are softer than the metal being joined. Thirdly, the metal being joined, i.e. the cleat, can fail in bearing as the bolts cut into it (*Fig. 4.32(c)*). This is the converse of the above situation and can only happen when the bolts are harder than the metal being joined.

It follows, therefore, that the design shear strength of the connection should be taken as the least of:

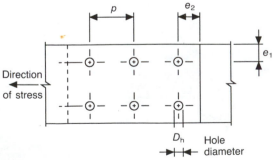

Direction of stress

Fig. 4.31 Rules for fastener spacing and edge/end distances to fasteners.

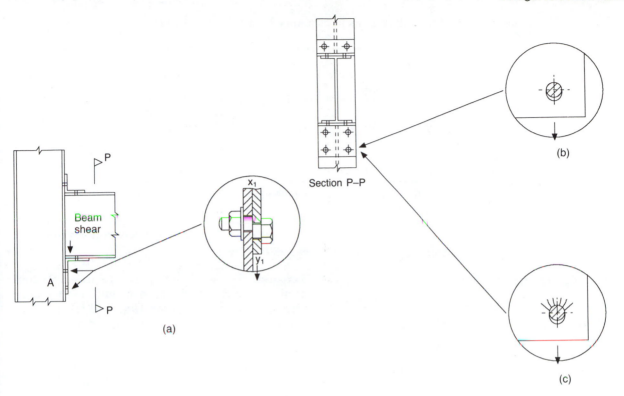

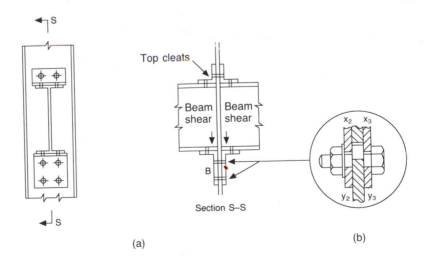

Fig. 4.32 *Failure modes of a beam-to-column connection: (a) single shear failure of bolt; (b) bearing failure of bolt; (c) bearing failure of cleat.*

Fig. 4.33 *Double shear failure.*

Table 4.21 Strength of bolts in clearance holes (Table 32 of BS 5950)

	Bolt grade (N mm^{-2})		Other grades of fasteners (N mm^{-2})
	4.6	8.8	
Shear strength, p_s	160	375	$0.48U_f$ but $\leqslant 0.69Y_f$
Bearing strength, p_{bb} (but see *Table 4.22*)	460	1035	$0.72(U_f + Y_f)$
Tension strength, p_t	195	450	$0.58U_f$ but $\leqslant 0.83 Y_f$

Note. Y_f is the specified minimum yield strength of the fastener and U_f is the specified minimum ultimate tensile strength of the fastener.

1. Shear capacity of the bolt,

$$P_s = p_sA_s \tag{4.35}$$

2. Bearing capacity of bolt,

$$P_{bb} = dtp_{bb} \tag{4.36}$$

3. Bearing capacity of connected ply,

$$P_{bs} = dtp_{bs} \leqslant 1/2etp_{bs} \tag{4.37}$$

where p_s is the shear strength of the bolt (*Table 4.21*), p_{bb} the bearing strength of the bolt (*Table 4.21*), p_{bs} the bearing strength of the ply (*Table 4.22*), e is the end distance e_2 and A_s is the effective area for shear, normally taken as the tensile stress area, A_t (*Table 4.23*).

Double shear. If a column supports two beams in the manner indicated in *Fig. 4.33*, the failure modes essentially remain the same as for the previous case, except that the bolts (B) will be in 'double shear'. This means that failure of the bolts will only occur once surfaces x₂y₂ and x₃y₃ exceed the shear strength of the bolt (*Fig. 4.33 (b)*). The shear capacity of bolts in double shear is given by

$$\text{Shear capacity of bolts in double shear} = 2P_s \tag{4.38}$$

Thus, double shear effectively doubles the shear strength of the bolt.

Tension. Tension failure may arise in simple connections as a result of excessive tension in the bolts (*Fig. 4.34(a)*) or cover plates (*Fig. 4.34(b)*). The tension capacity of ordinary bolts, P_t, is given by

$$P_t = p_tA_t \tag{4.39}$$

where p_t is the tension strength of the bolt (*Table 4.21*) and A_t is the tensile stress area of bolt (*Table 4.23*). The tensile capacity of a flat plate is given by

$$P_t = A_ep_y \tag{4.40}$$

where effective area, A_e, is

$$A_e = K_eA_n < A_g \tag{4.41}$$

in which $K_e = 1.2$ for grade 43 steel plates, A_g is the gross area of plate $= bt$ (*Fig. 4.34(b)*) and A_n is the net area of plate $= A_g -$ allowance for bolt holes ($= D_ht$, *Fig. 4.34(b)*).

Combined shear and tension. Where ordinary bolts are subject to combined shear and tension (*Fig. 4.35*), in addition to checking their shear and tension capacities separately, the following relationship should also be satisfied:

Table 4.22 Bearing strength of connected parts for ordinary bolts in clearance holes, p_{bs} (Table 33, BS 5950)

Design grade of steel (N mm^{-2})			Other grades of steel (N mm^{-2})
43	50	55	
460	550	650	$0.65 (U_s + Y_s)$

Note. Y_s is the specified minimum yield strength, and U_s is the specified minimum ultimate tensile strength.

Table 4.23 Tensile stress area, A_t

Nominal size and thread diameter (mm)	Tensile stress area, A_t (mm^2)
12	84.3
16	157
20	245
22	303
24	353
27	459
30	561

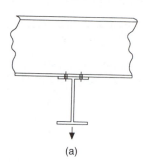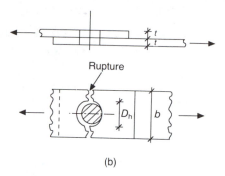

(a)   (b)

Fig. 4.34 *Typical tension failures: (a) bolts in tension; (b) cover plate in tension.*

$$\frac{F_s}{P_s} + \frac{F_t}{P_t} \leq 1.4 \qquad (4.42)$$

where F_s is the applied shear, F_t the applied tension, P_s the shear capacity (equation 4.35) and P_t the tension capacity (equation 4.39).

4.10.3.2 HSFG bolts

Shear and bearing. If HSFG bolts, rather than ordinary bolts, were used in the connection detail shown in *Fig. 4.32*, failure of the connection would principally be as a result of (a) slip between the connected members or (b) bearing failure of the connecting ply. Thus with connections using HSFG bolts, the design shear strength is taken as the lesser of (1) slip resistance, P_{sL}, given by

$$P_{sL} = 1.1 K_s \mu \, P_o \qquad (4.43)$$

and (2) bearing capacity of connected ply, P_{bg}, given by

$$P_{bg} = dt p_{bg} \leq 1/3 e t p_{bg} \qquad (4.44)$$

where P_o is the minimum shank tension (*Table 4.24*), K_s is 1.0 for clearance holes, μ the slip factor ≤ 0.45, p_{bg} the bearing strength of connected parts (*Table*

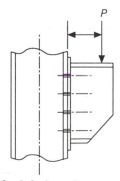

Fig. 4.35 *Bracket bolted to column.*

4.25), F_s the applied shear and F_t the externally applied tension.

Tension. The tension capacity, P_t, of a friction grip bolt is given by

$$P_t = 0.9 P_o \qquad (4.45)$$

Combined shear and tension. When friction grip bolts are subject to combined shear and tension, then the following additional check should be carried out:

$$\frac{F_s}{P_{sL}} + 0.8\frac{F_t}{P_t} \leq 1 \qquad (4.46)$$

Table 4.24   Proof load of HSFG bolts, P_o

Nominal size and thread diameter (mm)	Minimum shank tension or proof load (kN)
12	49.4
16	92.1
20	144
22	177
24	207
27	234
30	286

Table 4.25   Bearing strength of parts connected by parallel shank friction grip fasteners, p_{bg} (Table 34, BS 5950)

Design grade of steel (N mm^{-2})			Other grades of steel (N mm^{-2})
43	50	55	
825	1065	1210	$2.2U_s$ but $\leq 3.0Y_s$

Note. Y_s is the specified minimum yield strength of the steel and U_s is the specified minimum ultimate tensile strength.

Example 4.16 Analysis of a double angle web cleat beam-to-column connection

Show that the double angle web cleat beam-to-column connection detail shown below is suitable to resist the design shear force, V, of 400 kN. Assume the steel is grade 43 and the bolts are M20, grade 8.8 in 2 mm clearance holes.

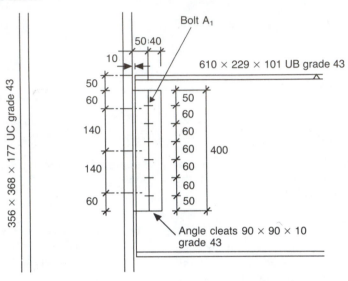

CHECK FASTENER SPACING AND EDGE/END DISTANCES

Diameter of bolt, d_b	= 20 mm
Diameter of bolt hole, D_h	= 22 mm
Pitch of bolt, p	= 140 mm and 60 mm
Edge distance, e_1	= 40 mm
End distance, e_2	= 60 mm and 50 mm
Thickness of angle cleat, t_p	= 10 mm

The following conditions need to be met:

Pitch	$\geqslant 2.5 d_b = 2.5 \times 20 = 50 < 140$ and 60	OK
Pitch	$\leqslant 14 t_p = 14 \times 10 = 140 \leqslant 140$ and 60	OK
Edge distance e_1	$\geqslant 1.4 D_h = 1.4 \times 22 = 30.8 < 40$	OK
End distance e_2	$\geqslant 1.4 D_h = 1.4 \times 22 = 30.8 < 60$ and 50	OK
e_1 and e_2	$\leqslant 11 t_p \varepsilon = 11 \times 10 \times 1 = 110 < 40$, 50 and 60	OK

(For grade 43 steel with $t_p = 10$ mm, $p_y = 275$ N mm^{-2}, $\varepsilon = 1$.) Hence all fastener spacings and edge/end distances to fasteners are satisfactory.

CHECK STRENGTH OF BOLTS CONNECTING CLEATS TO SUPPORTING COLUMN

Shear

6 No., M20 grade 8.8 bolts. Hence $A_s = 245$ mm^2 (*Table 4.23*) and $p_s = 375$ N mm^{-2} (*Table 4.21*). Shear capacity of single bolt, P_s, is

$$P_s = p_s A_s = 375 \times 245 = 91.9 \times 10^3 = 91.9 \text{ kN}$$

Shear capacity of bolt group is

$$6 P_s = 6 \times 91.9 = 551.4 \text{ kN} > V = 400 \text{ kN}$$

Hence bolts are adequate in shear.

Bearing

Bearing capacity of bolt, P_{bb}, is given by

$$P_{bb} = d_b t p_{bb} = 20 \times 10 \times 1035 = 207 \times 10^3 = 207 \text{ kN}$$

Since thickness of angle cleat (= 10 mm) < thickness of column flange (= 23.8 mm), bearing capacity of cleat is critical. Bearing capacity of cleat, P_{bs}, is given by $P_{bs} = d_b t p_{bs} = 20 \times 10 \times 460 = 92 \times 10^3$ N = 92 kN $\leqslant 1/2 e t p_{bs} = 0.5 \times 60 \times 10 \times 460 = 138 \times 10^3$ N
Hence total capacity of connection is

$$6 \times 92 = 552 \text{ kN} > V = 400 \text{ kN}$$

Therefore bolts are adequate in bearing.

CHECK STRENGTH OF BOLT GROUP CONNECTING CLEATS TO WEB OF SUPPORTED BEAM

Shear

6 No., M20 grade 8.8 bolts; from above, $A_s = 245$ mm^2 and $p_s = 375$ N mm^{-2}. Since bolts are in double shear, shear capacity of each bolt is $2P_s = 2 \times 91.9 = 183.8$ kN
Loads applied to the bolt group are vertical shear, $V = 400$ kN and moment, $M = 400 \times 50 \times 10^{-3} = 20$ kN m.
Outermost bolt (A_1) subject to greatest shear force which is equal to the resultant of the load due to the moment, $M = 20$ kN m and vertical shear force, $V = 400$ kN. Load on the outermost bolt due to moment, F_{mb}, is given by

$$F_{mb} = \frac{M}{Z} A = \frac{20 \times 10^3}{420A} A = 47.6 \text{ kN}$$

where A is the area of bolt, Z the modulus of the bolt group is

$$\frac{I}{y} = \frac{63\ 000}{150} A = 420A \text{ mm}^3$$

and I the inertia of the bolt group is

$$2A\ (30^2 + 90^2 + 150^2) = 63\ 000A \text{ mm}^4$$

Load on outermost bolt due to shear, F_{vb}, is given by

$$F_{vb} = V/\text{No. of bolts} = 400/6 = 66.7 \text{ kN}$$

Resultant shear force of bolt, F_s, is

$$F_s = (F_{vb}{}^2 + F_{mb}{}^2)^{1/2} = (66.7^2 + 47.6^2)^{1/2} = 82 \text{ kN}$$

Since F_s (= 82 kN) < $2P_s$ (= 183.8 kN) the bolts are adequate in shear.

Bearing

Bearing capacity of bolt, P_{bb}, is

$$P_{bb} = 207 \text{ kN (from above)} > F_s \quad \text{OK}$$

Bearing capacity of each cleat, P_{bs}, is

$$P_{bs} = 92 \text{ kN} \quad \text{(from above)}$$

Bearing capacity of both cleats is

$$2 \times 92 = 184 \text{ kN} > F_s \quad \text{OK}$$

Bearing capacity of the web, P_{bs}, is

$$P_{bs} = d_b t_w p_{bs} = 20 \times 10.6 \times 460 \times 10^{-3}$$

$$= 97.52 \text{ kN} > F_s \quad \text{OK}$$

Hence bolts, cleats and beam web are adequate in bearing.

SHEAR STRENGTH OF CLEATS

Shear capacity of a single angle cleat P_v, is

$$P_v = 0.6 p_y A_v = 0.6 \times 275 \times 3600 \times 10^{-3} = 594 \text{ kN}$$

where $A_v = 0.9 A_0$ (clause 4.2.3 of BS 5950) $= 0.9 \times$ thickness of cleat (t_p) $\times$ length of cleat (l_p) $0.9 \times 10 \times 400 = 3600 \text{ mm}^2$. Since shear force $V/2$ (= 200 kN) < P_v (= 594 kN) the angle is adequate in shear.

BENDING STRENGTH OF CLEATS

$$M = \frac{V}{2} \times 50 \times 10^{-3} = \frac{400}{2} \times 50 \times 10^{-3} = 10 \text{ kN m}$$

Assume moment capacity of one angle of cleat, M_c, is

$$M_c = p_y Z = 275 \times 266.7 \times 10^3 = 73.3 \times 10^6 \text{ N mm} = 73.3 \text{ kN m} > M$$

where $Z = \dfrac{t_p l_p^2}{6} = \dfrac{10 \times 400^2}{6} = 266\,667 \text{ mm}^3$. Angle cleat is adequate in bending.

LOCAL SHEAR STRENGTH OF THE BEAM

Shear capacity of the supported beam, P_v, is $P_v = 0.6 p_y A_v = 0.6 \times 275 \times 6383.3 = 1053.2 \times 10^3 \text{ N} > V$ (= 400 kN)

where $A_v = t_w D = 10.6 \times 602.2 = 6383.3 \text{ mm}^2$. Hence supported beam at the end is adequate in shear.

Example 4.17 Analysis of a bracket-to-column connection

Show that the bolts in the bracket-to-column connection below are suitable to resist the design shear force of 200 kN. Assume the bolts are all M16, grade 8.8.

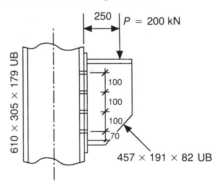

Since the bolts are subject to combined shear and tension, the bolts should be checked for shear, tension and combined shear and tension separately.

SHEAR

Design shear force, P	= 200 kN
Number of bolts, N	= 8
Shear force/bolt, F_s	= P/N = 200/8 = 25 kN

Shear capacity of bolt, P_s, is

$$P_s = p_s A_s = 375 \times 157 = 58.9 \text{ kN} > F_s \quad \text{OK}$$

TENSILE CAPACITY
Maximum bolt tension, F_t, is

$$F_t = P e y_1 / 2\Sigma\, y^2 = 200 \times 250 \times 370/2(70^2 + 170^2 + 270^2 + 370^2)$$
$$= 38 \text{ kN}$$

Tension capacity, P_t, is

$$P_t = p_t A_t = 157 \times 450 = 70.6 \times 10^3 \text{ kN} = 70.6 \text{ kN} > F_t \quad \text{OK}$$

COMBINED SHEAR AND TENSION
Combined check:

$$\frac{F_s}{P_s} + \frac{F_t}{P_t} \leq 1.4$$

$$\frac{25}{58.9} + \frac{38}{70.6} = 0.96 \leq 1.4 \quad \text{OK}$$

Hence the M16, grade 8.8 bolts are satisfactory.

Example 4.18 Analysis of beam splice connection

Show that the splice connection shown below is suitable to resist a design bending moment, M, and shear force, F, of 270 kN m and 300 kN respectively. Assume the steel grade is 43 and the bolts are general grade, M22, HSFG bolts.

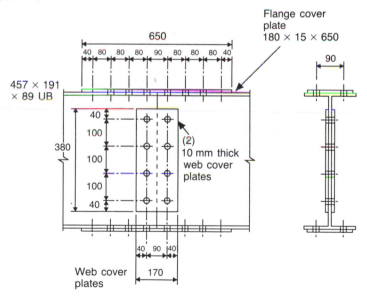

Assume that (1) flange cover plates resist the design bending moment, M and (2) web cover plates resist the design shear force, F, and the torsional moment ($= Fe$), where e is half the distance between the centroids of the bolt groups either side of the joint.

FLANGE SPLICE

Check bolts in flange cover plate
Single cover plate on each flange. Hence the bolts in the flanges are subject to single shear and must resist a shear force equal to

$$\frac{\text{Applied moment}}{\text{Overall depth of beam}} = \frac{M}{D} = \frac{270 \times 10^3}{463.6} = 582.4 \text{ kN}$$

Slip resistance of single bolt in single shear, P_{sL}, is

$$P_{sL} = 1.1K_s\mu P_o = 1.1 \times 1.0 \times 0.45 \times 177 = 87.6 \text{ kN} \quad (P_o = 177 \text{ kN}, \text{ Table 4.24})$$

Since thickness of beam flange (= 17.7 mm) > thickness of cover plate (= 15 mm), bearing in cover plate is critical. Bearing capacity, P_{bg}, of cover plate is

$$dtp_{bg} \leqslant 1/3etp_{bg}$$

$$dtp_{bg} = 22 \times 15 \times 825 = 272.2 \times 10^3 \text{ N} \quad (p_{bg} = 825 \text{ N/mm}^2, \text{ Table 4.25})$$
$$1/3etp_{bg} = 1/3 \times 40 \times 15 \times 825 = 165 \times 10^3 \text{ N}$$

Hence shear strength based on slip resistance of bolts. Slip resistance of 8 No., M22 bolts = $8P_{sL}$ = $8 \times 87.6 = 700.8$ kN > shear force in bolts = 582.4 kN. Therefore the 8 No., M22 HSFG bolts provided are adequate in shear.

Check tension capacity of flange cover plate
Gross area of plate, A_g, is

$$A_g = 180 \times 15 = 2700 \text{ mm}^2$$

Net area of plate, A_n, is

$$A_n = A_g - \text{area of bolt holes}$$
$$= 2700 - 2(15 \times 24) = 1980 \text{ mm}^2$$

Force in cover plate, F_t, is

$$F_t = \frac{M}{D + T_{fp}} = \frac{270 \times 10^3}{478.6} = 564.1 \text{ kN}$$

where T_{fp} is the thickness of flange cover plate = 15 mm. Tension capacity of plate, P_t, is

$$P_t = A_e p_y = 2376 \times 275 = 653.4 \times 10^3 \text{ N}$$

where $A_e = K_e A_n = 1.2 \times 1980 = 2376$ mm^2. Since $P_t > F_t$, cover plate is OK in tension.

Check compressive capacity of flange cover plate
Top cover plate in compression will also be adequate provided the compression flange has sufficient lateral restraint.

WEB SPLICE

Check bolts in web splice
Vertical shear $F_v = 300$ kN
Torsional moment = $F_v e_o$ (where e_o is half the distance between the centroids of the bolt groups either side of the joint in accordance with assumption (2) above)

$$= 300 \times 45 = 13\,500 \text{ kN mm}$$

Maximum resultant force F_R occurs on outermost bolts, e.g. bolt (A) and is given by the following expression:

$$F_R = \sqrt{(F_{vs}^2 + F_{tm}^2)}$$

where F_{vs} is the force due to vertical shear and F_{tm} the force due to torsional moment.

$$F_{vs} = \frac{F_v}{N} = \frac{300}{4} = 75 \text{ kN}$$

$$F_{tm} = \frac{\text{Torsional moment} \times A_b}{Z_b}$$

where A_b is the area of single bolt and Z_b the modulus of the bolt group which is

$$\frac{\text{Intertia of bolt group}}{\text{Distance to furthest bolt}} = \frac{2A_b(50^2 + 150^2)}{150} = 333.33\,A_b$$

Hence

$$F_{tm} = \frac{13\,500A_b}{333.33A_b} = 40.5 \text{ kN}$$

and

$$F_R = \sqrt{(40.5^2 + 75^2)} = 85.2 \text{ kN}$$

Slip resistance of single bolt in single shear, $P_{sL} = 87.6$ kN and slip resistance of single bolt in double shear is

$$2P_{sL} = 2 \times 87.6 = 175.2 \text{ kN} > F_R$$

Therefore, shear strength of bolts in web splice is adequate.

Check web of beam
The forces on the edge of the holes in the web may give rise to bearing failure and must therefore be checked. Here e = edge distance for web and

$$\tan \theta = \frac{40.5}{75}$$

$$\Rightarrow \theta = 28.37°$$

$$e = \frac{45}{\sin \theta} = \frac{45}{0.475} = 94.7$$

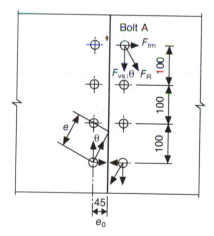

Bearing capacity of web is

$$= dtp_{bg} \leqslant 1/3etp_{bg} = 1/3 \times 94.7 \times 10.6 \times 825 = 276 \times 10^3$$
$$= 22 \times 10.6 \times 825 = 192.4 \times 10^3 \text{ N} > F_R$$

Hence web of beam is adequate in bearing.

Check web cover plates
The forces on the edge of the holes in the web cover plates may give rise to bearing failure and must therefore also be checked. Here e = edge distance for web cover plates and $\theta = 28.37°$. Thus

$$e = \frac{40}{\cos \theta} = \frac{40}{0.88} = 45.5$$

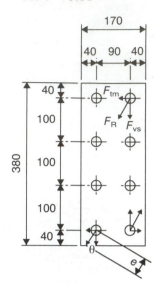

Bearing capacity of one plate is

$$= dtp_{bg} \leq 1/3 etp_{bg} = 1/3 \times 45.5 \times 10 \times 825 = 125 \times 10^3$$
$$= 22 \times 10 \times 825 = 181.5 \times 10^3 \text{ N}$$

Thus bearing capacity of single web plate = 125 kN and bearing capacity of two web plates = 2 × 125 = 250 kN > F_R. Hence web cover plates are adequate in bearing.

Check web cover plates for shear and bending

Check shear capacity. Shear force applied to one plate is

$$F_v/2 = 300/2 = 150 \text{ kN}$$

Torsional moment applied to one plate is

$$150 \times 45 = 6750 \text{ kN mm} = 6.75 \text{ kN mm}$$

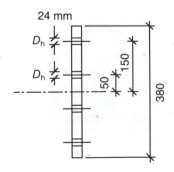

Shear capacity of one plate, P_v, is

$$P_v = 0.6 \, p_y A_v = 0.6 \times 275 \times 0.9 \times 10 \, (380 - 4 \times 24)$$
$$= 421.7 \text{ kN} > 150 \text{ kN}$$

where $A_v = 0.9 A_n$. Hence cover plate is adequate for shear.

Check moment capacity. Since applied shear force (= 150 kN) < $0.6 P_v$ = 253 kN, moment capacity, $M_c = p_y Z$ where $Z = I/y$. Referring to the above diagram, I of plate = $10 \times 380^3/12 - 2(10 \times 24)50^2 - 2(10 \times 24)150^2 = 337 \times 10^5$ mm^2

$$Z = I/y = 337 \times 10^5/190 = 177 \times 10^3 \text{ mm}^2$$

$$M_c = p_y Z = 275 \times 177 \times 10^3 = 48.7 \times 10^6 \text{ N mm} = 48.7 \text{ kN m} > 6.75 \text{ kN m}$$

Hence web cover plates are also adequate for bending.

Example 4.19 Shear resistance of a welded end plate beam-to-column connection

Calculate the design shear resistance of the connection shown below, assuming that the steel is grade 43 and the bolts are M20, grade 8.8 in 2 mm clearance holes.

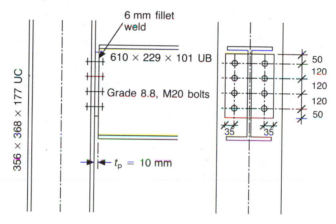

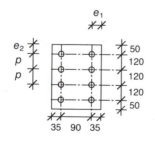

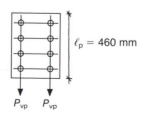

CHECK FASTENER SPACINGS AND END/EDGE DISTANCES

Diameter of bolt, d_b = 20 mm
Diameter of bolt hole, D_h = 22 mm
Pitch of bolt, p = 120 mm
Edge distance, e_1 = 35 mm
End distance, e_2 = 50 mm

Thickness of end plate, $t_p = 10$ mm
The following conditions need to be met:

$$\text{Pitch } ñ\ 2.5d_b = 2.5 \times 20 = 50 < 120 \quad \text{OK}$$
$$\text{Pitch} < 14t_p = 14 \times 10 = 140 > 120 \quad \text{OK}$$
$$\text{Edge distance } e_1 > 1.4D_h = 1.4 \times 22 = 30.8 < 35 \quad \text{OK}$$
$$\text{End distance } e_2 > 1.4D_h = 1.4 \times 22 = 30.8 < 50 \quad \text{OK}$$
$$e_1 \text{ and } e_2 < 11t_p\varepsilon = 11 \times 10 \times 1 = 110 > 35, 50 \quad \text{OK}$$

For grade 43 steel with $t_p = 10$ mm, $p_y = 275$ N mm^{-2} *(Table 4.3)*, $\varepsilon = 1$ (equation 4.4).

BOLT GROUP STRENGTH

Shear
8 No., M20, grade 8.8 bolts; $A_s = 245$ mm^2 *(Table 4.23)*, $p_s = 375$ N mm^{-2}. Shear capacity of single bolt, P_s, is

$$P_s = p_sA_s = 375 \times 245 = 91.9 \times 10^3 \text{ N}$$

Shear capacity of bolt group $= 8 \times 91.9 = 735$ kN.

Bearing
Bearing capacity of bolt, P_{bb}, is

$$P_{bb} = d_bt_p \ P_{bb} = 20 \times 10 \times 1035 = 207 \times 10^3 \text{ N} = 207 \text{ kN}$$

End plate is thinner than column flange and will therefore be critical. Bearing capacity of end plate, P_{bs}, is

$$P_{bs} = d_bt_pp_{bs} = 20 \times 10 \times 460 = 92 \times 10^3 \text{ N} = 92 \text{ kN } ß \frac{1}{2} \times e_2t_pp_{bs} = \frac{1}{2} \times 50 \times 10 \times 460 = 115 \times 10^3 \text{ N}$$

Hence bearing capacity of connection $= 8 \times 92 = 736$ kN.

END PLATE SHEAR STRENGTH

$$A_v = 0.9A_n = 0.9t_p\ (l_p - 4D_h) = 0.9 \times 10 \times (460 - 4 \times 22) = 3348 \text{ mm}^2$$

Shear capacity assuming single plane of failure, P_{vp}, is $P_{vp} = 0.6p_yA_v = 0.6 \times 275 \times 3348 = 552.4 \times 10^3 = 552.4$ kN
Shear capacity assuming two failure planes is

$$2P_{vp} = 2 \times 552.4 = 1104.8 \text{ kN}$$

WELD STRENGTH
(Readers should refer to *section 4.10.4* before performing this check.) Fillet weld of 6 mm provided. Hence

Leg length, $s = 6$ mm   Design strength, $p_w = 215$ N mm^{-2} (E43 electrode)
Throat size, $a = 0.7s = 0.7 \times 6 = 4.2$ mm

Effective length of weld, l_w, is

$$l_w = 2(l_p - 2s) = 2(460 - 2 \times 6) = 896 \text{ mm}$$

Hence design shear strength of weld, V_w, is

$$V_w = p_wal_w = 215 \times 4.2 \times 896 = 809 \times 10^3 \text{ N} = 809 \text{ kN}$$

LOCAL SHEAR STRENGTH OF BEAM WEB AT THE END PLATE

$$P_{vb} = 0.6p_yA_v = 0.6 \times 275(0.9 \times 10.6 \times 460) = 724 \text{ kN}$$

Hence, strength and connection is controlled by shear strength of beam web and is equal to 724 kN.

4.10.4 WELDED CONNECTIONS

The two main types of welded joints are **fillet welds** and **butt welds**. Varieties of each type are shown in *Fig. 4.36*. Essentially the process of welding consists of heating and melting steel in and/or around the gap between the pieces of steel that are being welded together. Welding rods consist of a steel rod surrounded by a flux which helps the metal to melt and flow into the joint. Welding can be accomplished using oxyacetylene equipment, but the easiest method is electric arc welding.

4.10.4.1 Strength of welds

If welding is expertly carried out using the correct grade of welding rod, the resulting weld should be considerably stronger than the pieces held together. However, to allow for some variation in the quality of welds, it is assumed that the weld strength for fillet welds is as given in Table 36 of BS 5950, reproduced as *Table 4.26*. However, the design strength of the weld can be taken as the same as that for the parent metal if the joint is a butt weld, or alternatively a fillet weld satisfying the following conditions:

1. The weld is symmetrical as shown in *Fig. 4.37*.
2. It is made with suitable electrodes which will produce specimens at least as strong as the parent metal.
3. The sum of throat sizes (*Fig. 4.37*) is not less than the connected plate thickness.
4. The weld is principally subject to direct tension or compression (*Fig. 4.37*).

Table 4.26 Design, strength, p_w (Table 36, BS 5950)

Design grade of steel	Electrode strength to BS 639 (N mm^{-2})			Other types (N mm^{-2})
	E43	E51	E51[a]	
40 or 43	215	215	–	
WR50 and 50	215	255	–	0.5U_e but $\leqslant$ 0.54U_s
55	–	255	275[a]	

[a] Only applies to electrodes having a minimum tensile strength of 550 N mm^{-2} and a minimum yield strength of 450 N mm^{-2}. *Note* U_e is the minimum tensile strength of the electrode based on all weld tensile tests as specified in BS 709; U_s is the specified minimum ultimate tensile strength of the steel.

4.10.4.2 Design details

Figure 4.36 also indicates what is meant by the weld **leg length**, s, and the **effective throat size**, a, which should not be taken as greater than 0.7s. The **effective length** of a run of weld should be taken as the actual length, less one leg length for each end of the weld. Where the weld ends at a corner of the metal, it should be continued around the corner for a distance greater than 2s. In a lap joint, the minimum lap length should be not less than 4t, where t is the thinner of the pieces to be joined. For fillet welds, the 'vector sum of the design stresses due to all forces and moments transmitted by the weld should not exceed the design strength, p_w'.

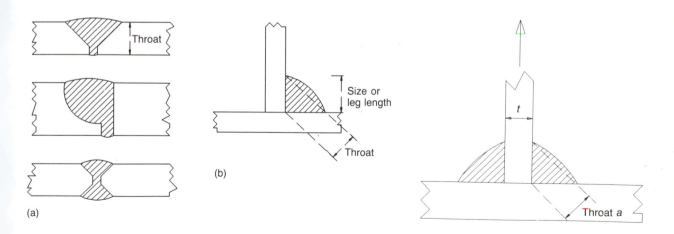

Fig. 4.36 *Types of weld: (a) butt welds; (b) fillet weld.*

Fig. 4.37 *Special fillet weld.*

Example 4.20 Analysis of a welded beam-to-column connection

A grade 43 steel $610 \times 229 \times 101$ UB is to be connected, via a welded end plate, on to a $356 \times 368 \times 177$ UC. The connection is to be designed to transmit a bending moment of 500 kN m and a shear force of 300 kN. Show that the proposed welding scheme for this connection is adequate.

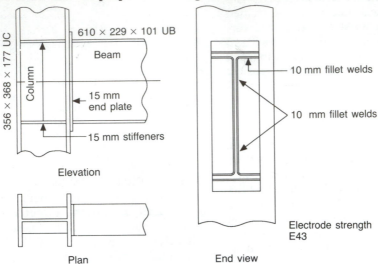

Elevation

Plan

End view

Electrode strength E43

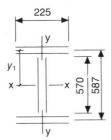

Leg length of weld $s = 10$ mm. Effective length of weld is

$$4(l_w - 2s) + 2(h_w - 2s) = 4(225 - 2 \times 10) + 2(570 - 2 \times 10)$$
$$= 1920 \text{ mm}$$

Weld force = shear force / effective length of weld = 300/1920 = 0.16 kN mm^{-1}. Weld second moment of area, I_{xx}, is

$$I_{xx} = 4[\, 1 \times (l_w - 2s)y_1^2] + 2\,[1 \times (h_w - 2s)^3/12$$

$$= 4[(205)293.5^2] + 2(550^3/12)$$

$$= 98\,365\,811 \text{ mm}^4$$

Weld shear (moment) is

$$My/I_{xx} = 500 \times 10^3 \times 293.5/98\,365\,811$$
$$= 1.49 \text{ kN mm}^{-1}$$
$$\text{where } y = 587/2 = 293.5$$
$$\text{Resultant} = \sqrt{(1.49^2 + 0.16^2)} = 1.50 \text{ kN mm}^{-1}$$

Steel grade 43, Electrode strength E43. Hence from *Table 4.26*, design strength $p_w = 215$ N mm^{-2}. Weld capacity of 10 mm fillet weld is

$ap_w = 0.7 \times 10 \times 215 \times 10^{-3} = 1.51$ kN mm^{-1} > required where the throat thickness of weld, $a = 0.7s = 0.7 \times 10 = 7$ mm. Hence proposed welding scheme is just adequate.

4.11 Summary

This chapter has considered the design of a number of structural steelwork elements including beams, columns and connections to BS 5950: Part 1: *Code of practice for Design in Simple and Continuous Construction: Hot Rolled Sections*. The ultimate limit state of strength and the serviceability limit state of deflection principally influence the design of steel elements. Many steel structures are still analysed by assuming that individual elements are simply supported at their ends. Steel sections are classified as compact, semi-compact or slender depending on how the sections perform in bending. The design of flexural members generally involves considering the limit states of bending, shear, web bearing/buckling, deflection and if the compression flanges are not fully restrained, lateral torsional buckling. Columns subject to axial load and bending are normally checked for local instability and overall buckling.

The two principal methods of connecting steel elements of a structure are bolting and welding. Design of bolted connections, using ordinary (or black) bolts, usually involves checking that neither the bolt nor the elements being joined exceed their shear, bearing or tension capacities. Where HSFG bolts are used, the slip resistance must also be determined. Welded connections are most often used to weld end plates or cleats to members, a task which is normally performed in the fabrication yard rather than on site.

Questions

1. A simply supported beam spanning 8 m has central point dead and imposed loads of 200 and 100 kN respectively. Assuming the beam is fully laterally restrained select and check suitable UB sections in (a) grade 43 and (b) grade 55 steel.

2. A simply supported beam spanning 8 m has uniformly distributed dead and imposed loads of 20 and 10 kN m^{-1} respectively. Assuming the beam is fully laterally restrained select and check suitable UB sections in (a) grade 43 and (b) grade 55 steel.

3. For the fully laterally restrained beam shown in *Fig. 4.38*, select and check a suitable UB section in grade 43 steel to satisfy shear, bending and deflection criteria.

4. If the beam in *Fig. 4.38* is laterally unrestrained, select and check a suitable section in grade 43 steel to additionally satisfy lateral torsional buckling, web bearing and buckling criteria. Assume that each support is 50 mm long and lateral restraint conditions at supports are as follows:
   (a) compression flange laterally restrained;
   (b) beam fully restrained against torsion;
   (c) both flanges free to rotate on plan;
   (d) destabilizing load conditions.

5. If two discrete lateral restraints, one at mid-span and one at the cantilever tip, are used to stabilize the beam in *Fig. 4.38*, select and check a suitable section in grade 43 steel to satisfy all the criteria in question 4.

6. Carry out designs for the beams shown in *Fig. 4.39*.

7. Select a suitable short column section in grade 43 steel to support a factored axial concentric load of 1000 kN and factored bending moments of 400 kN m about the major axis, and 100 kN m about the minor axis.

8. Select a suitable column section in grade 43 steel to support a factored axial concentric load of 1000 kN and factored bending moments of 400 kN m

m about the major axis, and 100 kN m about the minor axis, both applied at each end. The column is 10 m long and is fully fixed against rotation at top and bottom, and the floors it supports are braced against sway.

9. Design a splice connection for a 686 × 254 × 140 UB section in grade 43 steel to cater for half bending strength and half the shear capacity of the section.

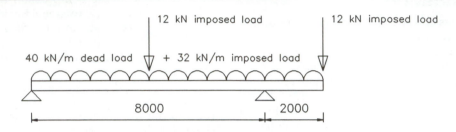

Fig. 4.38

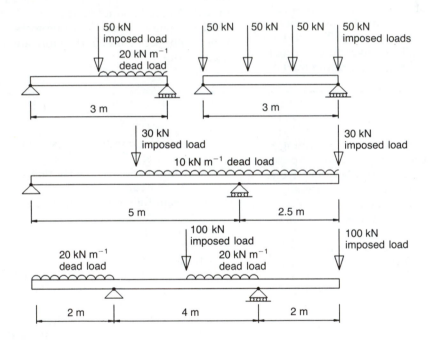

Fig. 4.39

Chapter 5

Design in unreinforced masonry to BS 5628

This chapter is concerned with the design of unreinforced masonry walls to BS 5628: Part 1. The chapter describes the composition and properties of the three main materials used in masonry construction: bricks, blocks and mortars. Masonry is primarily used these days in the construction of loadbearing and non-loadbearing walls. The primary aim of this chapter is to give guidance on the design of single leaf and cavity walls, with and without stiffening piers, subject to either vertical or lateral loading.

5.1 Introduction

Structural masonry was traditionally very widely used in civil and structural works including tunnels, bridges, retaining walls and sewerage systems (*Fig. 5.1*). However, the introduction of steel and concrete with their superior strength and cost characteristics led to a sharp decline in the use of masonry for these applications.

Over the past two decades or so, masonry has recaptured some of the market lost to steel and concrete due largely to the research and marketing work sponsored in particular by the Brick Development Association. For instance, everybody now knows that 'brick is beautiful'. Less well appreciated, perhaps, is the fact that masonry has excellent structural, thermal and acoustic properties. Furthermore, it displays good resistance to fire and the weather.

Fig. 5.1 *Traditional application of masonry in construction: (a) brick bridge; (b) brick sewer; (c) brick retaining wall.*

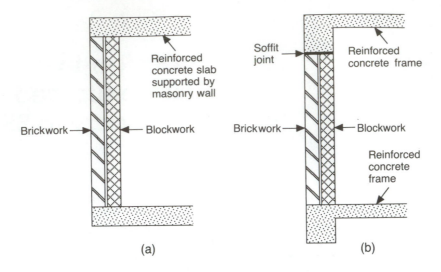

Fig. 5.2 (a) Loadbearing and (b) non-loadbearing masonry walls.

On this basis it has been argued that it is faster and cheaper to build certain types of buildings using only masonry rather than using a combination of materials to provide these properties.

Brick manufacturers have also pointed out that, unlike steelwork, masonry does not require regular maintenance nor, indeed, suffers from the durability problems which have plagued concrete. The application of reinforcing and prestressing techniques to masonry design will, it is believed, considerably improve its structural properties and hence its appeal to designers.

Despite the above, masonry is primarily used nowadays for the construction of loadbearing and non-loadbearing walls (*Fig. 5.2*). These structures are principally designed to resist lateral and vertical loading. The lateral loading arises mainly from the wind pressure acting on the wall. The vertical loading is attributable to dead plus imposed loading from any supported floors, roofs, etc. and/or self-weight of the wall.

Design of masonry structures in the UK is carried out in accordance with the recommendations given in BS 5628: *Code of Practice for Use of Masonry*. This code is divided into three parts:

Part 1: *Structural Use of Unreinforced Masonry.*
Part 2: *Structural Use of Reinforced and Prestressed Masonry.*
Part 3: *Materials and Components, Design and Workmanship.*

Part 1 was originally published in 1978 and Parts 2 and 3 were issued in 1985. This book deals only with the design of unreinforced masonry walls, subject to vertical or lateral loading. It should, therefore, be assumed that all references to BS 5628 refer to Part 1, unless otherwise noted.

Before looking in detail at the design of masonry walls, the composition and properties of the component materials are considered in the following sections.

5.2 Materials

Structural masonry basically consists of bricks or blocks bonded together using mortar or grout. In cavity walls, wall ties complying with BS 1243 or DD 140 are also used to tie together the two skins of masonry (*Fig. 5.3*). In external walls, damp-proof courses are also necesssary to prevent moisture ingress to the building fabric (*Fig. 5.4*).

The following discussion will concentrate on the composition and properties of the three main components of structural masonry, namely (i) bricks, (ii) blocks and (iii) mortar. The reader is referred to BS 5628: Parts 1 and 3 and BRE Digest 380 for guidance on the design and specification of wall ties and damp-proof systems for normal applications.

5.2.1 BRICKS

Bricks are manufactured from a variety of materials such as clay, lime and sand/flint, concrete and natural stone. Of these, clay bricks are by far the most commonly used variety in the UK.

Clay bricks are manufactured by shaping suitable

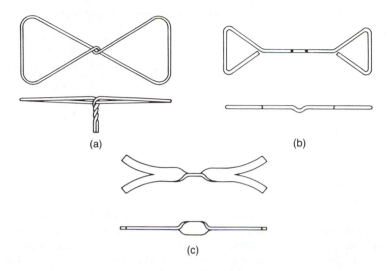

Fig. 5.3 *Wall ties to BS 1243: (a) butterfly tie; (b) double triangle tie; (c) vertical twist tie.*

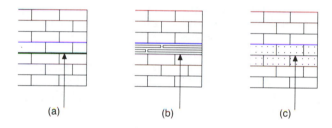

Fig. 5.4 *Damp-proof courses (a) bitumen, pitch and bitumen, polymer, lead, copper, polyethylene d.p.c.; (b) two courses of bonded slate d.p.c.; (c) brick d.p.c. (usually two courses) (based on Table 12, BS 5628: Part 3).*

clays to units of standard size, normally taken to be $215 \times 102.5 \times 65$ mm (*Fig. 5.5*). Sand facings and face textures may then be applied to the 'green' clay. Alternatively, the clay units may be perforated or frogged in order to reduce the self-weight of the unit. Thereafter, the clay units are fired in kilns to a temperature in the range 900–1500 °C in order to produce a brick suitable for structural use. It is because of the firing process that bricks have excellent fire-resistant properties.

In design it is normal to refer to the coordinating size of bricks. This is usually taken to be $225 \times 112.5 \times 75$ mm and is based on the actual or work size of the brick, i.e. $215 \times 102.5 \times 65$ mm, plus an allowance

Fig. 5.5 *Typical bricks: (a) solid; (b) perforated; (c) frogged.*

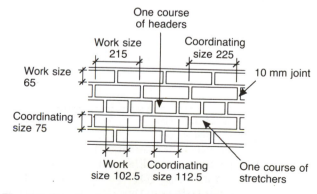

Fig. 5.6 *Coordinating and work size of bricks.*

165

Table 5.1 Classification of bricks by compressive strength and water absorption (Table 4, BS 3921)

Type	Average compressive strength $(N\ mm^{-2})$ not less than	Average absorption (% by weight) not greater than
Engineering A	70	4.5
Engineering B	50	7.0
Damp-proof course 1	5	4.5
Damp-proof course 2	5	7.0
All others	5	No limit

of 10 mm for the mortar joint (*Fig. 5.6*). Clay bricks are also manufactured in metric modular format having a coordinating size of $200 \times 100 \times 75$ mm. Other cuboid and special shapes are also available (BS 4729).

Bricks are normally available in three categories, namely common, facing and engineering. Common bricks are those suitable for general construction work, with no special claim to give an attractive appearance. Facing bricks are specially made or selected to give an attractive appearance on the basis of colour and texture. Engineering bricks tend to be dense and strong and are designated class A and class B on the basis of strength and water absorption (*Table 5.1*).

BS 3921 also classifies clay bricks with respect to their resistance to frost and the maximum soluble salt content, using the designations shown in *Table 5.2*. The reader is referred to BS 3921 for definitions of these designations. Guidance on the selection of clay bricks most appropriate for particular situations as regards frost resistance and soluble salt content can be found in Table 13 of BS 5628: Part 3.

5.2.2 BLOCKS
Blocks are walling units but, unlike bricks, are normally made from concrete. They are available in two basic types: aerated concrete and aggregate concrete. The aerated blocks are made from a mixture of sand, pulverized fuel ash, cement and aluminium powder. The aggregate blocks have a composition similar to that of normal concrete, consisting chiefly of sand, coarse and fine aggregate and cement plus extenders. Aerated blocks tend to have lower densities than aggregate blocks, which accounts for the former's superior thermal properties, lower unit weights and lower strengths. Generally, aerated blocks are more expensive than aggregate blocks.

Blocks are manufactured in three basic forms: solid, cellular and hollow (*Fig. 5.7*). Solid blocks have no formed holes or cavities other than those inherent in the material. Cellular blocks have one or more formed holes or cavities which do not pass through the block. Hollow blocks are similar to cellular blocks except that the holes or cavities pass through the block.

For structural design, the two most important properties of blocks are their size and compressive strength. *Tables 5.3* and *5.4* give the most commonly available sizes and compressive strengths of concrete blocks. The most frequently used block has a work face of 440×215 mm, thickness 100 mm and compressive strength 3.5 N mm^{-2}; 2.8 N mm^{-2} is a popular strength for aerated concrete blocks, and 7.0 N mm^{-2} for aggregate concrete blocks as it can be used below ground. Guidance on the selection and specification of concrete blocks in masonry construction can be found in BS 5628: Part 3 and BS 6073 respectively.

Table 5.2 Frost resistance and soluble salt content designations for clay bricks (BS 3921)

Designation	Frost resistance	Soluble salt content
FL	Frost resistant (F)	Low (L)
FN	Frost resistant (F)	Normal (N)
ML	Moderately frost resistant (M)	Low (L)
MN	Moderately frost resistant (M)	Normal (N)
OL	Not frost resistant (O)	Low (L)
ON	Not frost resistant (O)	Normal (N)

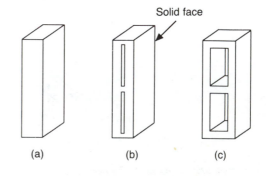

Fig. 5.7 *Concrete blocks: (a) solid; (b) cellular; (c) hollow.*

Table 5.3 Work sizes of concrete blocks (Table 1, BS 6073: Part 2)

Length (mm)	Height (mm)	Thickness (mm)														
		60	75	90	100	115	125	140	150	175	190	200	215	220	225	250
390	190	x	x	x	x	x		x	x		x	x				
440	140	x	x	x	x			x	x		x	x			x	
440	190	x	x	x	x			x	x		x		x	x		
440	215	x	x	x	x	x	x	x	x	x	x	x	x	x	x	x
440	290	x	x	x	x			x	x		x	x	x			
590	140		x	x	x			x	x		x	x	x			
590	190		x	x	x			x	x		x	x	x			
590	215		x	x	x		x	x	x	x		x	x		x	x

5.2.3 MORTARS

The primary function of the mortar is to bind together the individual brick or block units, thereby allowing the transfer of compression, shear and tensile stresses between adjacent units. However, there are several other properties which the mortar must possess for ease of construction and maintenance. For instance, the mortar should be easy to spread and remain plastic for a sufficient length of time in order that the units can be accurately positioned before setting occurs. On the other hand, the setting time should not be too excessive otherwise the mortar may be squeezed out as successive courses of units are laid. Additionally, the mortar should be able to resist water uptake by the absorbent bricks, otherwise hydration and hence full development of the mortar strength may be prevented.

In its most basic form, mortar simply consists of a mixture of sand and ordinary Portland cement (OPC). However, such a mix is generally unsuitable for use in masonry since it will tend to be too strong in comparison with the strength of the bricks/blocks. It is generally desirable to provide the lowest grade of mortar possible, taking into account the strength and durability of the bricks/blocks to be used. This is to ensure that any cracking which occurs in the masonry due to settlement of the foundations, caused by thermal/moisture movements, etc., will occur in the mortar rather than the masonry units themselves. Cracks in the mortar tend to be smaller because the mortar is more flexible than the masonry units and, hence, are easier to repair.

In order to produce a more practicable mix, it is normal to add lime to the cement mortar. This has the effect of reducing the strength of the mix, but at the same time increasing its workability and bonding properties. For most forms of masonry construction a 1:1:6 cement: lime: sand mortar is suitable. Similar properties can also be imparted to the mortar by adding a plasticizer to the mix. Alternatively, the OPC can be substituted with masonry cement which is a mixture of approximately 75% OPC, an inert filler and a plasticizer. Such a mix will have intermediate properties to those displayed by the cement, lime, sand mortar and cement, sand, plasticizer mortar.

By varying the proportions of the constituents referred to above, mortars of differing compressive and bond strengths, plasticity and frost susceptibility can be produced as shown in *Table 5.5*. As will be noted from the table there are three basic types of mortars and four mortar strengths associated with each mortar type.

Table 5.6, which is based on Table 13 of BS 5628: Part 3, gives guidance on the selection of masonry units and mortars for particular applications as regards durability.

Table 5.4 Compressive strength of concrete blocks (Clause 5.3, BS 6073: Part 2)

Compressive strengths (N mm^{-2})
2.8
3.5
5.0
7.0
10.0
15.0
20.0
35.0

5.3 Masonry design

The foregoing has summarized some of the more important properties of the component materials which are used in masonry construction. As noted at the beginning of this chapter, masonry is mostly used these days in the construction of loadbearing and panel walls. Loadbearing walls primarily resist vertical load-

Table 5.5 Types of mortars (Table 1, BS 5628)

	Mortar designa-tion	Type of mortar (proportion by volume)			Mean compressive strength at 28 days (N mm^{-2})	
		Cement : lime : sand	Masonry cement : sand	Cement : sand with plasticizer	Preliminary (laboratory) tests	Site tests
Increasing strength	Increasing ability to accommodate (i)	$1 : 0$ to $\frac{1}{4} : 3$	–	–	16.0	11.0
	movement, (ii)	$1 : \frac{1}{2} : 4$ to $4\frac{1}{2}$	$1 : 2\frac{1}{2}$ to $3\frac{1}{2}$	$1 : 3$ to 4	6.5	4.5
e.g. due to	(iii)	$1 : 1 : 5$ to 6	$1 : 4$ to 5	$1 : 5$ to 6	3.6	2.5
settlement,	(iv)	$1 : 2 : 8$ to 9	$1 : 5\frac{1}{2}$ to $6\frac{1}{2}$	$1 : 7$ to 8	1.5	1.0
temperature and moisture changes						

Direction of change in properties is shown by the arrows

Increasing resistance to frost attack during construction

Improvement in bond and consequent resistance to rain penetration

ing while panel walls primarily resist lateral loading. *Figure 5.8* shows some commonly used wall types.

The remainder of this chapter considers the design of (1) loadbearing walls, with and without stiffening piers, resisting vertical compression loading (*section 5.5*) and (2) panel walls resisting lateral loading (*section 5.6*).

5.4 Symbols

For the purposes of this chapter, the following symbols have been used. These have largely been taken from BS 5628.

GEOMETRIC PROPERTIES

t	actual thickness of wall or leaf
h	height of panel between restraints
L	length of wall between restraints
A	cross-sectional area of loaded wall
Z	sectional modulus
t_{ef}	effective thickness of wall
h_{ef}	effective height of wall
K	effective thickness coefficient
e_x	eccentricity of loading at top of wall
λ	slenderness ratio
β	capacity reduction factor

COMPRESSION

G_k characteristic dead load

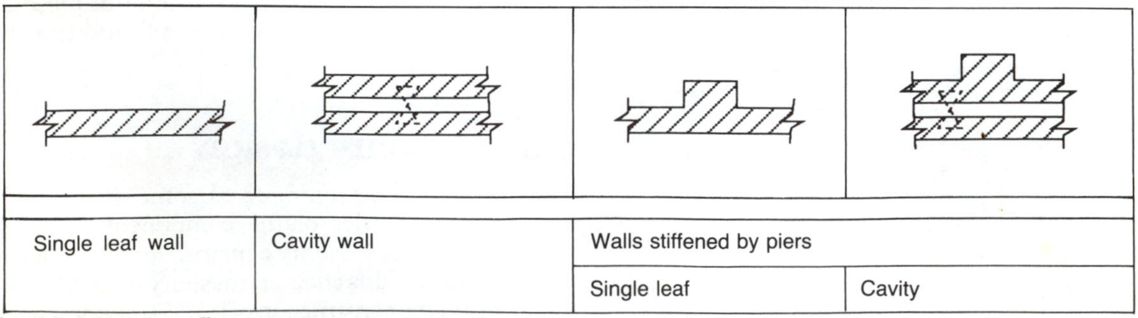

Single leaf wall	Cavity wall	Walls stiffened by piers	
		Single leaf	Cavity

Fig. 5.8 Masonry walls.

Q_k	characteristic imposed load	M	ultimate design moment
W_k	characteristic wind load	M_R	design moment of resistance
γ_f	partial safety factor for load	M_{par}	design moment with plane of failure parallel to bed joint
γ_m	partial safety factor for materials		
f_k	characteristic compressive strength of masonry	M_{perp}	design moment with plane of failure perpendicular to bed joint
N	ultimate design vertical load	$M_{k\,par}$	design moment of resistance with plane of failure parallel to bed joint
N_R	ultimate design vertical load resistance of wall		
		$M_{k\,perp}$	design moment of resistance with plane of failure perpendicular to bed joint

FLEXURE

$f_{kx\,par}$	characteristic flexural strength of masonry with plane of failure parallel to bed joint
$f_{kx\,perp}$	characteristic flexural strength of masonry with plane of failure perpendicular to bed joint
α	bending moment coefficient
μ	orthogonal ratio

5.5 Design of vertically loaded masonry walls

In common with most modern codes of practice dealing with structural design, BS 5628: *Code of Practice for Use of Masonry* is based on the limit state philosophy (*Chapter 1*). This code states that the

Table 5.6 Desirable minimum quality of brick, block and mortar for durability (based on Table 13, BS 5628: Part 3)

Location	Clay bricks	Concrete blocks (type)[a]	Mortar designation	
			NF[b]	PF[b]
Internal walls, inner leaf of external cavity walls, backing (inner face) of external solid walls	NF[b] (O) PF[b] (M)	T	(iv)	(iii)
External walls, outer leaf of external cavity walls	(1)[c] (M) (2) (M) (3) (F)	T R R[g]	(iv)[d] (iii)[e] (ii)[e]	(iii)[d] (iii)[e,f] (ii)[e,f]
Facing (outer face) of external solid construction	(F) —	– (see External walls)	(ii)[h] (see External walls)	(ii)[h]
Damp-proof course (d.p.c.)	d.p.c.	(Not applicable	(i)	(i)
External freestanding walls (with coping)	(F)	S	(iii)	(iii)
Earth retaining walls (back filled with free-draining material)	(F)	R[g]	(ii)[e,i]	(ii)[e,i]
Sewerage work (foul water)	A	–	(ii)[h,i]	(ii)[h,i]

[a] R: minimum density 1500 kg m^{-3} *and* average compressive strength $\geqslant$ 7 N mm^{-2}. S: minimum density 1500 kg m^{-3} *and* average compressive strength $\geqslant$ 3.5 N mm^{-2} or average compressive strength $\geqslant$ 7 N mm^{-2}. T: all types except that blocks less than 75 mm thick generally only suitable for internal partitions.
[b] NF, no risk of frost during construction; PF, possibility of frost during construction; (O) not frost resistant; (F) frost resistant; (M) moderately frost resistant.
[c] (1) Above d.p.c. (2) Below d.p.c. but not closer than 150 mm to finished ground level. (3) Below 150 mm above finished ground level.
[d] For clay brickwork, not less than mortar designation (iii) should be used; if this brickwork is to be rendered, see footnote f.
[e] Where sulphates are present in the soil or ground water, see footnote h.
[f] Where clay bricks inadvertently become wet, the use of plasticized mortars is desirable.
[g] Concrete blocks are generally not suitable for use in contact with ground from which there is any danger of sulphate attack.
[h] Cement mortars with sulphate-resisting cement in place of ordinary Portland cement should be used.

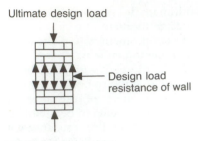

Ultimate design load

Design load resistance of wall

primary aim of design is to ensure an adequate margin of safety against the ultimate limit state being reached. In the case of vertically loaded walls this is achieved by ensuring that the ultimate design load (N) does not exceed the design load resistance of the wall (N_R):

$$N \leqslant N_R \qquad (5.1)$$

The ultimate design load is a function of the actual loads bearing down on the wall. The design load resistance is related to the design strength of the masonry wall. The following subsections discuss the procedures for estimating the:

1. ultimate design load
2. design strength of masonry walls and
3. design load resistance of masonry walls.

5.5.1 ULTIMATE DESIGN LOADS, N

As discussed in *section 2.3*, the loads acting on a structure are divided into three basic types, namely dead loads, imposed (or live) loads and wind loads. Generally, the ultimate design load is obtained by multiplying the characteristic (dead/imposed/wind) loads (F_k) by the appropriate partial safety factor for loads (γ_f)

$$N = \gamma_f F_k \qquad (5.2)$$

5.5.1.1 Characteristic loads (clause 21, BS 5628)

The characteristic dead loads (G_k) and imposed loads (Q_k) are obtained from the following: (i) BS 648: *Schedule of Weights for Building Materials* and (ii) BS 6399: *Design Loadings for Buildings*, Part 1: *Code*

of *Practice for Dead and Imposed Loads*. The characteristic wind load (W_k) is calculated in accordance with CP 3: Chapter V: *Wind Loads* (which is to be superseded in due course by Part 2 of BS 6399).

5.5.1.2 Partial safety factors for loads (γ_f) (clause 22, BS 5628)

As discussed in *section 2.3.2*, the applied loads may be greater than anticipated for a number of reasons. Therefore, it is normal practice to factor up the characteristic loads. *Table 5.7* shows the partial safety factors on loading for various load combinations. Thus, with structures subject to only dead and imposed loads the partial safety factors for the ultimate limit state are usually taken to be 1.4 and 1.6 respectively. The ultimate design load for this load combination is given by

$$N = 1.4G_k + 1.6Q_k \qquad (5.3)$$

In assessing the effect of dead, imposed and wind loads for the ultimate limit state, the partial safety factor is generally taken to be 1.2 for all the load types. Hence

$$N = 1.2(G_k + Q_k + W_k) \qquad (5.4)$$

5.5.2 DESIGN STRENGTH

The design strength of masonry is given by

$$\text{Design strength} = \frac{\beta f_k}{\gamma_m} \qquad (5.5)$$

where f_k is the characteristic compressive strength of masonry, γ_m the partial safety factor for materials and β the capacity reduction factor. Each of these factors are discussed in the following sections.

5.5.2.1 Characteristic compressive strength of masonry, f_k (clause 23.1, BS 5628)

The basic characteristic compressive strengths of normally bonded masonry constructed under laboratory conditions and tested at an age of 28 days are given in *Table 5.8*. As will be noted from the table,

Table 5.7 Partial factors of safety on loading, γ_f, for various load combinations

Load combinations	Ultimate limit state		
	Dead	*Imposed*	*Wind*
Dead and imposed	$1.4G_k$ or $0.9G_k$	$1.6Q_k$	
Dead and wind	$1.4G_k$ or $0.9G_k$		The larger of $1.4W_k$ or $0.015G_k$
Dead, imposed and wind	$1.2G_k$	$1.2Q_k$	The larger of $1.2W_k$ or $0.015G_k$

Table 5.8 Characteristic compressive strength of masonry, f_k (Table 2, BS 5628)
(a) Constructed with standard format bricks

Mortar designation	Compressive strength of unit (N mm^{-2})								
	5	10	15	20	27.5	35	50	70	100
(i)	2.5	4.4	6.0	7.4	9.2	11.4	15.0	19.2	24.0
(ii)	2.5	4.2	5.3	6.4	7.9	9.4	12.2	15.1	18.2
(iii)	2.5	4.1	5.0	5.8	7.1	8.5	10.6	13.1	15.5
(iv)	2.2	3.5	4.4	5.2	6.2	7.3	9.0	10.8	12.7

(b) Constructed with blocks having a shape factor of 0.6

Mortar designation	Compressive strength of unit (N mm^{-2})							35 or greater
	2.8	3.5	5.0	7.0	10	15	20	
(i)	1.4	1.7	2.5	3.4	4.4	6.0	7.4	11.4
(ii)	1.4	1.7	2.5	3.2	4.2	5.3	6.4	9.4
(iii)	1.4	1.7	2.5	3.2	4.1	5.0	5.8	8.5
(iv)	1.4	1.7	2.2	2.8	3.5	4.4	5.2	7.3

(c) Constructed with hollow blocks having a shape factor of between 2 and 4

Mortar designation	Compressive strength of unit (N mm^{-2})							35 or greater
	2.8	3.5	5.0	7.0	10	15	20	
(i)	2.8	3.5	5.0	5.7	6.1	6.8	7.5	11.4
(ii)	2.8	3.5	5.0	5.5	5.7	6.1	6.5	9.4
(iii)	2.8	3.5	5.0	5.4	5.5	5.7	5.9	8.5
(iv)	2.8	3.5	4.4	4.8	4.9	5.1	5.3	7.3

(d) Constructed with solid concrete blocks having a shape factor of between 2 and 4

Mortar designation	Compressive strength of unit (N mm^{-2})							35 or greater
	2.8	3.5	5.0	7.0	10	15	20	
(i)	2.8	3.5	5.0	6.8	8.8	12.0	14.8	22.8
(ii)	2.8	3.5	5.0	6.4	8.4	10.6	12.8	18.8
(iii)	2.8	3.5	5.0	6.4	8.2	10.0	11.6	17.0
(iv)	2.8	3.5	4.4	5.6	7.0	8.8	10.4	14.6

the compressive strength is a function of several factors including the type and size of masonry unit, compressive strength of the unit and quality of the mortar used. The following discusses how the characteristic strength of masonry constructed from (a) brickwork and (b) blockwork is determined using *Table 5.8*.

Brickwork. The basic characteristic strengths of masonry built with standard format bricks and mortar designations (i)–(iv) (*Table 5.5*) are obtained from *Table 5.8(a)*. The values in the table may be modified where the following conditions, among others, apply:

1. If the horizontal cross-sectional area of the loaded wall (A) is less than 0.2 m^2, the basic compressive strength should be multiplied by the factor (0.7 + 1.5A) (clause 23.1.1, BS 5628).
2. When brick walls are constructed so that the thickness of the wall or loaded inner leaf of a cavity wall is equal to the width of a standard format brick, the basic compressive strength may be multiplied by 1.15 (clause 23.1.2, BS 5628).

Blockwork. *Tables 5.8(b)*, (c) and (d) are used, singly or in combination, to determine the basic char-

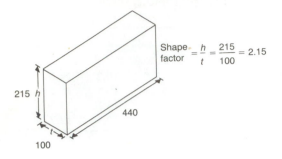

$$\text{Shape factor} = \frac{h}{t} = \frac{215}{100} = 2.15$$

Fig. 5.9 *Definition of shape factor.*

Table 5.10 Capacity reduction factor, β (Table 7, BS 5628)

Slenderness	Eccentricity at top of wall, e_x, ratio			
h_{ef}/t_{ef}	0.05t	0.1t	0.2t	0.3t
0	1.00	0.88	0.66	0.44
6	1.00	0.88	0.66	0.44
8	1.00	0.88	0.66	0.44
10	0.97	0.88	0.66	0.44
12	0.93	0.87	0.66	0.44
14	0.89	0.83	0.66	0.44
16	0.83	0.77	0.64	0.44
18	0.77	0.70	0.57	0.44
20	0.70	0.64	0.51	0.37
22	0.62	0.56	0.43	0.30
24	0.53	0.47	0.34	
26	0.45	0.38		
27	0.40	0.33		

acteristic compressive strength of blockwork masonry. As in the case of brick masonry, the values given in these tables may be modified where the cross-sectional area of loaded wall is less than 0.2 m^2 (see above).

Table 5.8(b) applies to masonry constructed using (solid or hollow) blocks having a shape factor of 0.6. The shape factor is the ratio of the height to least horizontal length of the block (*Fig. 5.9*). *Table 5.8(c)* applies to masonry constructed using hollow blocks having a shape factor of between 2.0 and 4.0. *Table 5.8(d)* applies to masonry constructed using solid blocks having a shape factor of between 2.0 and 4.0.

When walls are constructed using hollow blocks having a shape factor of between 0.6 and 2.0, the characteristic strengths should be determined by interpolating between the values in *Tables 5.8(b)* and (*c*). When walls are constructed using solid blocks having a shape factor of between 0.6 and 2.0, the characteristic strengths should be determined by interpolating between the values in *Tables 5.8(b)* and (*d*).

5.5.2.2 Partial safety factor for materials (γm)
The quality control exercised both during the manufacture of individual structural units and during construction influences the design strength of the masonry element. Two degrees of quality control are

Table 5.9 Partial safety factors for material strengths, γm (Table 4, BS 5628)

		Category of construction control	
		Special	Normal
Category of manufacturing control of structural units	Special	2.5	3.1
	Normal	2.8	3.5

recognized: normal and special. *Table 5.9* shows the values of the partial safety factor for material strengths for various combinations of categories of manufacturing control of structural units and construction control.

The degree of quality control for manufacture and construction will usually be assumed to be 'normal' unless the provisions outlined in clauses 27.2.1.2 and 27.2.2.2 (BS 5628) relating to the 'special' category can clearly be met.

5.5.2.3 Capacity reduction factor (β)
Masonry walls which are tall and slender are likely to be less stable under compressive loading than walls which are short and stocky. Similarly, eccentric loading will reduce the compressive load capacity of the wall. In both cases, the reduction in load capacity reflects the increased risk of failure due to instability rather than crushing of the materials. These two effects are taken into account by means of the capacity reduction factor, β (*Table 5.10*). This factor generally reduces the design strength of the member, in some cases by as much as 70%, and is a function of the following factors which are discussed below: (a) slenderness ratio and (b) eccentricity of loading.

Slenderness ratio (SR). The SR indicates the type of failure which may arise when a member is subject to compressive loading. Thus, walls which are short and stocky will have a low SR and tend to fail by crushing. Walls which are tall and slender will have higher SRs and will also fail by crushing. How-

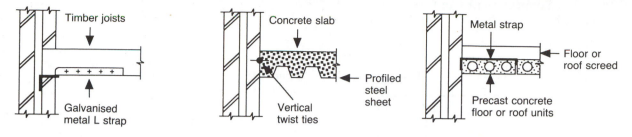

Fig. 5.10 *Details of horizontal supports providing simple resistance.*

ever, in this case, failure will arise as a result of excessive bending of the wall rather than direct crushing of the materials. Similarly, walls which are rigidly fixed at their ends will tend to have lower SRs and hence higher design strengths than walls which are partially fixed or unrestrained.

As can be appreciated from the above, the SR depends upon the height of the member, the cross-sectional area of loaded wall and the restraints at the member ends. BS 5628 defines SR as

$$SR = \frac{\text{effective height}}{\text{effective thickness}} = \frac{h_{ef}}{t_{ef}} \quad (5.6)$$

The following discusses how the (i) effective height and (ii) effective thickness of masonry walls are determined.

EFFECTIVE HEIGHT (h_{ef}). The effective height of a loadbearing wall depends on the actual height of the member and the type of restraint at the wall ends. Such restraint that exists is normally provided by any supported floors, roofs, etc. and may be designated 'simple' (i.e. pin-ended) or 'enhanced' (i.e. partially fixed), depending on the actual construction details used. *Figures 5.10* and *5.11* show typical examples of horizontal restraints which provide simple and enhanced resistance respectively.

According to clause 28.3.2 of BS 5628 the effective height of a wall may be taken as:

1. 0.75 times the clear distance between lateral supports which provide enhanced resistance to lateral movements; or
2. the clear distance between lateral supports which provide simple resistance to lateral movement.

EFFECTIVE THICKNESS (t_{ef}). For single leaf walls the effective thickness is taken as the actual thickness (t) of the wall (*Fig. 5.12(a)*):

$$t_{ef} = t \quad \text{(single leaf wall)}$$

If the wall was stiffened with piers (*Fig. 5.12(b)*), in order to increase its load capacity, the effective thickness is given by $t_{ef} = tK$ (single leaf wall stiffened with piers) where K is obtained from *Table 5.11*.

For cavity walls the effective thickness should be taken as the greater of (i) two-thirds the sum of the actual thickness of the two leaves, i.e. $\frac{2}{3}(t_1 + t_2)$ or (ii) the actual thickness of the thicker leaf, i.e. t_1 or t_2 (*Fig. 5.13(a)*). Where the cavity wall is stiffened by piers, the effective thickness of the wall is taken as the greatest of (i) $\frac{2}{3}(t_1 + Kt_2)$ or (ii) t_1 or (iii) Kt_2 (*Fig. 5.13(b)*).

Eccentricity of vertical loading. In addition to the slenderness ratio, the capacity reduction factor is also a function of the eccentricity of loading. When considering the design of walls it is not realistic to assume that the load will be applied

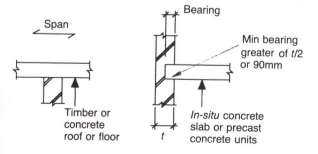

Fig. 5.11 *Details of horizontal supports providing enhanced resistance.*

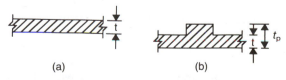

Fig. 5.12 *Effective thickness: (a) single leaf wall; (b) single leaf wall stiffened with piers (based on Fig. 3 of BS 5628).*

Table 5.11 Stiffness coefficient for walls stiffened by piers (Table 5, BS 5628)

Ratio of pier spacing (centre to centre) to pier width	Ratio t_p/t of pier thickness to actual thickness of wall to which it is bonded		
	1	2	3
6	1.0	1.4	2.0
10	1.0	1.2	1.4
20	1.0	1.0	1.0

Note. Linear interpolation between the values given in the table is permissible, but not extrapolation outside the limits given.

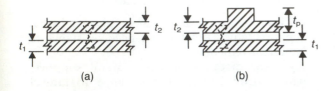

Fig. 5.13 *Effective thickness: (a) cavity wall; (b) cavity wall stiffened with piers (based on Fig. 3 of BS 5628).*

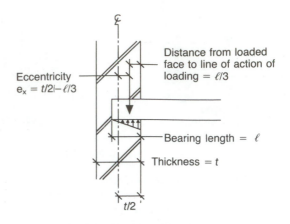

Fig. 5.14 *Eccentricity for wall supporting single floor.*

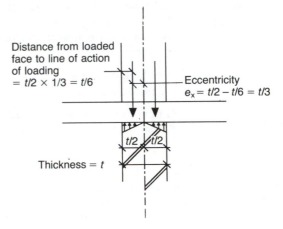

Fig. 5.15 *Eccentricity for wall supporting continuous floor.*

truly axially, but rather that it will occur at some eccentricity to the centroid of the wall. This eccentricity is normally expressed as a fraction of the wall thickness.

Clause 31 of BS 5628 recommends that the loads transmitted to a wall by a single floor or roof act at one-third of the length of the bearing surface from the loaded edge (*Fig. 5.14*). This is based on the assumption that the stress distribution under the bearing surface is triangular in shape as shown in *Fig. 5.14*. Where a uniform floor is continuous over a wall, each span of the floor should be taken as being supported individually on half the total bearing area (*Fig. 5.15*).

5.5.3 DESIGN VERTICAL LOAD RESISTANCE OF WALLS (N_R)

The foregoing has discussed how to evaluate the design strength of a masonry wall being equal to the characteristic strength (f_k) multiplied by the capacity reduction factor (β) and divided by the appropriate safety factor for materials (γ_m). The characteristic strengths and safety factors for materials are obtained from *Tables 5.8* and *5.9* respectively. The effective thickness and effective height of the member are used to determine the slenderness ratio and thence, together with the eccentricity of loading, the capacity reduction factor via *Table 5.10*.

The design strength is used to estimate the vertical load resistance of a wall, N_R, which is given by

$$N_R = \text{stress} \times \text{area} = \frac{\beta f_k}{\gamma_m} \times t \times 1 = \frac{\beta t f_k}{\gamma_m} \quad (5.7)$$

As stated earlier, the primary aim of design is to ensure that the ultimate design load does not exceed the design load resistance of the wall. Thus from equations 5.1, 5.2 and 5.7

$$N \leqslant N_R$$

$$\gamma_f (G_k, Q_k, \text{ or } W_k) \leqslant \frac{\beta t f_k}{\gamma_m} \quad (5.8)$$

Equation 5.8 provides the basis for the design of vertically loaded walls. The full design procedure is summarized in *Fig. 5.16*.

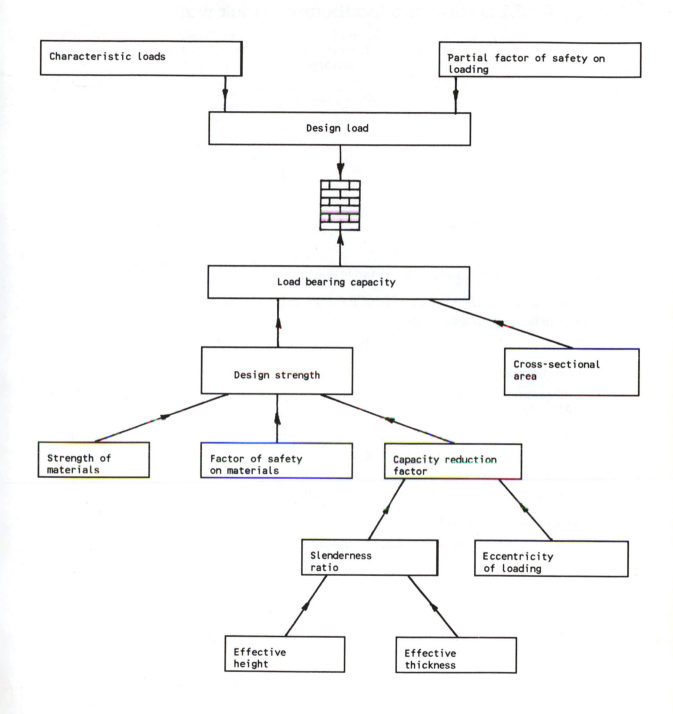

Fig. 5.16 *Design procedure for vertically loaded walls.*

Example 5.1 Design of a loadbearing brick wall

The internal loadbearing brick wall shown in *Fig. 5.17* supports an ultimate axial load of 140 kN m^{-1} run including self-weight of the wall. The wall is 102.5 mm thick and 4 m long. Assuming that normal manufacturing and construction controls apply, design the wall.

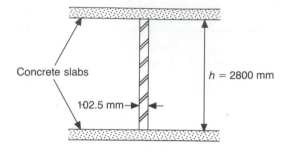

Concrete slabs

102.5 mm

$h = 2800$ mm

Fig. 5.17

LOADING
Ultimate design load, $N = 140$ kN m$^{-1} = 140$ N mm^{-1}

DESIGN VERTICAL LOAD RESISTANCE OF WALL

Characteristic compressive strength

$$\text{Basic value} = f_k$$

Check modification factor

SMALL PLAN AREA. Modification factor does not apply since horizontal cross-sectional area of wall, $A = 0.1025 \times 4.0 = 0.41$ m$^2 \nless 0.2$ m^2.

NARROW BRICK WALL. Modification factor is 1.15 since wall is one brick thick. Hence modified characteristic compressive strength is 1.15f_k.

Safety factor for materials (γ_m)
Manufacturing and construction controls are 'normal'. Hence from *Table 5.9*, $\gamma_m = 3.5$.

Capacity reduction factor (β)

Eccentricity. Since wall is axially loaded assume eccentricity of loading, $e_x < 0.05t$.

Slenderness ratio (SR). Concrete slab provides 'enhanced' resistance to wall:
$$h_{ef} = 0.75 \times \text{height} = 0.75 \times 2800 = 2100 \text{ mm}$$
$$t_{ef} = \text{actual thickness (single leaf)} = 102.5 \text{ mm}$$

$$\text{SR} = \frac{h_{ef}}{t_{ef}} = \frac{2100}{102.5} = 20.5 < \text{permissible} = 27$$

Hence, from *Table 5.10*, $\beta = 0.68$.

Design vertical load resistance of wall (N_R)

$$N_R = \frac{\beta \times \text{modified characteristic strength} \times t}{\gamma_m}$$

$$= \frac{0.68 \times (1.15 f_k)\, 102.5}{3.5}$$

DETERMINATION OF f_k
For structural stability

$$N_R \geqslant N$$

$$\frac{0.68 \times 1.15 f_k \times 102.5}{3.5} \geqslant 140$$

$$\text{Hence } f_k \geqslant \frac{140}{22.9} = 6.1 \text{ N mm}^{-2}$$

SELECTION OF BRICK AND MORTAR TYPE
From *Table 5.8(a)*, the following brick/mortar combinations would be appropriate:

Compressive strength of bricks ($N \, mm^{-2}$)	Mortar designation	f_k ($N \, mm^{-2}$)
27.5	(iii)	7.1
20	(ii)	6.4

The actual brick type that will be specified on the working drawings will depend not only upon the structural requirements but also on aesthetics, durability, buildability and cost considerations. In this particular case the designer may specify the minimum requirements as brick strength: $\geqslant 27.5$ N mm^{-2}, mortar mix: 1:1:6, frost resistance/soluble salt content: ON. Assuming that the wall will be plastered on both sides, appearance of the brick will not need to be specified.

Example 5.2 Design of a brick wall with a 'small' plan area

Redesign the wall in *Example 5.1* assuming that it is only 1.5 m long.
The calculations for this case are essentially the same as for the 4 m long wall except for the fact that the plan area of the wall, A, is now less than 0.2 m^2, being equal to $0.1025 \times 1.5 = 0.1538$ m^2. Hence, the 'plan area' modification factor is equal to

$$(0.70 + 1.5A) = 0.7 + 1.5 \times 0.1538 = 0.93$$

The 1.15 factor applicable to narrow brick walls still applies, and therefore the modified characteristic compressive strength of masonry is given by

$$0.93 \times 1.15 f_k = 1.07 f_k$$

From the above $\gamma_m = 3.5$ and $\beta = 0.68$. Hence, the required characteristic compressive strength of masonry, f_k, is given by

$$\frac{0.68 \times 1.07 f_k \times 102.5}{3.5} \geqslant 140 \text{ kN m}^{-1} \Rightarrow f_k \geqslant \frac{140}{21.3} = 6.6 \text{ N mm}^{-2}$$

From *Table 5.8(a)*, any of the following brick/mortar combinations would be appropriate:

Compressive strength of bricks ($N \, mm^{-2}$)	Mortar designation	f_k ($N \, mm^{-2}$)
35	(iv)	7.3
27.5	(iii)	7.1
20	(i)	7.4

Hence it can be immediately seen that walls having similar construction details but a plan area of < 0.2 m^2 will have lower load-carrying capacities.

Example 5.3 Analysis of brick walls stiffened with piers

A 3.5 m high wall shown in cross-section in *Fig. 5.18* is constructed from clay bricks having a compressive strength of 20.5 N mm^{-2} (i.e. Class 3) laid in a 1:1:6 mortar. Calculate the ultimate loadbearing capacity of the wall assuming the partial safety factor for materials is 3.5 and the resistance to lateral loading is (a) 'enhanced' and (b) 'simple'.

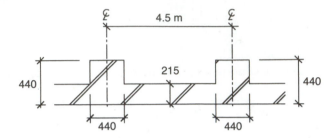

Fig. 5.18.

ASSUMING 'ENHANCED' RESISTANCE

Characteristic compressive strength (f_k). Assume compressive strength of bricks is 20 N mm^{-2} bricks. From *Table 5.5*, 1:1:6 mix corresponds to a grade (iii) mortar. This implies that $f_k = 5.8$ N mm^{-2} (*Table 5.8*).

Safety factor for materials (γ_m)

$$\gamma_m = 3.5$$

Capacity reduction factor (β)

ECCENTRICITY. Assume wall is axially loaded. Hence $e_x < 0.05t$.

SLENDERNESS RATIO (SR). With 'enhanced' resistance

$$h_{ef} = 0.75 \times \text{height} = 0.75 \times 3500 = 2625 \text{ mm}$$

$$\frac{\text{Pier spacing}}{\text{Pier width}} = \frac{4500}{440} = 10.2$$

$$\frac{\text{Pier thickness}}{\text{Thickness of wall}} = \frac{440}{215} = 2.0$$

Hence from *Table 5.11*, $K = 1.2$. The effective thickness of the wall, t_{ef}, is equal to

$$t_{ef} = tK = 215 \times 1.2 = 258 \text{ mm}$$

$$\text{SR} = \frac{h_{ef}}{t_{ef}} = \frac{2625}{258} = 10.2 < \text{permissible} = 27$$

Hence, from *Table 5.10*, $\beta = 0.96$.

Design vertical load resistance of wall (N_R)

$$N_R = \frac{\beta f_k t}{\gamma_m} = \frac{0.96 \times 5.8 \times 215}{3.5}$$

$$= 342 \text{ N mm}^{-1} \text{ run of wall}$$

$$= 342 \text{ kN m}^{-1} \text{ run of wall}$$

Hence the ultimate load capacity of wall, assuming enhanced resistance, is 342 kN m^{-1} run of wall.

ASSUMING 'SIMPLE' RESISTANCE

The calculation for this case is essentially the same as for 'enhanced' resistance except

$$h_{ef} = \text{actual height} = 3500 \text{ mm}$$
$$t_{ef} = 258 \text{ mm} \quad \text{(as above)}$$

Hence

$$\text{SR} = \frac{h_{ef}}{t_{ef}} = \frac{3500}{258} = 13.6.$$

Assuming $e_x < 0.05t$ (as above), this implies that $\beta = 0.9$ (*Table 5.10*). The design vertical load resistance of the wall, N_R, is given by

$$N_R = \frac{\beta f_k t}{\gamma_m} = \frac{0.9 \times 5.8 \times 215}{3.5}$$

$$= 320 \text{ N mm}^{-1} \text{ run of wall}$$

$$= 320 \text{ kN m}^{-1} \text{ run of wall}$$

Hence it can be immediately seen that, all other factors being equal, walls having simple resistance have a lower resistance to failure.

Example 5.4 Design of single leaf brick and block walls

Design the single leaf loadbearing wall shown in Fig. 5.19 using mortar designation (iii) and either (a) standard format bricks or (b) solid concrete blocks of length 390 mm, height 190 mm and thickness 100 mm. Assume that the manufacturing and construction controls are special and normal respectively.

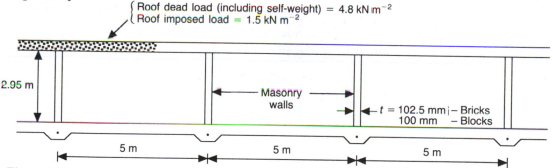

Fig. 5.19.

DESIGN FOR STANDARD FORMAT BRICKS

Loading

Characteristic dead load g_k. Assuming that the bricks are made from clay, the density of the brickwork can be assumed to be 55 kg m^{-2} per 25 mm thickness (*Table 2.1*) or $(55 \times (102.5/25)9.81 \times 10^{-3} =) 2.2$ kN m^{-2}. Therefore self-weight of wall is

$$\text{SW} = 2.2 \times 2.95 \times 1 = 6.49 \text{ kN m}^{-1} \text{ run of wall}.$$

$$\text{Dead load due to roof} = \frac{5+5}{2} \times 4.8 = 24 \text{ kN m}^{-1} \text{ run}$$

$$g_k = \text{SW} + \text{roof load}$$

$$= 6.49 + 24 = 30.49 \text{ kN m}^{-1} \text{ run of wall}$$

Characteristic imposed load, q_k

$$q_k = \text{roof load}$$

$$= \frac{5+5}{2} \times 1.5 = 7.5 \text{ kN m}^{-1} \text{ run of wall}$$

Ultimate design load, N

$$N = 1.4g_k + 1.6q_k$$

$$= 1.4 \times 30.49 + 1.6 \times 7.5 = 54.7 \text{ kN m}^{-1} \text{ run of wall}$$

$$= 54.7 \text{ N mm}^{-1} \text{ run of wall}$$

Design vertical load resistance of wall

Characteristic compressive strength

$$\text{Basic value} = f_k$$

MODIFICATION FACTORS

1. Small plan area – modification factor will not apply provided loaded plan area $A < 0.2$ m^2, i.e. provided that the wall length exceeds 2 m ≈ 0.2 m^2/0.1025 m.
2. Narrow brick wall – since wall is one brick thick, modification factor = 1.15.

Hence, modified characteristic compressive strength = $1.15f_k$.

Safety factor for materials (γ_m). Manufacturing control is special; construction control is normal. Hence from *Table 5.9* $\gamma_m = 3.1$.

Capacity reduction factor (β)

ECCENTRICITY. Assuming the wall is symmetrically loaded, eccentricity of loading, $e_x < 0.05t$.

SLENDERNESS RATIO (SR). Concrete slab provides 'enhanced' resistance to wall:

$$h_{ef} = 0.75 \times \text{height} = 0.75 \times 2950 = 2212.5 \text{ mm}$$

$$t_{ef} = \text{actual thickness (single leaf)} = 102.5 \text{ mm}$$

$$\text{SR} = \frac{h_{ef}}{t_{ef}} = \frac{2212.5}{102.5} = 21.6$$

Hence from *Table 5.10* $\beta = 0.63$.

Design vertical load resistance of wall (N_R)

$$N_R = \frac{\beta \times (\text{modified characteristic strength}) \times t}{\gamma_m}$$

$$= \frac{0.63 \times (1.15f_k)102.5}{3.1}$$

Determination of f_k
For structural stability

$$N_R \geqslant N$$

$$\frac{0.63 \times 1.15f_k \times 102.5}{3.1} \geqslant 54.7$$

$$f_k \geqslant \frac{54.7}{23.95} = 2.3 \text{ N mm}^{-2}$$

Selection of brick and mortar type

From *Table 5.8(a)*, any of the following brick/mortar combinations would be appropriate:

Compressive strength of bricks (N mm^{-2})	Mortar designation	f_k (N mm^{-2})
10	(iv)	3.5
5	(iii)	2.5

DESIGN FOR SOLID CONCRETE BLOCKS

Loading

Assuming that the density of the blockwork is the same as that for brickwork, i.e. 55 kg m^{-2} per 25 mm thickness (*Table 2.1*), self-weight of wall is 6.49 kN m^{-1} run of wall.

$$\text{Dead load due to roof} = 24 \text{ kN m}^{-1} \text{ run of wall}$$
$$\text{Imposed load from roof} = 7.5 \text{ kN m}^{-1} \text{ run of wall}$$
$$\text{Ultimate design load, } N = 1.4(6.49 + 24) + 1.6 \times 7.5$$
$$= 54.7 \text{ kN m}^{-1} \text{ run of wall}$$

Design vertical load resistance of wall

Characteristic compressive strength. Characteristic strength is f_k assuming that plan area of wall is greater than 0.2 m^2. Note that the narrow wall modification factor only applies to brick walls.

Safety factor for materials

$$\gamma_m = 3.1$$

Capacity reduction factor

ECCENTRICITY

$$e_x < 0.05t$$

SLENDERNESS RATIO (SR).

Concrete slab provides 'enhanced' resistance to wall:

$$h_{ef} = 0.75 \times \text{height} = 0.75 \times 2950 = 2212.5 \text{ mm}$$

$$t_{ef} = \text{actual thickness (single leaf)} = 100 \text{ mm}$$

$$SR = \frac{h_{ef}}{t_{ef}} = \frac{2212.5}{100} = 22.1$$

Hence from *Table 5.10*, $\beta = 0.61$.

Design vertical load resistance of wall (N_R)

$$N_R = \frac{\beta \times (\text{modified characteristic strength}) \times t}{\gamma_m}$$

$$= \frac{0.61 \times (f_k) \, 100}{3.1}$$

Determination of f_k

For structural stability

$$N_R \geqslant N$$

$$\frac{0.61 \times (f_k)\,100}{3.1} \geqslant 54.7$$

$$f_k \geqslant \frac{54.7}{19.68} = 2.8 \text{ N mm}^{-2}$$

Selection of block and mortar type

$$\text{Shape factor for block} = \frac{\text{height}}{\text{thickness}} = \frac{190}{100} = 1.9$$

Interpolating between *Tables 5.8(b)* and *(d)* a solid block of compressive strength 3.5 N mm^{-2} used with mortar designation (iv) would produce masonry of compressive strength 3.3 N mm^{-2} which would be appropriate here.

Example 5.5 Design of a cavity wall

A cavity wall of length 6 m supports the loads shown in Fig. 5.20. The inner loadbearing leaf is built using concrete blocks of length 440 mm, height 215 mm, thickness 100 mm and faced with plaster, and the outer leaf from standard format clay bricks. Design the wall assuming that the manufacturing control is special and construction control is normal. The self-weight of the blocks and plaster can be taken to be 2.4 kN m^{-2}.

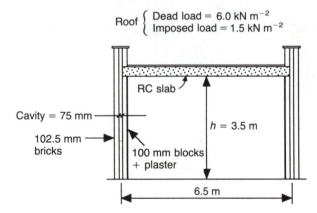

Fig. 5.20.

Clause 29.1 of BS 5628 on cavity walls states that 'where the load is carried by one leaf only, the loadbearing capacity of the wall should be based on the horizontal cross-sectional area of that leaf alone, although the stiffening effect of the other leaf can be taken into account when calculating the slenderness ratio'. Thus, it should be noted that the following calculations relate to the design of the inner loadbearing leaf.

LOADING

Characteristic dead load g_k

$$\begin{aligned}
g_k &= \text{roof load} + \text{self-weight of wall} \\
&= \frac{(6.5 \times 1)6}{2} + (3.5 \times 1)2.4 \\
&= 19.5 + 8.4 = 27.9 \text{ kN m}^{-1} \text{ run of wall}
\end{aligned}$$

Characteristic imposed load, q_k

$$q_k = \text{roof load}$$

$$= \frac{(6.5 \times 1)1.5}{2} = 4.9 \text{ kN m}^{-1} \text{ run of wall}$$

Ultimate design load, N

$$N = 1.4g_k + 1.6q_k$$
$$= 1.4 \times 27.9 + 1.6 \times 4.9 = 46.9 \text{ kN m}^{-1} \text{ run of wall}$$

DESIGN VERTICAL LOAD RESISTANCE OF WALL

Characteristic compressive strength
Basic value = f_k

Check modification factors

1. Small plan area – modification factor does not apply since loaded plan area, $A > 0.2 \text{ m}^2$ where $A = 6 \times 0.1 = 0.6 \text{ m}^2$.
2. Narrow brick wall – modification factor does not apply either since wall is of cavity construction.

Safety factor for materials
Manufacturing control is special; construction control is normal. Hence, from *Table 5.9*, $\gamma_m = 3.1$.

Capacity reduction factor

Eccentricity. The load from the concrete roof will be applied eccentrically as shown in the figure. The eccentricity is given by

$$e_x = \frac{t}{2} - \frac{t}{3} = \frac{t}{6} = 0.167t$$

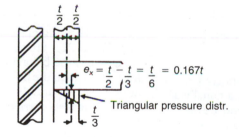

Triangular pressure distr.

Slenderness ratio (SR). Concrete slab provides 'enhanced' resistance to wall:

$$h_{ef} = 0.75 \times \text{height} = 0.75 \times 3500 = 2625 \text{ mm}$$
$$t_{ef} = 135 \text{ mm}$$

since t_{ef} is the greater of

$$2/3(t_1 + t_2) \, [= 2/3(102.5 + 100) = 135 \text{ mm}]$$

and $t_1 \, [= 102.5 \text{ mm}]$ or $t_2 \, [= 100 \text{ mm}]$.

$$\text{SR} = \frac{h_{ef}}{t_{ef}} = \frac{2625}{135} = 19.4 < \text{permissible} = 27$$

Slenderness ratio h_{ef}/t_{ef}	Eccentricity at top of wall, e_x		
	$0.1t$	$0.167t$	$0.2t$
18	0.70		0.57
19.4	0.658	0.570	0.528
20	0.64		0.51

Hence by interpolating between the values in *Table 5.10*, $\beta = 0.57$.

Design vertical load resistance of wall (N_R)

$$N_R = \frac{\beta f_k t}{\gamma_m} = \frac{0.57 \times f_k \times 100}{3.1}$$

DETERMINATION OF f_k
For structural stability

$$N_R \geqslant N$$

$$\frac{0.57 f_k 100}{3.1} \geqslant 46.9$$

$$f_k \geqslant \frac{46.9}{18.39} = 2.6 \text{ N mm}^{-2}$$

SELECTION OF BRICK AND MORTAR TYPE

$$\text{Shape factor} = \frac{215}{100} = 2.15$$

From *Table 5.8(d)*, a solid concrete block with a compressive strength of 2.8 N mm^{-2} used with mortar type (iv) would be appropriate. The bricks and mortar for the outer leaf would be selected on the basis of appearance and durability.

5.6 Design of laterally loaded wall panels

A non-loadbearing wall which is supported on a number of sides is usually referred to as a panel wall. Panel walls are very common in the UK and are mainly used to clad framed buildings. These walls are primarily designed to resist lateral loading from the wind.

The purpose of this section is to describe the design of laterally loaded panel walls. This requires an understanding of the following factors which are discussed below:

1. characteristic flexural strength
2. orthogonal ratio
3. support conditions
4. limiting dimensions
5. basis of design.

5.6.1 CHARACTERISTIC FLEXURAL STRENGTH OF MASONRY (f_{kx})

The majority of panel walls tend to behave as a two-way bending plate when subject to lateral loading (*Fig. 5.21*). However, it is more convenient to understand the behaviour of such walls if the bending processes in the horizontal and vertical directions are studied separately.

Bending tests carried out on panel walls show that the flexural strength of masonry is significantly greater when the plane of failure occurs perpendicular to the bed joint rather than when failure occurs parallel to the bed joint (*Fig. 5.22*). This is because, in the former case, failure is a complex mix of shear, etc. and end bearing, while in the latter case cracks need only form at the mortar/brick interface.

Table 5.12 shows the characteristic flexural strengths of masonry constructed using a range of brick/block types and mortar designations for the two failure modes. Note that for masonry built with clay bricks, the flexural strengths are primarily related to the water-absorption properties of the brick, but for masonry built with concrete blocks the flexural strengths are related to the compressive strengths and thicknesses of the blocks.

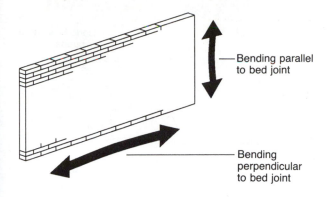

— Bending parallel to bed joint

— Bending perpendicular to bed joint

Fig. 5.21 *Panel wall subject to two-way bending.*

5.6.2 ORTHOGONAL RATIO, μ

The orthogonal ratio is the ratio of the characteristic flexural strength of masonry when failure occurs parallel to the bed joints ($f_{kx\,par}$) to that when failure occurs perpendicular to the bed joint ($f_{kx\,perp}$):

$$\mu = \frac{f_{kx\,par}}{f_{kx\,perp}} \qquad (5.9)$$

As will be discussed later, the orthogonal ratio is used to calculate the size of the bending moment in panel walls.

For masonry constructed using clay, calcium silicate or concrete bricks a value of 0.35 for μ can normally be assumed in design. No such unique value exists for masonry walls built with concrete blocks and must,

Table 5.12 Characteristic flexural strength of masonry, f_{kx} in N mm^{-2} (Table 3, BS 5628)

	Plane of failure parallel to bed joints			Plane of failure perpendicular to bed joints		
Mortar designation	(i)	(ii) and (iii)	(iv)	(i)	(ii) and (iii)	(iv)
Clay bricks having a water absorption						
less than 7%	0.7	0.5	0.4	2.0	1.5	1.2
between 7% and 12%	0.5	0.4	0.35	1.5	1.1	1.0
over 12%	0.4	0.3	0.25	1.1	0.9	0.8
Calcium silicate bricks	0.3	0.3	0.2	0.9	0.9	0.6
Concrete bricks	0.3	0.3	0.2	0.9	0.9	0.6
Concrete blocks of compressive strength in N/mm^2:						
2.8 used in walls of					0.40	0.4
3.5 thickness*	0.25		0.2		0.45	0.4
7.0 up to 100mm					0.60	0.5
2.8 used in walls					0.25	0.2
3.5 of thickness*	0.15		0.1		0.25	0.2
7.0 250mm					0.35	0.3
10.5 used in walls					0.75	0.6
14.0 of any thickness*	0.25		0.2		0.90†	0.7†
and over }						

*The thickness should be taken to be the thickness of the wall, for a single-leaf wall, or the thickness of the leaf, for a cavity wall.
†When used with flexural strength in parallel direction, assume the orthogonal ratio μ = 0.3.

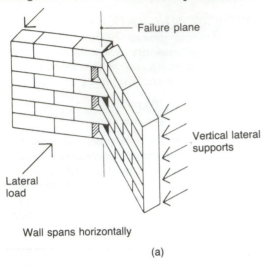

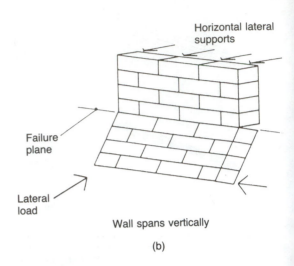

Fig. 5.22 *Failure modes of walls subject to lateral loading: (a) failure perpendicular to bed joint; (b) failure parallel to bed joint.*

therefore, be determined for individual block types.

5.6.3 SUPPORT CONDITIONS

In order to assess the lateral resistance of masonry panels it is necessary to take into account the support conditions at the edges. Three edge conditions are possible:

1. a free edge
2. a simply supported edge
3. a restrained edge.

A free edge is one that is unsupported (*Fig. 5.23*). A restrained edge is one that, when the panel is loaded, will fail in flexure before rotating. All other supported edges are assumed to be simply supported. This is despite the fact that some degree of fixity may actually exist in practice. *Figures 5.24* and *5.25* show typical detail providing simple and restrained support conditions respectively.

5.6.4 LIMITING DIMENSIONS

Clause 36.3 of BS 5628 lays down various limits on

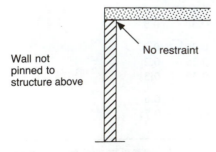

Wall not pinned to structure above

No restraint

Fig. 5.23 *Detail providing free edge.*

the dimensions of laterally loaded panel walls depending upon the support conditions. These are chiefly in respect of (i) the area of the panel (ii) the height and length of the panel. Specifically, the code mentions the following limits:

1. *Panel supported on three edges.*
    (a) *two or more sides continuous*: height × length equal to $1500t_{ef}^2$ or less;
    (b) *all other cases*: height × length equal to $1350t_{ef}^2$ or less.
2. *Panel supported on four edges.*
    (a) *three or more sides continuous*: height × length equal to $2250t_{ef}^2$ or less;
    (b) *all other cases*: height × length equal to $2025t_{ef}^2$ or less.
3. *Panel simply supported at top and bottom.* Height equal to $40t_{ef}$ or less.

Figure 5.26 illustrates the above limits on panel sizes. The height and length restrictions referred to in (ii) above only apply to panels which are supported on three or four edges, i.e. (1) and (2). In such cases the code states that no dimension should exceed 50 times the effective thickness of the wall (t_{ef}).

5.6.5 BASIS OF DESIGN
(CLAUSE 36.4, BS 5628)

The preceding sections have summarized much of the background material needed to design laterally loaded panel walls. This section now considers the design procedure in detail.

The principal aim of design is to ensure that the ultimate design moment (M) does not exceed the design moment of resistance of the panel (M_d):

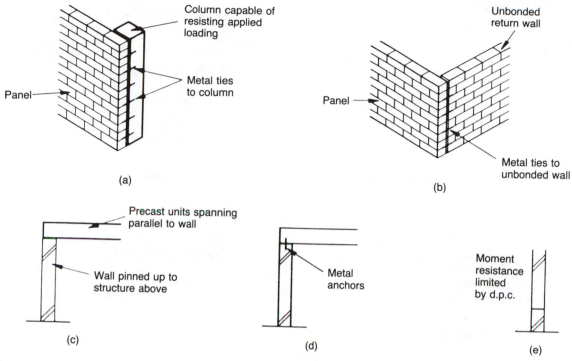

Fig. 5.24 *Details providing simple support conditions: (a) and (b) vertical support; (c)–(e) horizontal support (based on Figs 7 and 8, BS 5628).*

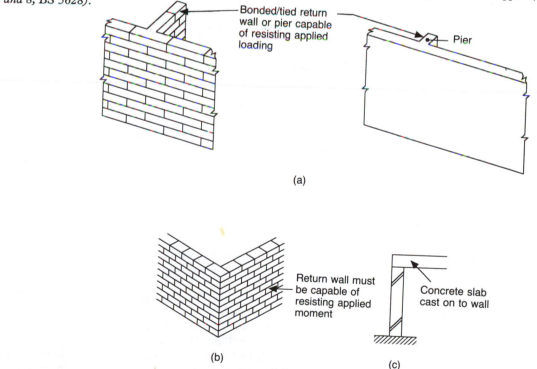

Fig. 5.25 *Details providing restrained support conditions: (a) and (b) vertical support; (c) horizontal support at head of wall (based on Figs 7 and 8, BS 5628).*

$h \times L \leqslant 1500t_{ef}^2$
where t_{ef} is the effective thickness
but neither h nor L to be greater than
$50 \times t_{ef}$

Two or more sides continuous

$h \times L \leqslant 2250t_{ef}^2$
neither h nor L to be greater than
$50 \times t_{ef}$

Three or more sides continuous

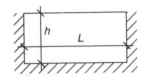

$h \times L \leqslant 1350t_{ef}^2$
neither h nor L to be greater than
$50 \times t_{ef}$

All other cases

(a)

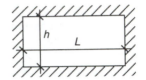

$h \times L \leqslant 2025t_{ef}^2$
neither h nor L to be greater than
$50 \times t_{ef}$

All other cases

(b)

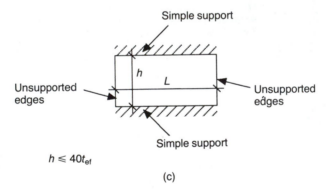

$h \leqslant 40t_{ef}$

(c)

Fig. 5.26 *Panel sizes: (a) panels supported on three sides; (b) panels supported on four edges; (c) panel simply supported top and bottom.*

$$M \leqslant M_d \qquad (5.10)$$

Since failure may take place about either axis (*Fig. 5.21*), for a given panel, there will be two design moments and two corresponding moments of resistance. The ultimate design moment per unit height of a panel when the plane of failure is perpendicular to the bed joint, M_{perp} (*Fig. 5.22*), is given by

$$M_{perp} = \alpha W_k \gamma_f L^2 \qquad (5.11)$$

The ultimate design moment per unit height of a panel when the plane of failure is parallel to the bed joint, M_{par}, is given by

$$M_{par} = \mu \alpha W_k \gamma_f L^2 \qquad (5.12)$$

where μ is the orthogonal ratio, α the bending moment coefficient taken from *Table 5.13*, γ_f the partial safety factor for loads (*Table 5.7*), L the length of the panel between supports and W_k the characteristic wind load per unit area.

The corresponding design moments of resistance

when the plane of bending is perpendicular, $M_{k\,perp}$, or parallel, $M_{k\,par}$, to the bed joint is given by equations 5.13 and 5.14 respectively:

$$M_{k\,perp} = \frac{f_{kx\,perp}Z}{\gamma_m} \qquad (5.13)$$

$$M_{k\,par} = \frac{f_{kx\,par}Z}{\gamma_m} \qquad (5.14)$$

where $f_{kx\,perp}$ is the characteristic flexural strength perpendicular to the plane of bending, $f_{kx\,par}$ the characteristic flexural strength parallel to the plane

of bending (*Table 5.12*), Z the section modulus and γ_m the partial safety factor for materials (*Table 5.9*).

Equations 5.10–5.14 form the basis for the design of laterally loaded panel walls. It should be noted, however, since by definition $\mu = f_{kx\,par} / f_{kx\,perp}$, $M_{k\,par} = \mu M_{k\,perp}$ (by dividing equation 5.14 by equation 5.13) and $M_{par} = \mu M_{perp}$ (by dividing equations 5.11 by equation 5.12) that either equations 5.11 and 5.13 or equations 5.12 and 5.14 can be used in design. The full design procedure is summarized in *Fig. 5.27* and illustrated by means of the following examples.

Table 5.13 Bending moment coefficients in laterally loaded wall panels (based on Table 9, BS 5628)

Key to support conditions

——— denotes free edge
∠∠∠ simply supported edge
⋁⋁⋁ an edge over which full continuity exists

	μ	Values of α h/L						
		0.30	0.50	0.75	1.00	1.25	1.50	1.75
	1.00	0.031	0.045	0.059	0.071	0.079	0.085	0.090
	0.90	0.032	0.047	0.061	0.073	0.081	0.087	0.092
	0.80	0.034	0.049	0.064	0.075	0.083	0.089	0.093
	0.70	0.035	0.051	0.066	0.077	0.085	0.091	0.095
	0.60	0.038	0.053	0.069	0.080	0.088	0.093	0.097
	0.50	0.040	0.056	0.073	0.083	0.090	0.095	0.099
A	0.40	0.043	0.061	0.077	0.087	0.093	0.098	0.101
	0.35	0.045	0.064	0.080	0.089	0.095	0.100	0.103
	0.30	0.048	0.067	0.082	0.091	0.097	0.101	0.104
	1.00	0.020	0.028	0.037	0.042	0.045	0.048	0.050
	0.90	0.021	0.029	0.038	0.043	0.046	0.048	0.050
	0.80	0.022	0.031	0.039	0.043	0.047	0.049	0.051
	0.70	0.023	0.032	0.040	0.044	0.048	0.050	0.051
C	0.60	0.024	0.034	0.041	0.046	0.049	0.051	0.052
	0.50	0.025	0.035	0.043	0.047	0.050	0.052	0.053
	0.40	0.027	0.038	0.044	0.048	0.051	0.053	0.054
	0.35	0.029	0.039	0.045	0.049	0.052	0.053	0.054
	0.30	0.030	0.040	0.046	0.050	0.052	0.054	0.054
	1.00	0.008	0.018	0.030	0.042	0.051	0.059	0.066
	0.90	0.009	0.019	0.032	0.044	0.054	0.062	0.068
	0.80	0.010	0.021	0.035	0.046	0.056	0.064	0.071
	0.70	0.011	0.023	0.037	0.049	0.059	0.067	0.073
E	0.60	0.012	0.025	0.040	0.053	0.062	0.070	0.076
	0.50	0.014	0.028	0.044	0.057	0.066	0.074	0.080
	0.40	0.017	0.032	0.049	0.062	0.071	0.078	0.084
	0.35	0.018	0.035	0.052	0.064	0.074	0.081	0.086
	0.30	0.020	0.038	0.055	0.068	0.077	0.083	0.089

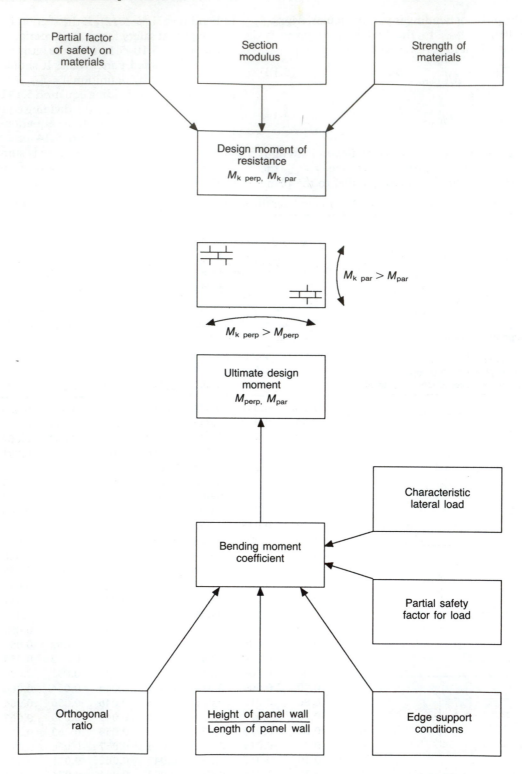

Fig. 5.27 *Design procedure for laterally loaded panel walls.*

Example 5.6 Analysis of a one-way spanning wall panel

Estimate the design wind pressure that the cladding panel shown below can resist given that it is constructed using clay bricks having a water absorption of < 7% and mortar designation (iii). Assume $\gamma_f = 1.2$ and $\gamma_m = 3.5$.

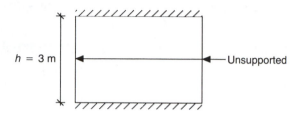

ULTIMATE DESIGN MOMENT (M)

Since the vertical edges are unsupported, the panel must span vertically and, therefore, equations 5.11 and 5.12 cannot be used to determine the design moment here. The critical plane of bending will be parallel to the bed joint and the ultimate design moment at mid-height of the panel, M, is given by (clause 36.4.2 of BS 5628)

$$M = \frac{\text{ultimate load} \times \text{height}}{8}$$

Ultimate load on the panel per unit length of wall = wind pressure $\times$ area
$$= (\gamma_f W_k) \, (\text{height} \times \text{length of panel})$$
$$= 1.2 W_k 3000 \times 1000 = 3.6 W_k 10^6 \, \text{N m}^{-1} \text{ length of wall}$$

Hence

$$M = \frac{3.6 W_k \times 10^6 \times 3000}{8} = 1.35 \times 10^9 W_k \, \text{m}^{-1} \text{ length of wall}$$

MOMENT OF RESISTANCE (M_d)

Section modulus (Z)

$$Z = \frac{bd^2}{6} = \frac{10^3 \times 102.5^2}{6} = 1.75 \times 10^6 \, \text{mm}^3 \, \text{m}^{-1} \text{ length of wall}$$

Moment of resistance

The design moment of resistance, M_d, is equal to the moment of resistance when the plane of failure is parallel to the bed joint, $M_{k \, par}$. Hence

$$M_d = M_{k \, par} = \frac{f_{kx \, par} Z}{\gamma_m} = \frac{0.5 \times 1.75 \times 10^6}{3.5} = 0.25 \times 10^6 \, \text{N mm m}^{-1}$$

where $f_{kx \, par} = 0.5 \, \text{N mm}^{-2}$ from *Table 5.12*, since water absorption of bricks < 7% and the mortar is designation (iii)

For structural stability:

$$M \leqslant M_d$$
$$1.35 \times 10^9 W_k \leqslant 0.25 \times 10^6$$
$$W_k \leqslant 0.185 \times 10^{-3} \, \text{N mm}^{-2}$$

Hence the maximum wind pressure that the panel can resist is $0.185 \times 10^{-3} \, \text{N mm}^{-2}$ or $185 \, \text{N m}^{-2}$

Example 5.7 Analysis of a two-way spanning panel wall

The panel wall shown in *Fig. 5.28* is constructed using clay bricks having a water absorption of greater than 12% and mortar designation (ii). If the manufacturing and construction controls are both normal, calculate the design wind pressure, W_k, the wall can withstand. Assume that the wall is simply supported on all four edges.

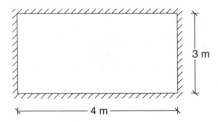

3 m

4 m

Fig. 5.28

ULTIMATE DESIGN MOMENT (M)

Orthogonal ratio (μ)

$$\mu = \frac{f_{kx} \, \text{par}}{f_{kx} \, \text{perp}} = \frac{0.3}{0.9} = 0.33 \, (Table \, 5.12)$$

Bending moment coefficient (α)

$$\frac{h}{L} = \frac{3}{4} = 0.75$$

Hence, from *Table 5.13(E)*, $\alpha = 0.053$.

Ultimate design moment (M)

$$M = M_{\text{perp}} = \alpha \gamma_f W_k L^2 = 0.053 \times 1.2 W_k \times 4^2 \, \text{kN m m}^{-1} \, \text{run}$$

$$= 1.0176 \times 10^6 W_k \, \text{N mm m}^{-1} \, \text{run}$$

MOMENT OF RESISTANCE (M_d)

Safety factor for materials (γ_m)

$$\gamma_m = 3.5 \quad (Table \, 5.9)$$

Section modulus (Z)

$$Z = \frac{bd^2}{6} = \frac{10^3 \times 102.5^2}{6} = 1.75 \times 10^6 \text{ mm}^3$$

Moment of resistance (M_d)

$$M_d = M_{k\,perp} = \frac{f_{kx\,perp}Z}{\gamma_m} = \frac{0.9 \times 1.75 \times 10^6}{3.5} = 0.45 \times 10^6 \text{ N mm m}^{-1} \text{ run}$$

DETERMINATION OF MAXIMUM WIND PRESSURE (W_k)
For structural stability

$$M \leqslant M_d$$
$$1.0176 \times 10^6 W_k \leqslant 0.45 \times 10^6$$
$$\Rightarrow W_k \leqslant 0.44 \text{ kN m}^{-2}$$

Hence the panel is able to resist a wind pressure of 0.44 kN m^{-2}.

Example 5.8 Design of a two-way spanning single-leaf panel wall

A 102.5 mm brick wall is to be designed to withstand a wind pressure, W_k, of 0.65 kN m^{-2}. Assuming that the edge support conditions are as shown in *Fig. 5.29* and that both the manufacturing and construction controls are normal, determine suitable brick/mortar combination(s) for the wall.

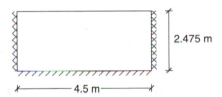

Fig. 5.29

ULTIMATE DESIGN MOMENT (M)

Orthogonal ratio (μ)
Assume $\mu = 0.35$.

Bending moment coefficient (α)

$$\frac{h}{L} = \frac{2475}{4500} = 0.55$$

Hence, from *Table 5.13(C)*, $\alpha = 0.040$.

Ultimate design moment (M)
$M = M_{perp} = \alpha\gamma_f W_k L^2 = 0.040 \times 1.2 \times 0.65 \times 4.5^2 \text{ kN m m}^{-1} \text{ run}$
   $= 0.632 \text{ kN m m}^{-1} \text{ run} = 0.632 \times 10^6 \text{ N mm m}^{-1} \text{ run}$

MOMENT OF RESISTANCE (M_d)

Safety factor for materials (γ_m)

$$\gamma_m = 3.5 \quad (Table\ 5.9)$$

Section modulus (Z)

$$Z = \frac{bd^2}{6} = \frac{10^3 \times 102.5^2}{6} = 1.75 \times 10^6 \text{ mm}^3$$

Moment of resistance (M_d)

$$M_d = M_{k\,perp} = \frac{f_{kx\,perp}Z}{\gamma_m} = \frac{1.75 \times 10^6 f_{kx\,perp}}{3.5}$$
$$= 0.5 \times 10^6 f_{kx\,perp} \text{N mm m}^{-1} \text{ run}$$

DETERMINATION OF $f_{kx\,perp}$
For structural stability:

$$M \leqslant M_d$$
$$0.632 \times 10^6 \leqslant 0.5 \times 10^6 f_{kx\,perp}$$
$$\Rightarrow f_{kx\,perp} \leqslant 1.26 \text{ N mm}^{-2}$$

SELECTION OF BRICK AND MORTAR TYPE
From *Table 5.12*, any of the following brick/mortar combinations would be appropriate:

Clay brick having a water absorption:	Mortar designation	$f_{kx\,perp}$ $(N\,mm^{-2})$
< 7%	(iii)	1.5
7%–12%	(i)	1.5

Note that the actual value of μ for the brick/mortar combinations above is 0.33 and not 0.35 as assumed. However, this difference is insignificant and will not affect the choice of the brick/mortar combinations shown in the above table.

Example 5.9 Analysis of a two-way spanning cavity panel wall

Determine the maximum design wind pressure, W_k, which can be resisted by the cavity wall shown in *Fig. 5.30* if the construction details are as follows:

1. *Outer leaf*: clay brick having a water absorption < 7% laid in a 1 : 1 : 6 mortar (i.e. designation (iii)).
2. *Inner leaf*: solid concrete blocks of compressive strength 3.5 N mm^{-2} and length 390 mm, height 190 mm and thickness 100 mm also laid in a 1: 1 : 6 mortar.

Assume that the top edge of the wall is unsupported, but that the base and vertical edges are simply supported. The manufacturing and construction controls can both be assumed to be normal.

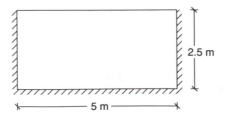

2.5 m

5 m

Fig. 5.30

The design wind pressure, W_k, depends upon the available capacities of both leaves of the wall. Therefore, it is normal practice to work the capacities of the outer and inner leaf separately, and to then sum them in order to determine W_k.

OUTER LEAF
Ultimate design moment (M)

Orthogonal ratio (μ)

$$\mu = \frac{f_{kx\ par}}{f_{kx\ perp}} = \frac{0.5}{1.5} = 0.33 \ (Table\ 5.12)$$

Bending moment coefficient (α)

$$\frac{h}{L} = \frac{2.5}{5.0} = 0.5$$

Hence from *Table 5.13(A)*, $\alpha = 0.065$.

Ultimate design moment (M)

$$M = M_{perp} = \alpha \gamma_f W_k L^2 = 0.065 \times 1.2 \ (W_k)_{outer} \times 5^2 \ \text{kN m m}^{-1} \ \text{run}$$
$$= 1.95 \times 10^6 (W_k)_{outer} \ \text{N mm m}^{-1} \ \text{run}$$

Moment of resistance (M_d)

Safety factor for materials (γ_m)

$$\gamma_m = 3.5 \quad (Table\ 5.9)$$

Section modulus (Z)

$$Z = \frac{bd^2}{6} = \frac{10^3 \times 102.5^2}{6} = 1.75 \times 10^6 \ \text{mm}^3$$

Moment of resistance (M_d)

$$M_d = M_{k\ perp} = \frac{f_{kx\ perp} Z}{\gamma_m} = \frac{1.5 \times 1.75 \times 10^6}{3.5} = 0.75 \times 10^6 \ \text{N mm m}^{-1}$$

Maximum permissible wind pressure on outer leaf ($W_k)_{outer}$
For structural stability:

$$M \leq M_d$$
$$1.95 \ (W_k)_{outer} \times 10^6 \leq 0.75 \times 10^6$$
$$(W_k)_{outer} \leq 0.38 \ \text{kN m}^{-2}$$

INNER LEAF
Ultimate design moment (M)

Orthogonal ratio (μ)

$$\mu = \frac{f_{kx\ par}}{f_{kx\ perp}} = \frac{0.25}{0.45} = 0.55 \quad (Table\ 5.12)$$

Bending moment coefficient (α)

$$\frac{h}{L} = \frac{2.5}{5.0} = 0.5$$

Hence from *Table 5.13(A)*, $\alpha = 0.055$.

Ultimate design moment

$$M = M_{\text{perp}} = \alpha\gamma_f W_k L^2 = 0.055 \times 1.2 \, (W_k)_{\text{inner}} \times 5^2 \text{ kN m m}^{-1} \text{ run}$$
$$= 1.65 \times 10^6 \, (W_k)_{\text{inner}} \text{ Nmm m}^{-1} \text{ run}$$

Moment of resistance (M_d)

Safety factor for materials (γ_m)

$$\gamma_m = 3.5 \, (\textit{Table 5.9})$$

Section modulus (Z)

$$Z = \frac{bd^2}{6} = \frac{10^3 \times 100^2}{6} = 1.67 \times 10^6 \text{ mm}^3$$

Moment of resistance (M_d)

$$M_d = M_{k \text{ perp}} = \frac{f_{kx \text{ perp}} \, Z}{\gamma_m} = \frac{0.45 \times 1.67 \times 10^6}{3.5} = 0.21 \times 10^6 \text{ Nmm m}^{-1}$$

Maximum permissible wind pressure on inner leaf $(W_k)_{\text{inner}}$

For structural stability:

$$M \leqslant M_d$$
$$1.65 \, W_{k \text{ inner}} \times 10^6 \leqslant 0.21 \times 10^6 \text{ N mm m}^{-1} \text{ run}$$
$$(W_k)_{\text{inner}} \leqslant 0.13 \text{ kN m}^{-2}$$

Hence maximum wind pressure that the panel can withstand, W_k, is given by

$$W_k = (W_k)_{\text{inner}} + (W_k)_{\text{outer}} = 0.13 + 0.38 = 0.51 \text{ kN m}^{-2}$$

5.7 Summary

This chapter has explained the basic properties of the materials used in unreinforced masonry design and the design procedures involved in the design of (a) single leaf and cavity walls, with and without stiffening piers, subject to vertical loading and (b) panel walls resisting lateral loading. The design procedures outlined are in accordance with BS 5628: Part 1: *Structural Use of Unreinforced Masonry* which

is based on the limit state principles. In the case of vertically loaded walls it was found that the design process simply involves ensuring that the ultimate design load does not exceed the design strength of the wall. Panel walls, on the other hand, are normally subject to two-way bending. This means that, firstly, the designer must evaluate the design moment acting on the panel about one axis and then ensure that this does not exceed the corresponding moment of resistance of the panel.

Questions

1. (a) Discuss why there was a decline and what factors have lead to a revival in the use of masonry in construction over recent years.

   (b) The 4.0 m high wall shown below is constructed from clay bricks having a compressive strength of 20 N mm^{-2} laid in a 1 : 1 : 6 mortar. Calculate the design load resistance of the wall assuming the partial safety factor for materials is 3.5 and the resistance to lateral loading is simple.

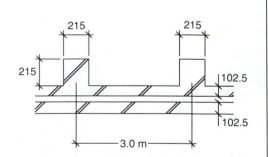

2. (a) Discuss the factors to be considered in the selection of brick, block and mortar types for particular applications.
(b) Design the external cavity wall and internal single leaf loadbearing wall for the single-storey building shown adjacent. Assume that the internal wall is made from solid concrete blocks of length 390 mm, height 190 mm and thickness 100 mm

Roof dead load (including self-weight) = 5.5 kN m^{-2}
Roof imposed load = 1.5 kN m^{-2}

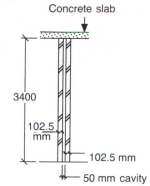

and that the external wall consists of 102.5 mm thick brick outer leaf with $390 \times 190 \times 100$ mm concrete block. Assume that the manufacturing and construction controls are special and normal respectively and that the self-weight of the brick and blockwork is 2.2 kN m^{-2}.

3. (a) Explain the difference between simple and enhanced resistance and sketch typical construction details showing examples of both types of restraint.
(b) The brick cavity wall shown adjacent supports an ultimate axial load of 200 kN m^{-1}. Assuming that the load is equally shared by both leaves and the manufacturing and construction controls are normal and special respectively, design the wall.

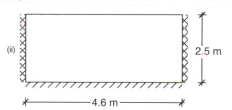

4. (a) Explain with the use of sketches the limiting dimensions of laterally loaded panel walls.
(b) A 102.5 mm brick wall is to be designed to withstand a wind pressure of 550 N m^{-2}. For the support conditions shown in (i) and (ii) below, determine suitable brick/mortar combinations for the two cases. Assume that both the manufacturing and construction controls are normal.

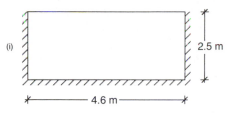

5. (a) In all the examples on the design of laterally loaded panel wall discussed in this chapter the self-weight of the wall has been ignored. What effect do you think that including the self-weight would have on the design compressive strength of masonry?
(b) Determine the maximum design wind pressure which can be resisted by the cavity wall shown adjacent if the construction details are as follows: (1) outer leaf: clay bricks having a water absorption 7–12% laid in mortar designation (ii); (2) inner leaf: solid concrete blocks of compressive strength 3.5 N mm^{-2} and length 390 mm,

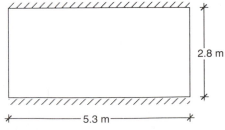

height 190 mm and thickness 100 mm also laid in mortar designation (ii). Assume that the manufacturing and construction controls are both special and normal respectively and the vertical edges are unsupported.

Chapter 6

Design in Timber to BS 5268

198 - 200
216 - 220

This chapter is concerned with the design of timber elements to BS 5268: Part 2, which is based on the permissible stress philosophy. The chapter describes how timber is specified for structural purposes and discusses some of the basic concepts involved such as stress grading, grade stresses and strength classes. The primary aim of this chapter is to give guidance on the design of flexural members, e.g. beams and joists, compression members, e.g. posts and columns, and load-sharing systems, e.g. stud walling.

6.1 Introduction

Wood is a very versatile raw material and is widely used in construction, especially in countries such as Canada, Sweden, Finland, Norway and Poland, where there is an abundance of good-quality timber. Timber can be used in a range of structural applications including marine works: construction of wharves, piers, cofferdams; heavy civil works: bridges, piles, shoring, pylons; domestic housing: roofs, floors, partitions; shuttering for precast and *in situ* concrete; falsework for brick or stone construction.

Of all the construction materials which have been discussed in this book, only timber is naturally occurring. This makes it a very difficult material to characterize and partly accounts for the wide variation in the strength of timber, not only between different species but also between timber of the same species and even from the same log. Quite naturally, this led to uneconomical use of timber which was costly for individuals and the nation as a whole. However, this problem has now been largely overcome by specifying stress graded timber (*section 6.2*).

There is an enormous variety of timber species. They are divided into softwoods and hardwoods, a botanical distinction, not on the basis of mechanical strength. Softwoods are derived from trees with needle-shaped leaves and are usually evergreen, e.g. fir, larch, spruce, hemlock, pine. Hardwoods are derived from trees with broad leaves and are usually deciduous,

e.g. ash, elm, oak, teak, iroko, ekki, greenheart. Obviously the suitability of a particular timber type for any given purpose will depend upon various factors such as performance, cost, appearance and availability. This makes specification very difficult. The task of the structural engineer has been simplified, however, by grouping timber species into nine strength classes for which typical design parameters, e.g. grade stresses and moduli of elasticity, have been produced (*section 6.3*). Most standard design in the UK is with softwoods.

Design of timber elements is normally carried out in accordance with BS 5268: *Structural Use of Timber*. This is divided into the following parts:

Part 1: *Limit State Design, Materials and Workmanship.*
Part 2: *Code of Practice for Permissible Stress Design, Materials and Workmanship.*
Part 3: *Code of Practice for Trussed Rafter Roofs.*
Part 4: *Fire Resistance of Timber Structures.*
Part 5: *Code of Practice for the Preservative Treatment of Structural Timber.*
Part 6: *Code of Practice for Timber Frame Walls.*
Part 7: *Recommendations for the Calculation Basis for Span Tables.*

The design principles which will be outlined in this chapter are based on the contents of Part 2 of the code. It should therefore be assumed that all future references to BS 5268 refer exclusively to Part 2. As pointed out in *Chapter 1* of this book, BS 5268 is based on permissible stress design rather than limit state design. This means in practice that a partial safety factor is applied only to material properties, i.e. the permissible stresses (*section 6.4*) and not the loading.

Specifically, this chapter gives guidance on the design of timber beams, joists, columns and stud walling. The design of timber formwork is not covered here as it was considered to be rather too specialized a topic and, therefore, inappropriate for a book of this nature. However, before discussing the design process in detail, the following sections will expand on the more general aspects mentioned above, namely:

1. stress grading
2. grade stress and strength class
3. permissible stress.

6.2 Stress grading

The strength of timber is a function of several parameters including the moisture content, density, size of specimen and the presence of various strength-reducing characteristics such as knots, slope of grain, fissures and wane. Prior to the introduction of BS 5268 the strength of timber was determined by carrying out short-term loading tests on small timber specimens free from all defects. The data was used to estimate the **minimum strength** which was taken as the value below which not more than 1% of the test results fell. These strengths were multiplied by a reduction factor to give **basic stresses.** The reduction factor made an allowance for the reduction in strength due to duration of loading, size of specimen and other effects normally associated with a safety factor such as accidental overload, simplifying assumptions made during design and design inaccuracies, together with poor workmanship. Basic stress was defined as the stress which could safely be permanently sustained by timber free from any strength-reducing characteristics.

However, it was found that basic stress was not directly applicable to structural size timber since structural size timber invariably contains defects which further reduces its strength. To take account of this, timber was visually classified into one of four grades, namely 75, 65, 50 and 40, which indicated the percentage free from defects. The grade stress for structural size timber was finally obtained by multiplying the grade designations expressed as a percentage (e.g. 75%, 65% etc) by the basic stress for the timber.

With the introduction of BS 5268 the concept of basic stresses has been largely abandoned and the new approach for assessing the strength of timber appears to have a limit state 'feel'. The first step in this process involves grading structural size timber. Grading is still carried out visually, although it is now common practice to do this mechanically. The latter approach offers the advantage of greater economy in the use of timber since it takes into account the density of timber which significantly influences its strength.

Mechanical stress grading is based on the fact that there is a direct relationship between the modulus of elasticity measured over a relatively short span, i.e. stiffness, and bending strength. The stiffness is assessed non-destructively by feeding individual pieces of timber through a series of rollers on a machine which automatically applies small transverse loads over short successive lengths and measures the deflections. These are compared with permitted deflections appropriate to given stress grades and the machine automatically assesses the grade of the timber over its entire length.

The numbered grades (i.e. 75, 65, 50 and 40) have been withdrawn and replaced by two visual grades: General Structural (GS) and Special Structural (SS) and four machine grades: MGS, MSS, M75 and M50. The MGS and MSS grades correspond directly with the visual GS and SS grades respectively. The structural strength of the four machine grades occur in the order: M75 > MSS > M50 > MGS.

The (visually or mechanically) graded timber is then subject to short-term load tests. The results are used to determine the **characteristic stress**, which is taken to be the value below which not more than 5% of test results fall (*Fig. 6.1*). Finally, the grade stress is obtained by dividing the characteristic stress by a reduction factor which includes adjustments for a standard depth of specimen of 300 mm, duration of load and a factor of safety.

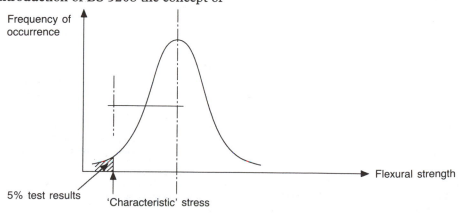

Fig. 6.1 *Frequency distribution curve for flexural strength of timber*

Table 6.1 Grade stresses for softwoods for the dry exposure condition (N mm^{-2}) (Table 10, BS 5268)

Standard name	Grade	Bending parallel to grain[a]	Tension parallel to grain[a]	Compression Parallel to grain	Compression Perpendicular to grain[b]	Shear parallel to grain	Modulus of elasticity Mean	Modulus of elasticity Minimum
Redwood/	SS/MSS	7.5	4.5	7.9	2.1	0.82	10 500	7 000
whitewood	GS/MGS	5.3	3.2	6.8	1.8	0.82	9 000	6 000
(imported)	M75	10.0	6.0	8.7	2.4	1.32	11 000	7 000
and Scots pine (British grown)	M50	6.6	4.0	7.3	2.1	0.82	9 000	6 000
Corsican	SS/MSS	7.5	4.5	7.9	2.1	0.82	9 500	6 500
pine	GS/MGS	5.3	3.2	6.8	1.8	0.82	8 000	5 000
(British	M75	10.0	6.0	8.7	2.4	1.33	10 500	7 000
grown)	M50	6.6	4.0	7.3	2.0	0.83	9 000	5 500
Sitka spruce	SS/MSS	5.7	3.4	6.1	1.6	0.64	8 000	5 000
and European	GS/MGS	4.1	2.5	5.2	1.4	0.64	6 500	4 500
spruce	M75	6.6	4.0	6.4	1.8	1.02	9 000	6 000
(British grown)	M50	4.5	2.7	5.5	1.6	0.64	7 500	5 000
Douglas fir	SS/MSS	6.2	3.7	6.6	2.4	0.88	11 000	7 000
(British	GS/MGS	4.4	2.6	5.6	2.1	0.88	9 500	6 000
grown)	M75	10.0	6.0	8.7	2.9	1.41	11 000	7 500
	M50	6.6	4.0	7.3	2.4	0.88	9 500	6 000
Larch	SS	7.5	4.5	7.9	2.1	0.82	10 500	7 000
(British grown)	GS	5.3	3.2	6.8	1.8	0.82	9 000	6 000
Parana pine	SS	9.0	5.4	9.5	2.4	1.03	11 000	7 500
(imported)	GS	6.4	3.8	8.1	2.2	1.03	9 500	6 000
Pitch pine	SS	10.5	6.3	11.0	3.2	1.16	13 500	9 000
(Caribbean)	GS	7.4	4.4	9.4	2.8	1.16	11 000	7 500
Western red	SS	5.7	3.4	6.1	1.7	0.63	8 500	5 500
cedar	GS	4.1	2.5	5.2	1.6	0.63	7 000	4 500
(imported)								
Douglas fir	SS	7.5	4.5	7.9	2.4	0.85	11 000	7 500
larch	GS	5.3	3.2	6.8	2.2	0.85	10 000	6 500
(Canada)								
Douglas fir	SS	7.5	4.5	7.9	2.4	0.85	11 000	7 500
larch	GS	5.3	3.2	6.8	2.2	0.85	9 500	6 000
(USA)								
Hem-fir	SS/MSS	7.5	4.5	7.9	1.9	0.68	11 000	7 500
(Canada)	GS/MGS	5.3	3.2	6.8	1.7	0.68	9 000	6 000
	M75	10.0	6.0	9.3	2.4	1.13	12 000	8 000
	M50	6.6	4.0	7.7	2.1	0.71	10 500	7 000
Hem-fir	SS	7.5	4.5	7.9	1.9	0.68	11 000	7 500
(USA)	GS	5.3	3.2	6.8	1.7	0.68	9 000	6 000
Spruce-pine-	SS/MSS	7.5	4.5	7.9	1.8	0.68	10 000	6 500
Fir	GS/MGS	5.3	3.2	6.8	1.6	0.68	8 500	5 500
(Canada)	M75	9.7	5.8	8.5	2.1	1.10	10 500	7 000
	M50	6.2	3.7	7.1	1.8	0.68	9 000	5 500
Western	SS	6.6	4.0	7.0	1.7	0.66	9 000	6 000
whitewoods	GS	4.7	2.8	6.0	1.5	0.66	7 500	5 000
(USA)								
Southern	SS	9.6	5.8	10.2	2.5	0.98	12 500	8 500
pine	GS	6.8	4.1	8.7	2.2	0.98	10 500	7 000
(USA)								

[a] Stresses applicable to timber 300 mm deep (or wide): for other section sizes see 14.6 and 16.2 of BS 5268.
[b] When the specifications specifically prohibit wane at bearing areas, the SS grade compression perpendicular to the grain stress may be multiplied by 1.33 and used for all grades.

6.3 Grade stress and strength class

Table 6.1 shows typical timber species/grade combinations and associated grade stresses and moduli of elasticity. This information would enable the designer to determine the size of a timber member given the intensity and distribution of the loads to be carried. However, it would mean that the contractor's choice of material would be limited to one particular species/grade combination, which could be difficult to obtain. It would obviously be better if a range of species/grade combinations could be specified and the contractor could then select the most economical one. Such an approach forms the basis of grouping timber species/grade combinations with similar strength characteristics into strength classes (Table 6.2).

In all there are nine strength classes, SC1–SC9, with SC1 having the lowest strength characteristics. Classes SC1–SC5 usually comprise the softwoods and classes SC6–SC9 the hardwoods. The grade stresses and moduli of elasticity associated with each strength class are shown in Table 6.3. In the UK structural timber design is normally based on strength classes SC3 to SC5. These classes cover a wide range of softwoods which display good struc-tural properties and are both plentiful and cheap.

Some time after the publication of this book the strength class system in BS 5628 will be changed to the European standard system used for Eurocode 5 (section 11.4.2). Apart from changes to the number of classes, their boundary values and the assignment of grades and species, the advantages, general principles and method of use described in this book will still be valid.

6.4 Permissible stresses

The grade stresses given in Tables 6.1 and 6.3 were derived assuming particular conditions of service and loading. In order to take account of the actual conditions that individual members will be subject to during their design life, the grade stresses are multiplied by modification factors known as K-factors. The modified stresses are termed permissible stresses.

BS 5268 lists over 70 K-factors. However, the following subsections describe only those modification factors relevant to the design of simple flexural and compression members, namely:

K_1:   Wet exposure geometrical properties factor
K_2:   Wet exposure strength characteristics factor
K_3:   Duration of loading factor

Table 6.2   Softwood species/grade combinations which satisfy the requirements for strength classes graded to BS 4978 (based on Table 3, BS 5268)

| Standard name | Strength class | | | | |
	SC1	SC2	SC3	SC4	SC5
Imported					
Parana pine			GS	SS	
Pitch pine (Caribbean)			GS		SS
Redwood			GS/M50	SS	M75
Whitewood			GS/M50	SS	M75
Western red cedar	GS	SS			
Douglas fir-larch (Canada)			GS	SS	
Douglas fir-larch (USA)			GS	SS	
Hem-fir (Canada)			GS/M50	SS	M75
Hem-fir (USA)			GS	SS	
Spruce-pine-fir (Canada)			GS/M50	SS/M75	
Sitka spruce (Canada)		GS	SS		
Western whitewoods (USA)	GS		SS		
Southern pine (USA)			GS	SS	
British grown					
Douglas fir		GS	M50/SS		M75
Larch			GS	SS	
Scots pine			GS/M50	SS	M75
Corsican pine		GS	M50	SS	M75
European spruce	GS	M50/SS	M75		
Sitka spruce	GS	M50/SS	M75		

Table 6.3 Grade stresses and moduli of elasticity for strength classes for the dry exposure condition (based on Table 9, BS 5268)

Strength class	Bending parallel to grain $(\sigma_{m,g,\parallel})$ $(N\,mm^{-2})$	Tension parallel to grain $(\sigma_{t,g,\parallel})$ (Nmm^{-2})	Compression parallel to grain $(\sigma_{c,g,\parallel})$ $(N\,mm^{-2})$	Compression perpendicular to grain[a] $(\sigma_{c,g,\perp})$ $(N\,mm^{-2})$		Shear parallel to grain (τ_g) $(N\,mm^{-2})$	Modulus of elasticity $(N\,mm^{-2})$ Mean (E_{mean})	Minimum (E_{min})	Approximate density[b] $(kg\,m^{-3})$
SC1	2.8	2.2	3.5	2.1	1.2	0.46	6 800	4 500	540
SC2	4.1	2.5	5.3	2.1	1.6	0.66	8 000	5 000	540
SC3	5.3	3.2	6.8	2.2	1.7	0.67	8 800	5 800	540
SC4	7.5	4.5	7.9	2.4	1.9	0.71	9 900	6 600	590
SC5	10.0	6.0	8.7	2.8	2.4	1.00	10 700	7 100	590/760
SC6[c]	12.5	7.5	12.5	3.8	2.8	1.50	14 100	11 800	840
SC7[c]	15.0	9.0	14.5	4.4	3.3	1.75	16 200	13 600	960
SC8[c]	17.5	10.5	16.5	5.2	3.9	2.00	18 700	15 600	1 080
SC9[c]	20.5	12.3	19.5	6.1	4.6	2.25	21 600	18 000	1 200

[a] When the specification specifically prohibits wane at bearing areas, the higher value may be used.
[b] Densities in the table may be considered only crude approximations.
[c] Classes SC6, SC7, SC8 and SC9 will usually comprise the denser hardwoods.

K_5: Notched end factor
K_7: Depth factor
K_8: Load-sharing system factor
K_{12}: Compression member stress factor.

6.4.1 GEOMETRICAL PROPERTIES, K_1

Variations in the moisture content of timber will cause it to shrink or swell. The geometrical properties of timber sections are normally quoted for the dry exposure condition (*Table 6.8*). This is assumed to be when the moisture content of timber is 18% or below. The geometrical properties for use in designing for the wet exposure conditions are obtained by multiplying the dry exposure value by a modification factor K_1 given in *Table 6.4*. Normally, dry exposure conditions can be assumed to exist where the timber members are either indoors or are protected from contact with water. In all other cases wet exposure conditions should be assumed, including solid timber members more than 100 mm thick, irrespective of exposure conditions, because it is impractical to dry timber more than 75 mm thick artificially.

6.4.2 STRENGTH CHARACTERISTICS, K_2

The strength characteristics of timber also vary with the moisture content. The grade stresses and moduli of elasticity shown in *Tables 6.1* and *6.3* apply to timber exposed to dry conditions. Where wet conditions exist, the values in *Tables 6.1* and *6.3* are multiplied by a modification factor K_2 given in *Table 6.5*.

Apart from its effect on structural properties, the moisture content of timber significantly influences its durability properties and clause 5.4 of BS 5268 points out that wood is less prone to decay if its moisture content is below 25% and may be considered immune below 20%. Furthermore, care should be taken on site to ensure that material supplied in a dry condition is adequately protected from the weather.

6.4.3 DURATION OF LOADING, K_3

The stresses given in *Tables 6.1* and *6.3* apply to long-term loading. Where the applied loads will act for shorter durations, e.g. snow and wind, the grade stresses can be increased. *Table 6.6* gives the modification factor K_3 by which these values should be multiplied for various load combinations.

Table 6.4 Wet exposure geometrical properties modification factor K_1 (Table 2, BS 5268)

Geometrical property	Value of K_1
Thickness, width, radius of gyration	1.02
Cross-sectional area	1.04
First moment of area, section modulus	1.06
Second moment of area	1.08

Table 6.5 Wet exposure strength characteristics modification factor (Table 16, BS 5268)

Property	Values of K_2
Bending parallel to grain	0.8
Tension parallel to grain	0.8
Compression parallel to grain	0.6
Compression perpendicular to grain	0.6
Shear parallel to grain	0.9
Mean and minimum modulus of elasticity	0.8

Table 6.6 Modification factor K_3 for duration of loading (Table 17, BS 5268)

Duration of loading	Value of K_3
Long term (e.g. dead + permanent imposed[a])	1.00
Medium term (e.g. dead + snow, dead + temporary imposed)	1.25
Short term (e.g. dead + imposed + wind[b], dead + imposed + snow + wind[b])	1.50
Very short term (e.g. dead + imposed + wind[c])	1.75

[a] For imposed floor loads $K_3 = 1.00$.
[b] For wind, short-term category applies to class C (15 s gust) as defined in CP 3: Chapter V: Part 2.
[c] For wind, very short-term category applies to classes A and B (3 or 5 s gust) as defined in CP3: Chapter V: Part 2.

6.4.4 NOTCHED ENDS, K_5

Notches at the ends of flexural members will result in high shear concentrations which may cause structural failure and must, therefore, be taken into account during design (*Fig. 6.2*). In notched members the grade shear stresses parallel to the grain (*Tables 6.1* and *6.3*) are multiplied by a modification factor K_5 calculated as follows:

1. For a notch on the top edge (*Fig. 6.2(a)*):

$$K_5 = \frac{h(h_e - a) + ah_e}{h_e^2} \quad \text{for } a \leqslant h_e \qquad (6.1)$$

$$K_5 = 1.0 \qquad \text{for } a > h_e \qquad (6.2)$$

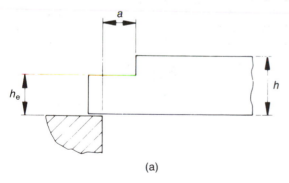

(a)

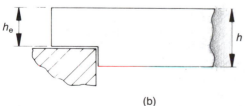

(b)

Fig. 6.2 *Notched beams: (a) beam with notch on top edge; (b) beam with notch on underside (Fig. 2, BS 5268).*

2. For a notch on the underside (*Fig. 6.2(b)*):

$$K_5 = \frac{h_e}{h} \qquad (6.3)$$

Clause 14.4 of BS 5268 also notes that the effective depth, h_e, should not be less than $0.5h$, i.e. $K_5 \geqslant 0.5$.

6.4.5 DEPTH FACTOR, K_7

The grade bending stresses given in *Table 6.3* only apply to timber sections having a depth h of 300 mm. For other depths of beams, the grade bending stresses are multiplied by the depth factor K_7, defined in clause 14.6 of BS 5268 as follows:
$K_7 = 1.17$ for solid beams having a depth < 72 mm
$K_7 = (300/h)^{0.11}$ for solid beams   (6.4)

$$\text{with } 72 \text{ mm} < h < 300 \text{ mm}$$

$K_7 = 0.81(h^2 + 92\,300)/(h^2 + 56\,800)$ for solid beams with $h > 300$ mm

6.4.6 LOAD-SHARING SYSTEMS, K_8

The grade stresses given in *Tables 6.1* and *6.3* apply to individual members, e.g. isolated beams and columns, rather than assemblies. When four or more members such as rafters, joists or wall studs, spaced a maximum of 610 mm centre to centre act together to resist a common load, the grade stress should be multiplied by a load-sharing factor K_8 which has a value of 1.1.

6.4.7 COMPRESSION MEMBERS, K_{12}

The grade compression stresses parallel to the grain given in *Tables 6.1* and *6.3* are used to design struts and columns. These values apply to compression members with slenderness ratios less than 5 which would fail by crushing. Where the slenderness ratio of the member is equal to or greater than 5 the grade stresses should be multiplied by the modification factor K_{12} given in *Table 6.7*. Alternatively Appendix C of BS 5268 gives a formula for K_{12} which could be used. This is based on the Perry–Robertson equation which has also been used to develop the steel column design Tables 27(a)–(d) in BS 5950 (*section 4.9*).

The factor K_{12} takes into account the tendency of the member to fail by buckling and allows for imperfections such as out of straightness and accidental load eccentricities. The factor K_{12} is based on the minimum modulus of elasticity, E_{min}, irrespective of whether the compression member acts alone or forms part of a load-sharing system and the compression stress, $\sigma_{c,||}$, is given by

$$\sigma_{c,||} = \sigma_{c,g,||}K_3 \qquad (6.5)$$

Table 6.7 Modification factor K_{12} for compression members (Table 22, BS 5268)

Values of K_{12}

$\dfrac{E}{\sigma_{c,\parallel}}$	Values of slenderness ratio $\lambda\ (= L_e/i)$																			
	<5	5	10	20	30	40	50	60	70	80	90	100	120	140	160	180	200	220	240	250
	Equivalent L_e/b (for rectangular sections)																			
	<1.4	1.4	2.9	5.8	8.7	11.6	14.5	17.3	20.2	23.1	26.0	28.9	34.7	40.5	46.2	52.0	57.8	63.6	69.4	72.3
400	1.000	0.975	0.951	0.986	0.827	0.735	0.621	0.506	0.408	0.330	0.271	0.225	0.162	0.121	0.094	0.075	0.061	0.051	0.043	0.040
500	1.000	0.975	0.951	0.899	0.837	0.759	0.664	0.562	0.466	0.385	0.320	0.269	0.195	0.148	0.115	0.092	0.076	0.063	0.053	0.049
600	1.000	0.975	0.951	0.901	0.843	0.774	0.692	0.601	0.511	0.430	0.363	0.307	0.226	0.172	0.135	0.109	0.089	0.074	0.063	0.058
700	1.000	0.975	0.951	0.902	0.848	0.784	0.711	0.629	0.545	0.467	0.399	0.341	0.254	0.195	0.154	0.124	0.102	0.085	0.072	0.067
800	1.000	0.975	0.952	0.903	0.851	0.792	0.724	0.649	0.572	0.497	0.430	0.371	0.280	0.217	0.172	0.139	0.115	0.096	0.082	0.076
900	1.000	0.976	0.952	0.904	0.853	0.797	0.734	0.665	0.593	0.522	0.456	0.397	0.304	0.237	0.188	0.153	0.127	0.106	0.091	0.084
1000	1.000	0.976	0.952	0.904	0.855	0.801	0.742	0.677	0.609	0.542	0.478	0.420	0.325	0.255	0.204	0.167	0.138	0.116	0.099	0.092
1100	1.000	0.976	0.952	0.905	0.856	0.804	0.748	0.687	0.623	0.559	0.497	0.440	0.344	0.272	0.219	0.179	0.149	0.126	0.107	0.100
1200	1.000	0.976	0.952	0.905	0.857	0.807	0.753	0.695	0.634	0.573	0.513	0.457	0.362	0.288	0.233	0.192	0.160	0.135	0.116	0.107
1300	1.000	0.976	0.952	0.905	0.858	0.809	0.757	0.701	0.643	0.584	0.527	0.472	0.378	0.303	0.247	0.203	0.170	0.144	0.123	0.115
1400	1.000	0.976	0.952	0.906	0.859	0.811	0.760	0.707	0.651	0.595	0.539	0.486	0.392	0.317	0.259	0.214	0.180	0.153	0.131	0.122
1500	1.000	0.976	0.952	0.906	0.860	0.813	0.763	0.712	0.658	0.603	0.550	0.498	0.405	0.330	0.271	0.225	0.189	0.161	0.138	0.129
1600	1.000	0.976	0.952	0.906	0.861	0.814	0.766	0.716	0.664	0.611	0.559	0.508	0.417	0.342	0.282	0.235	0.198	0.169	0.145	0.135
1700	1.000	0.976	0.952	0.906	0.861	0.815	0.768	0.719	0.669	0.618	0.567	0.518	0.428	0.353	0.292	0.245	0.207	0.177	0.152	0.142
1800	1.000	0.976	0.952	0.906	0.862	0.816	0.770	0.722	0.673	0.624	0.574	0.526	0.438	0.363	0.302	0.254	0.215	0.184	0.159	0.148
1900	1.000	0.976	0.952	0.907	0.862	0.817	0.772	0.725	0.677	0.629	0.581	0.534	0.447	0.373	0.312	0.262	0.223	0.191	0.165	0.154
2000	1.000	0.976	0.952	0.907	0.863	0.818	0.773	0.728	0.681	0.634	0.587	0.541	0.455	0.382	0.320	0.271	0.230	0.198	0.172	0.160

6.5 Timber design

Having discussed some of the more general aspects, the following sections will consider in detail the design of:

1. flexural members
2. compression members
3. stud walling.

6.6 Symbols

For the purposes of this chapter, the following symbols have been used. These have largely been taken from BS 5268.

GEOMETRICAL PROPERTIES

b	breadth of beam
h	depth of beam
A	total cross-sectional area
i	radius of gyration
I	second moment of area
Z	section modulus

BENDING

L	effective span
M	design moment
M_R	moment of resistance
$\sigma_{m,a,\|\|}$	applied bending stress parallel to grain
$\sigma_{m,g,\|\|}$	grade bending stress parallel to grain
$\sigma_{m,adm,\|\|}$	permissible bending stress parallel to grain

DEFLECTION

δ_t	total deflection
δ_m	bending deflection
δ_v	shear deflection
δ_p	permissible deflection
E	modulus of elasticity
E_{mean}	mean modulus of elasticity
E_{min}	minimum modulus of elasticity
G	shear modulus

SHEAR

F_v	design shear force
τ_a	applied shear stress parallel to grain
τ_g	grade shear stress parallel to grain
τ_{adm}	permissible shear stress parallel to grain

BEARING

F	bearing force
l_b	length of bearing
$\sigma_{c,a,\perp}$	applied compression stress perpendicular to grain
$\sigma_{c,g,\perp}$	grade compression stress perpendicular to grain
$\sigma_{c,adm,\perp}$	permissible bending stress perpendicular to grain

COMPRESSION

L_e	effective length of a column
λ	slenderness ratio
N	axial load
$\sigma_{c,a,\|\|}$	applied compression stress parallel to grain
$\sigma_{c,g,\|\|}$	grade compression stress parallel to grain
$\sigma_{c,adm,\|\|}$	permissible compression stress parallel to grain
$\sigma_{c,\|\|}$	compression stress $= \sigma_{c,g,\|\|} K_3$
σ_e	Euler critical stress

6.7 Flexural members

Beams, rafters and joists are examples of flexural members. All calculations relating to their design are based on the effective span and principally involves consideration of the following aspects which are discussed below:

1. bending
2. deflection
3. lateral torsional buckling
4. shear
5. bearing.

Generally, for medium span beams the design process follows the procedure outlined above. However, deflection is usually critical for long-span beams and shear for heavily loaded short-span beams.

6.7.1 EFFECTIVE SPAN

According to clause 14.3 of BS 5268, for simply supported beams, the effective span is normally taken as the distance between the centres of bearings (*Fig. 6.3*).

6.7.2 BENDING

If flexural members are not to fail in bending, the

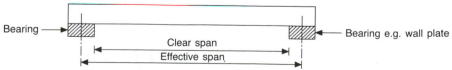

Fig. 6.3 *Effective span of simply supported beams.*

design moment (M) must not exceed the moment of resistance (M_R):

$$M \leqslant M_R \qquad (6.6)$$

The design moment is a function of the applied loads. The moment of resistance for a beam can be derived from the theory of bending (equation 2.5, *Chapter 2*) and is given by

$$M_R = \sigma_{m,adm,||} Z_{xx} \qquad (6.7)$$

where $\sigma_{m,adm,||}$ is the permissible bending stress parallel to the grain and Z_{xx} the section modulus. For rectangular sections

$$Z_{xx} = \frac{bd^2}{6} \quad (Fig.\ 6.4)$$

where b is the breadth of section and d the depth of section. The permissible bending stress is calculated by multiplying the grade bending stress, $\sigma_{m,g,||}$, by any relevant K-factors:

$$\sigma_{m,adm,||} = \sigma_{m,g,||} K_1 K_2 K_3 K_7 K_8 \quad \text{(as appropriate)} \qquad (6.8)$$

For a given design moment the minimum required section modulus, Z_{xx} req, can be calculated using equation 6.9, obtained by combining equations 6.6 and 6.7:

$$Z_{xx} \text{ req} > \frac{M}{\delta_{m,adm,||}} \qquad (6.9)$$

A suitable timber section can then be selected from Tables 98, 99 and 100 of BS 5268 for, respectively, sawn, planed all round and regularized softwoods. Table 98 is reproduced here as *Table 6.8*. (It should be noted that the new European standards to be used in BS 5268 will give two size classes.) Finally, the chosen section should be checked for deflection, lateral buckling, shear and bearing to assess its suitability as discussed below.

6.7.3 DEFLECTION

Excessive deflection of flexural members may result in damage to surfacing materials, ceilings, partitions and finishes, and to the functional needs as well as aesthetic requirements. Clause 14.7 of BS 5268 recommends that generally such damage can be avoided if the total deflection, δ_t, of the member when fully loaded does not exceed the permissible deflection, δ_p:

$$\delta_t \leqslant \delta_p \qquad (6.10)$$

The permissible deflection is generally given by

$$\delta_p = 0.003 \times \text{span} \qquad (6.11)$$

but for longer-span domestic floor joists, i.e. spans

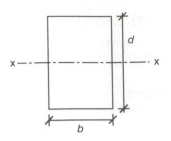

Fig. 6.4 *Section modulus.*

over 4.67 m, should not exceed 14 mm:

$$\delta_p \not> 14 \text{ mm} \qquad (6.12)$$

The total deflection, δ_t, is the summation of the bending deflection, δ_m, plus the shear deflection, δ_v:

$$\delta_t = \delta_m + \delta_v \qquad (6.13)$$

Table 6.9 gives the bending and shear deflection formulae for some common loading cases for beams of rectangular cross-section. The formulae have been derived by assuming that the shear modulus is equal to one-sixteenth of the permissible modulus of elasticity in accordance with clause 11 of BS 5268.

For solid timber members acting alone the deflections should be calculated using the minimum modulus of elasticity, but for load-sharing systems the deflections should be based on the mean modulus of elasticity.

6.7.4 LATERAL BUCKLING

If flexural members are not effectively laterally restrained, it is possible for the member to twist sideways before developing its full flexural strength (*Fig. 6.5*), thereby causing it to fail in bending, shear or deflection. This phenomenon is called lateral buckling and can be avoided by ensuring that the depth to breadth ratios given in *Table 6.10* are complied with.

6.7.5 SHEAR

If flexural members are not to fail in shear, the applied shear stress parallel to the grain, τ_a, should not exceed the permissible shear stress, τ_{adm}:

$$\tau_a \leqslant \tau_{adm} \qquad (6.14)$$

For a beam with a rectangular cross-section, the maximum applied shear stress occurs at the neutral axis and is given by

$$\tau_a = \frac{3F_v}{2A} \qquad (6.15)$$

where F_v is the applied maximum vertical shear force and A the cross-sectional area. The permissible shear stress is given by

Table 6.8 Geometrical properties of sawn softwoods (Table 98, BS5268)

Basic size[a] (mm)	Area (10^3 mm²)	Section modulus (10^3 mm³)		Second moment of area (10^6 mm⁴)		Radius of gyration (mm)	
		About x-x	About y-y	About x-x	About y-y	About x-x	About y-y
36 × 75	2.70	33.8	16.2	1.27	0.292	21.7	10.4
36 × 100	3.60	60.0	21.6	3.00	0.389	28.9	10.4
36 × 125	4.50	93.8	27.0	5.86	0.486	36.1	10.4
36 × 150	5.40	135	32.4	10.1	0.583	43.3	10.4
38 × 75	2.85	35.6	18.1	1.34	0.343	21.7	11.0
38 × 100	3.80	63.3	24.1	3.17	0.457	28.9	11.0
38 × 125	4.75	99.0	30.1	6.18	0.572	36.1	11.0
38 × 150	5.70	143	36.1	10.7	0.686	43.3	11.0
38 × 175	6.54	194	42.1	17.0	0.800	50.5	11.0
38 × 200	7.60	253	48.1	25.3	0.915	57.7	11.0
38 × 225	8.55	321	54.2	36.1	1.03	65.0	11.0
44 × 75	3.30	41.3	24.2	1.55	0.532	21.7	12.7
44 × 100	4.40	73.3	32.3	3.67	0.71	28.9	12.7
44 × 125	5.40	115	40.3	7.16	0.887	36.1	12.7
44 × 150	6.60	165	48.4	12.4	1.06	43.3	12.7
44 × 175	7.70	225	56.5	19.7	1.24	50.5	12.7
44 × 200	8.80	293	64.5	29.3	1.42	57.7	12.7
44 × 225	9.90	371	72.6	41.8	1.60	65.0	12.7
44 × 250	11.0	458	80.7	57.3	1.77	72.2	12.7
44 × 300	13.2	660	96.8	99.0	2.13	86.6	
47 × 75	3.53	44.1	27.6	1.65	0.649	21.7	13.6
47 × 100	4.70	78.3	36.8	3.92	0.865	28.9	13.6
47 × 125	5.88	122	46.0	7.65	1.08	36.1	13.6
47 × 150	7.05	176	55.2	13.2	1.30	43.3	13.6
47 × 175	8.23	240	64.4	21.0	1.51	50.5	13.6
47 × 200	9.40	313	73.6	31.3	1.73	57.7	13.6
47 × 225	10.6	397	82.8	44.6	1.95	65.0	13.6
47 × 250	11.8	490	92.0	61.2	2.16	72.2	13.6
47 × 300	14.1	705	110	106	2.60	86.6	13.6
50 × 75	3.75	46.9	31.3	1.76	0.781	21.7	14.4
50 × 100	5.00	83.3	41.7	4.17	1.04	28.9	14.4
50 × 125	6.25	130	52.1	8.14	1.30	36.1	14.4
50 × 150	7.50	188	62.5	14.1	1.56	43.3	14.4
50 × 175	8.75	255	72.9	22.3	1.82	50.5	14.4
50 × 200	10.0	333	83.3	33.3	2.08	57.7	14.4
50 × 225	11.3	422	93.8	47.5	2.34	65.0	14.4
50 × 250	12.5	521	104	65.1	2.60	72.2	14.4
50 × 300	15.0	750	125	113	3.13	86.6	14.4
63 × 100	6.30	105	66.2	5.25	2.08	28.9	18.2
63 × 125	7.88	164	82.7	10.3	2.60	36.1	18.2
63 × 150	9.45	236	99.2	17.7	3.13	43.3	18.2
63 × 175	11.0	322	116	28.1	3.65	50.5	18.2
63 × 200	12.6	420	132	42.0	4.17	57.7	18.2
63 × 225	14.2	532	149	59.8	4.69	65.0	18.2
75 × 100	7.50	125	93.8	6.25	3.52	28.9	21.7
75 × 125	9.38	195	117	12.2	4.39	36.1	21.7
75 × 150	11.3	281	141	21.1	5.27	43.3	21.7
75 × 175	13.1	383	164	33.5	6.15	50.5	21.7
75 × 200	15.0	500	188	50.0	7.03	57.7	21.7
75 × 225	16.9	633	211	71.2	7.91	65.0	21.7
75 × 250	18.8	781	234	97.7	8.79	72.2	21.7
75 × 300	22.5	1130	281	169	10.5	86.6	21.7
100 × 100	10.0	167	167	8.33	8.33	28.9	28.9
100 × 150	15.0	375	250	28.1	12.5	43.3	28.9
100 × 200		667	333	66.7	16.7	57.7	28.9
100 × 250		1040	417	130	20.8	72.2	28.9
100 × 300		1500	500	225	25.0	86.6	28.9
150 × 150	20.0	563	563	42.2	42.2	43.3	43.3
150 × 200	25.5	1000	750	100	56.3	57.7	43.3
150 × 300	30.0	2250	1130	338	84.4	86.6	43.3
200 × 200	40.0	1330	1330	133	133	57.7	57.7
250 × 250	62.5	2600	2600	326	326	72.2	72.2
300 × 300	90.0	4500	4500	675	675	86.6	86.6

[a] Basic size measured at 20% moisture content.

Table 6.9 Bending and shear deflections assuming $G = E/16$

Load distribution and supports	Deflection at Centre C or end E	
	Bending	Shear
(simply supported, UDL f, span L)	$\dfrac{5}{384} \times \dfrac{fL^4}{EI}$	$\dfrac{12}{5} \times \dfrac{fL^2}{EA}$
(simply supported, point load P at $L/2$)	$\dfrac{PL^3}{48EI}$	$\dfrac{24}{5} \times \dfrac{PL}{EA}$
(simply supported, two point loads P at a)	$\dfrac{Pa}{EI}\left[\dfrac{L^2}{8} - \dfrac{a^2}{6}\right]$	$\dfrac{96}{5} \times \dfrac{Pa}{EA}$
(fixed ends, UDL f, span L)	$\dfrac{fL^4}{384EI}$	$\dfrac{12}{5} \times \dfrac{fL^2}{EA}$
(fixed ends, point load P at $L/2$)	$\dfrac{PL^3}{192EI}$	$\dfrac{24}{5} \times \dfrac{PL}{EA}$
(cantilever, UDL f, end E)	$\dfrac{fL^4}{8EI}$	$\dfrac{48}{5} \times \dfrac{fL^2}{EA}$
(cantilever, point load P at end E)	$\dfrac{PL^3}{3EI}$	$\dfrac{96}{5} \times \dfrac{PL}{EA}$

$$\tau_{adm} = \tau_g K_1 K_2 K_3 K_5 K_8 \quad \text{(as appropriate)} \quad (6.16)$$

where τ_g is the grade shear parallel to grain (*Tables 6.1* and *6.3*).

6.7.6 BEARING PERPENDICULAR TO GRAIN

Bearing failure may arise in flexural members which are supported at their ends on narrow beams or wall plates. Such failures can be avoided by ensuring that the applied bearing stress, $\sigma_{c,a,\perp}$ never exceeds the permissible compression stress perpendicular to the grain, $\sigma_{c,adm,\perp}$:

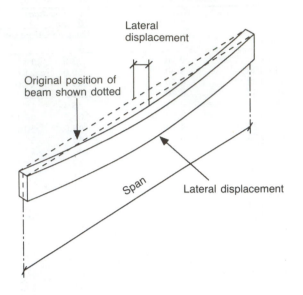

Fig. 6.5 *Lateral buckling.*

Table 6.10 Maximum depth to breadth ratios (Table 19, BS 5268)

Degree of lateral support	Maximum depth to breadth ratio
No lateral support	2
Ends held in position	3
Ends held in position and member held in line, as by purlins and tie-rods	4
Ends held in position and compression edge held in line, as by direct connection of sheathing, deck or joists	5
Ends held in position and compression edge held in line, as by direct connection of sheathing, deck or joists, together with adequate bridging or blocking spaced at intervals not exceeding six times the depth	6
Ends held in position and both edges held firmly in line	7

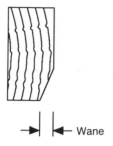

Fig. 6.6 *Wane.*

length. The permissible compression stress is obtained by multiplying the grade compression stress perpendicular to the grain, $\sigma_{c,g,\perp}$, by the K-factors for load duration (K_3) and load sharing (K_8) as appropriate:

$$\sigma_{c,adm,\perp} = \sigma_{c,g,\perp}K_3K_8 \qquad (6.19)$$

It should be noted that the grade compression stresses perpendicular to the grain given in *Tables 6.1* and *6.3* apply to (i) bearings of any length at the ends of members and (ii) bearings 150 mm or more in length at any position. Moreover, two values for the grade compression stress perpendicular to the grain are given for each strength class (*Table 6.3*). The lower value takes into account the amount of wane which is permitted within each stress grade (*Fig. 6.6*). If, however, the specification prohibits wane from occurring at bearing areas the higher value may be used.

$$\sigma_{c,a,\perp} \leqslant \sigma_{c,adm,\perp} \qquad (6.17)$$

The applied bearing stress is given by

$$\sigma_{c,a,\perp} = \frac{F}{bl_b} \qquad (6.18)$$

where F is the bearing force (usually maximum reaction), b the breadth of section and l_b the bearing

---

Example 6.1 Design of a timber beam

A timber beam with a clear span of 2.85 m supports a uniformly distributed load of 10 kN including self-weight of beam. Determine a suitable section for the beam using timber of strength class SC3. Assume that the bearing length is 150 mm and that the ends of the beam are held in position and compression edge held in line.

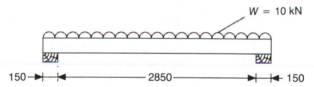

EFFECTIVE SPAN
Distance between centres of bearing (l) = 3000 mm

GRADE STRESS AND MODULUS OF ELASTICITY FOR SC3
Values in N mm^{-2} are as follows:

Bending parallel to grain ($\sigma_{m,g,\parallel}$)	Shear parallel to grain (τ_g)	Compression perpendicular to grain ($\sigma_{c,g,\perp}$)	Modulus of elasticity (E_{min})
5.3	0.67	1.7	5800

MODIFICATION FACTORS

$$K_3, \text{ duration of loading} = 1.0$$

$$K_7, \text{ depth factor} = \left(\frac{300}{h}\right)^{0.11}$$

Assume $h = 250$, $K_7 = 1.020$.

BENDING

$$M = \frac{Wl}{8} = \frac{10 \times 3}{8} = 3.75 \text{ kN m}$$

$$Z_{xx}\text{req (assuming } h = 250) \geqslant \frac{M}{\sigma_{m, adm, ||}} = \frac{3.75 \times 10^6}{5.3 \times 1.020}$$

$$= 694 \times 10^3 \text{ mm}^3$$

DEFLECTION

Permissible deflection $(\delta_p) = 0.003 \times$ span

The deflection due to shear (δ_s) is likely to be insignificant in comparison to the bending deflection (δ_b) and may be ignored in order to make a first estimate of the total deflection (δ_t):

$$\delta_t \text{ (ignoring shear deflection)} = \frac{5Wl^3}{384E_{min}I_{xx}} \quad (Table\ 6.9)$$

$$= \frac{5 \times 10^4 \times 3000^3}{384 \times 5800 \times I_{xx}}$$

Since $\delta_p \geqslant \delta_t$

$$0.003 \times 3000 \geqslant \frac{5 \times 10^4 \times 3000^3}{384 \times 5800 \times I_{xx}}$$

$$I_{xx} \text{ req} \geqslant 67.3 \times 10^6 \text{ mm}^4$$

From *Table 6.8*, section 75×250 provides

$$Z_{xx} = 781 \times 10^3 \text{ mm}^3 \qquad I_{xx} = 97.7 \times 10^6 \text{ mm}^4 \qquad A = 18.8 \times 10^3 \text{ mm}^2$$

Hence total deflection including shear deflection can now be calculated and is given by

$$\frac{5Wl^3}{384E_{min}I_{xx}} + \frac{12}{5}\frac{Wl}{E_{min}A} = \frac{5 \times 10^4 \times 3000^3}{384 \times 5800 \times 97.7 \times 10^6} +$$

$$\frac{12 \times 10^4 \times 3000}{5 \times 5800 \times 18.8 \times 10^3}$$

$$= 6.2 \text{ mm} + 0.7 \text{ mm}$$

$$= 6.9 \text{ mm} \leqslant \delta_p = 0.003 \times 3000 = 9 \text{ mm}$$

Therefore a beam with a 75×250 section is adequate for bending and deflection.

LATERAL BUCKLING

Permissible $\dfrac{d}{b} = 5 \quad (Table\ 6.10)$

Actual $\quad \dfrac{d}{b} = \dfrac{250}{75} = 3.3 <$ permissible

Hence the section is adequate for lateral buckling.

SHEAR

Permissible shear stress is

$$\tau_{adm} = \tau_g K_3 = 0.67 \times 1.0 = 0.67 \text{ N mm}^{-2}$$

Maximum shear force is

$$F_v = \frac{W}{2} = \frac{10 \times 10^3}{2} = 5 \times 10^3 \text{ N}$$

Maximum shear stress at neutral axis is

$$\tau_a = \frac{3}{2} \frac{F_v}{A} = \frac{3}{2} \times \frac{5 \times 10^3}{18.8 \times 10^3}$$
$$= 0.4 \text{ N mm}^{-2} < \text{permissible}$$

Therefore the section is adequate in shear.

BEARING
Permissible bearing stress is

$$\sigma_{c,adm,\perp} = \sigma_{c,g,\perp} K_3 = 1.7 \times 1.0 = 1.7 \text{ N mm}^{-2}$$

End reaction, F, is

$$\frac{W}{2} = \frac{10 \times 10^3}{2} = 5 \times 10^3 \text{ N}$$

$$\sigma_{c,a,\perp} = \frac{F}{b l_b} = \frac{5 \times 10^3}{75 \times 150} = 0.44 \text{ N mm}^{-2} < \text{permissible}$$

Therefore the section is adequate in bearing. Since all the checks are satisfactory, use 75 mm × 250 mm sawn SC3 beam.

Example 6.2 Design of timber floor joists

Design the timber floor joists for a domestic dwelling using timber of strength class SC2 given that:

1. the joists are spaced at 400 mm centres;
2. the floor has an effective span of 3.8 m;
3. the flooring is tongue and groove boarding with a self-weight of 0.1 kN m^{-2};
4. the ceiling is of plasterboard with a self-weight of 0.2 kN m^{-2}.

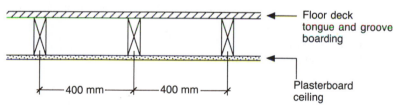

Floor deck tongue and groove boarding

Plasterboard ceiling

DESIGN LOADING

Tongue and groove boarding	= 0.10 kN m^{-2}
Ceiling	= 0.20 kN m^{-2}
Joists (say)	= 0.13 kN m^{-2}
Imposed floor load for domestic dwelling (*Table 2.2*)	= 1.50 kN m^{-2}
Total load	= 1.93 kN m^{-2}

Uniformly distributed load/joist (W) is

$$W = \text{joist spacing} \times \text{effective span} \times \text{load}$$
$$= 0.4 \times 3.8 \times 1.93 = 2.93 \text{ kN}$$

GRADE STRESSES AND MODULUS OF ELASTICITY FOR SC2
Values in N mm^{-2} are as follows:

Bending parallel to grain $(\sigma_{m,g,||})$	Compression perpendicular to grain $(\sigma_{c,g,\perp})$	Shear parallel to grain (τ_g)	Modulus of elasticity (E_{mean})
4.1	1.6	0.66	8000

MODIFICATION FACTORS

K_3, duration of loading $= 1.0$

K_8, load-sharing system $= 1.1$

K_7, depth factor $\quad = \left(\dfrac{300}{h}\right)^{0.11}$

where

$$\begin{aligned} h &= 225 & K_7 &= 1.032 \\ h &= 200 & K_7 &= 1.046 \\ h &= 175 & K_7 &= 1.061 \end{aligned}$$

BENDING

Bending moment $(M) = \dfrac{Wl}{8} = \dfrac{2.93 \times 3.8}{8} = 1.4$ kN m

Z_{xx} req (ignoring K_7) $= \dfrac{M}{\sigma_{m,g,||}\, K_3\, K_8} = \dfrac{1.4 \times 10^6}{4.1 \times 1.0 \times 1.1}$

$\qquad\qquad\qquad\qquad = 310 \times 10^3$ mm^3

From *Table 6.8* a 47×200 mm joist would be suitable ($Z_{xx} = 313 \times 10^3$ mm^3, $I_{xx} = 31.3 \times 10^6$ mm^4, $A = 9.4 \times 10^3$ mm^2).

Hence $K_7 = 1.046$. Therefore

$$Z_{xx}\ \text{req} = \frac{310 \times 10^3}{1.046} = 296 \times 10^3\ \text{mm}^3 < \text{provided} \quad \text{OK}$$

DEFLECTION

Permissible deflection $= 0.003 \times$ span

$\qquad\qquad\qquad\qquad = 0.003 \times 3800 = 11.4$ mm

Total deflection, δ_t, is

$$\delta_t = \text{bending deflection } (\delta_m) + \text{shear deflection } (\delta_v)$$

$$= \frac{5}{384} \times \frac{Wl^3}{E_{mean}I} + \frac{12}{5} \times \frac{Wl}{E_{mean}A}$$

$$= \frac{5}{384} \times \frac{2.93 \times 10^3 \times (3.8 \times 10^3)^3}{8 \times 10^3 \times 31.3 \times 10^6} + \frac{12}{5} \times \frac{2.93 \times 10^3 \times 3.8 \times 10^3}{8 \times 10^3 \times 9.4 \times 10^3}$$

$$= 8.4 + 0.4 = 8.8 < \text{permissible}$$

Therefore 47 mm $\times$ 200 mm joist is adequate in bending and deflection.

LATERAL BUCKLING

Permissible $\quad \dfrac{d}{b} = 5 \quad$ (*Table 6.10*)

Actual $\quad \dfrac{d}{b} = \dfrac{200}{47} = 4.3 < $ permissible

Therefore joist is satisfactory in lateral buckling.

SHEAR
Permissible shear is

$$\tau_{adm} = \tau_g K_3 K_8 = 0.66 \times 1.0 \times 1.1 = 0.726 \text{ N mm}^{-2}$$

Maximum shear force is

$$F_v = W/2 = \frac{2.93 \times 10^3}{2} = 1.47 \times 10^3 \text{ N}$$

Maximum shear stress at neutral axis is

$$\tau_a = \frac{3}{2} \frac{F_v}{A} = \frac{3}{2} \times \frac{1.47 \times 10^3}{9.4 \times 10^3} = 0.24 \text{ N mm}^{-2} < \text{permissible}$$

Therefore joist is adequate in shear.

BEARING
Permissible compression stress perpendicular to grain is

$$\sigma_{c,adm,\perp} = \sigma_{c,g,\perp} K_3 K_8 = 1.6 \times 1.0 \times 1.1 = 1.76 \text{ N mm}^{-2}$$

Maximum end reaction is

$$F = W/2 = \frac{2.93 \times 10^3}{2} = 1.47 \times 10^3 \text{ N}$$

Assuming that the floor joists span on to 100 mm wide wall plates the bearing stress is given by

$$\sigma_{c,a,\perp} = \frac{F}{bl_b} = \frac{1.47 \times 10^3}{47 \times 100} = 0.31 \text{ N mm}^{-2} < \text{permissible}$$

Therefore joist is adequate in bearing.

CHECK ASSUMED SELF-WEIGHT OF JOISTS
From *Table 6.3*, the density of timber of strength class SC2 is 540 kg m^{-3}. Hence, self-weight of the joists is

$$\frac{(47 \times 200 \times 10^{-6}) \times 540 \text{ kg m}^{-3} \times 9.81 \times 10^{-3}}{0.4 \text{ kg m}^{-3}} = 0.125 \text{ kN m}^{-2} < 0.13 \text{ kN m}^{-2} \quad \text{(assumed)}$$

Since all the checks are satisfactory use 47 mm × 200 mm SC2 sawn floor joists.

Example 6.3 Design of a notched floor joist
The joists in *Example 6.2* are to be notched at the bearings with a 75 mm deep notch as shown below. Check that the notched section is still adequate.

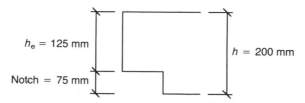

The presence of the notch affects only the shear stresses in the joists. For a notched member the permissible shear stress is given by

$$\tau_{adm} = \tau_g K_3 K_8 K_5,$$

where

$$K_5 = \frac{h_e}{h} = \frac{125}{200} = 0.625 > \text{min.} \ (= 0.5)$$

Hence

$$\tau_{adm} = 0.66 \times 1.0 \times 1.1 \times 0.625 = 0.45 \ \text{N mm}^{-2}$$

Applied shear parallel to grain, τ_a (from above) is

$$0.24 \ \text{N mm}^{-2} < \text{permissible}$$

Therefore the 47 mm × 200 mm sawn joists are also adequate when notched with a 75 mm deep bottom edge notch at the bearing.

Example 6.4 Analysis of a timber roof

A flat roof spanning 4.5 m is constructed using timber joists of grade GS whitewood with a section size of 47 × 225 mm and spaced at 450 mm centres. The total dead load due to the roof covering and ceiling including the self-weight of the joists is 1 kN m^{-2}. Calculate the maximum imposed load the roof can carry assuming that the duration of loading is (a) long term and (b) medium term.

DESIGN LOADING

$$\text{Dead load} = 1 \ \text{kN m}^{-2}$$
$$\text{Live load} = q \ \text{kN m}^{-2}$$

Uniformly distributed load/joist, W is

$$W = \text{joist spacing} \times \text{effective span} \times \text{(dead + live)}$$
$$= 0.45 \times 4.5 \ (1 + q)$$

GRADE STRESSES AND MODULUS OF ELASTICITY

Grade GS whitewood timber belongs to strength class SC3 (*Table 6.2*). Values in N mm^{-2} are as follows:

Bending parallel to grain $\sigma_{m,g,||}$	Compression perpendicular to grain $\sigma_{c,g,\perp}$	Shear parallel to grain τ_g	Modulus of elasticity E_{mean}
5.3	1.7	0.67	8800

MODIFICATION FACTORS

K_3, duration of loading (*Table 6.6*)   = 1.0 (long term) = 1.25 (medium term)

K_8, load-sharing system   = 1.1

K_7, depth factor $= \left(\dfrac{300}{h}\right)^{0.11}$

where $h = 225$, $K_7 = 1.032$.

GEOMETRICAL PROPERTIES

From *Table 6.8*, 47 × 225 section provides:

$$\text{Cross-sectional area} \ A = 10.6 \times 10^3 \ \text{mm}^2$$
$$\text{Section modulus about x–x,} \ Z_{xx} = 397 \times 10^3 \ \text{mm}^3$$
$$\text{Second moment of area about x–x,} \ I_{xx} = 44.6 \times 10^6 \ \text{mm}^4$$

BENDING

Long term

Permissible bending stress parallel to grain is

$$\sigma_{m,adm,||} = \sigma_{m,g,||}K_3K_7K_8$$

$$= 5.3 \times 1.0 \times 1.032 \times 1.1 = 6.02 \text{ N mm}^{-2}$$

Moment of resistance, M_R, is

$$M_R = \sigma_{m,adm,||} Z_{xx} = 6.02 \times 397 \times 10^3 \times 10^{-6} = 2.39 \text{ kN m}$$

Design moment, M, is

$$M = \frac{Wl}{8} = 0.45 \times 4.5 \,(1+q)\, \frac{4.5}{8} = 1.139\,(1+q)$$

Equating $M_R = M$,

$$2.39 = 1.139(1+q\;) \Rightarrow q = 1.09 \text{ kN m}^{-2}$$

Medium term

From above

$$\sigma_{m,adm,||} = 6.02K_3 \text{ (medium term)} = 7.52 \text{ N mm}^{-2}$$

$$M_R = 7.52 \times 397 \times 10^3 \times 10^{-6} = 2.98 \text{ kN m}$$

Equating $M_R = M$,

$$2.98 = 1.139(1+q\,) \Rightarrow q = 1.62 \text{ kN m}^{-2}$$

DEFLECTION

Max. total deflection = bending deflection + shear deflection

$$0.003L = \frac{5WL^3}{384E_{mean}I_{xx}} + \frac{12WL}{5E_{mean}A}$$

$$0.003 = W\left[\frac{5}{384} \times \frac{(4.5 \times 10^3)^2}{8800 \times 44.6 \times 10^6} + \frac{12}{5} \times \frac{1}{8800 \times 10.6 \times 10^3}\right]$$

$$= 6.975 \times 10^{-7}W$$

$$W = 4300 \text{ N/joist}$$

Load per unit area is

$$\frac{W}{\text{joist spacing} \times \text{span}} = \frac{4.3}{0.45 \times 4.5} = 2.12 \text{ kN m}^{-2}$$

Hence

$$q = 2.12 - \text{dead load} = 2.12 - 1 = 1.12 \text{ kN m}^{-2}$$

SHEAR

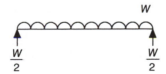

Permissible shear parallel to grain is

$$\tau_{adm} = \tau_g K_3 K_8 = 0.67 \times 1.0 \times 1.1 = 0.737 \text{ N mm}^{-2}$$

$$\text{Max. shear force } F_v = \frac{2}{3} \tau_{adm} A \quad \text{(equation 6.15)}$$

$$= \left(\frac{2}{3} \times 0.737 \times 10.6 \times 10^3 \right) \times 10^{-3} = 5.2 \text{ kN}$$

$$\text{Total load/joist} = 2F_v = 10.4 \text{ kN}$$

$$\text{Load per unit area} = \frac{10.4}{0.45 \times 4.5} = 5.13 \text{ kN m}^{-2}$$

Hence

$$q = 5.13 - 1 = 4.13 \text{ kN m}^{-2} \quad \text{(long term)}$$

and

$$q = 5.4 \text{ kN m}^{-2} \quad \text{(medium term, } K_3 = 1.25)$$

Hence the safe long-term imposed load that the roof can support is 1.09 kN m^{-2} (bending critical) and the safe medium-term imposed load is 1.12 kN m^{-2} (deflection critical).

6.8 Design of compression members

Struts and columns are examples of compression members. For design purposes BS 5268 divides compression members into two categories: (1) members subject to axial compression only and (2) members subject to combined bending and axial compression. The principal considerations in the design of compression members are:

1. slenderness ratio
2. axial compressive stress
3. permissible compressive stress.

The following subsections consider these more general aspects before describing in detail the design of the above two categories of compression members.

6.8.1 SLENDERNESS RATIO

The load-carrying capacity of compressive members is a function of the slenderness ratio, λ, which is given by

$$\lambda = \frac{L_e}{i} \quad (6.20)$$

where L_e is the effective length and i the radius of gyration. According to clause 15.4 of BS 5268, the slenderness ratio should not exceed 180 for compression members carrying dead and imposed loads other than loads resulting from wind in which case a slenderness ratio of 250 may be acceptable.

The radius of gyration, i, is given by

$$i = \sqrt{(I/A)} \quad (6.21)$$

where I is the moment of inertia and A the cross-section area. For rectangular sections

$$i = b/\sqrt{12} \quad (6.22)$$

where b is the least lateral dimension. The effective length of a column is obtained by multiplying the actual length, L, by a coefficient taken from *Table 6.11* which is a function of the fixity at the column ends.

$$L_e = L \times \text{coefficient} \quad (6.23)$$

In *Table 6.11* end condition (a) models the case of a column with both ends fully fixed and no relative horizontal motion possible between the column ends. End condition (c) models the case of a pin-ended column with no relative horizontal motion possible between column ends. End condition (e) models the case of a column with one end fully fixed and the other free. *Figure 6.7* illustrates all five combinations of end fixities.

6.8.2 AXIAL COMPRESSIVE STRESS

The axial compressive stress is given by

$$\sigma_{c, a, ||} = \frac{F}{A} \quad (6.24)$$

where F is the axial load and A the cross-sectional area.

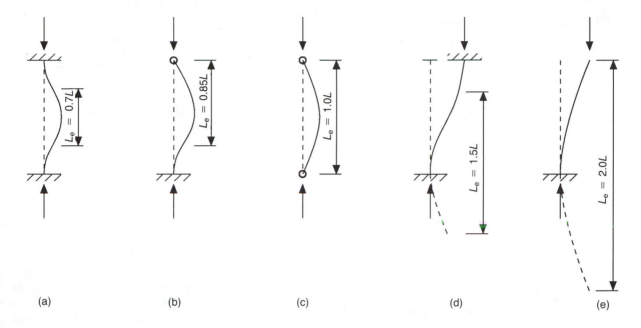

Fig. 6.7 *End conditions.*

6.8.3 PERMISSIBLE COMPRESSIVE STRESS

According to clause 15.5 of BS 5268, for compression members with slenderness ratios of less than 5, the permissible compressive stress should be taken as the grade compressive stresses parallel to the grain, $\sigma_{c,g,||}$, modified as appropriate for size, moisture content, duration and load sharing:

$$\sigma_{c,adm,||} = \sigma_{c,g||}, K_1 K_2 K_3 K_8 \text{ for } \lambda < 5 \quad (6.25)$$

For compression members with slenderness ratios equal to or greater than 5, the permissible compressive stress should be taken as the grade compressive stresses parallel to the grain, $\sigma_{c,g,||}$, modified as appropriate for size, moisture content, duration, load sharing and the modification factor K_{12}:

$$\sigma_{c,adm,||} = \sigma_{c,g,||} K_1 K_2 K_3 K_8 K_{12} \text{ for } \lambda \geqslant 5 \quad (6.26)$$

6.8.4 MEMBER DESIGN

Having discussed these common aspects it is now

Table 6.11 Effective length of compression members (Table 21, BS 5268)

End conditions		Effective length / Actual length (L_e/L)
(a)	Restrained at both ends in position and in direction	0.7
(b)	Restrained at both ends in position and one end in direction	0.85
(c)	Restrained at both ends in position but not in direction	1.0
(d)	Restrained at one end in position and in direction and at the other end in direction but not in position	1.5
(e)	Restrained at one end in position and in direction and free at the other end	2.0

possible to describe in detail the design of compression members. As pointed out earlier, BS 5268 distinguishes between two categories of members, that is, those subject to (a) axial compression only and (b) axial compression and bending.

6.8.4.1 Members subject to axial compression only

This category of compression member is designed so that the applied compressive stress, $\sigma_{c,a,||}$, does not exceed the permissible compressive stress parallel to the grain, $\sigma_{c,adm,||}$:

$$\sigma_{c,a,||} \leqslant \sigma_{c,adm,||} \qquad (6.27)$$

The applied compressive stress is calculated using equation 6.24 and the permissible compressive stress is given by equation 6.25 or 6.26 depending upon the slenderness ratio.

6.8.4.2 Members subject to axial compression and bending

This category includes compression members subject to eccentric loading which can be equated to an axial compression force and bending moment. Members which are restrained at both ends in position but not direction, which covers most real situations, should be so proportioned that

$$\frac{\sigma_{m,a,||}}{\sigma_{m,adm,||}\left(1 - \dfrac{1.5\sigma_{c,a,||}}{\sigma_e}K_{12}\right)} + \frac{\sigma_{c,a,||}}{\sigma_{c,adm,||}} \leqslant 1$$

$$(6.28)$$

where $\sigma_{m,a,||}$ is the applied bending stress, $\sigma_{m,adm,||}$ the permissible bending stress, $\sigma_{c,a,||}$ the applied compression stress, $\sigma_{c,adm,||}$ the permissible compression stress and σ_e the Euler critical

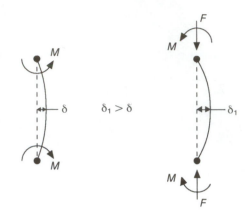

Fig. 6.8 *Bending in timber columns.*

stress $= \pi^2 E_{min}/(L_e/i)^2$.

Equation 6.28 is the normal interaction formula used to ensure that lateral instability does not arise in compression members subject to axial force and bending. Thus if the column was subject to compressive loading only, i.e. $M = 0$ and $\sigma_{m,a,||} = 0$, the designer would simply have to ensure that $\sigma_{c,a,||}/\sigma_{c,adm,||} \leqslant 1$. Alternatively, if the column was subject to bending only, i.e. $F = \sigma_{c,a,||} = 0$, the designer should ensure that $\sigma_{m,a,||}/\sigma_{m,adm,||} \leqslant 1$. However, if the column was subject to combined bending and axial compression, then the deflection as a result of the moment M would lead to additional bending due to the eccentricity of the force F as illustrated in *Fig. 6.8*. This is allowed for by the factor

$$1/[1 - (1.5\sigma_{c,a,||}\,K_{12})/\sigma_e]$$

in the above expression.

Example 6.5 Timber column resisting an axial load

A timber column of redwood GS grade consists of a 100 mm square section which is restrained at both ends in position but not in direction. Assuming that the actual height of the column is 3.75 m, calculate the maximum axial long-term load that the column can support.

SLENDERNESS RATIO

$$\lambda = \frac{L_e}{i} \qquad L_e = 1.0 \times h = 1.0 \times 3750 = 3750 \text{ mm}$$

$$i = \sqrt{(I/A)} = \sqrt{\left(\frac{db^3/12}{db}\right)} = \sqrt{\frac{b^2}{12}} = \frac{100}{\sqrt{12}} = 28.867$$

$$\lambda = \frac{3750}{28.867} = 129.9 \not> 180$$

GRADE STRESSES AND MODULUS OF ELASTICITY

Grade GS redwood belongs to strength class SC3 (*Table 6.2*). Values in N mm^{-2} are as follows:

Compression parallel to grain $\sigma_{c,g,||}$	Modulus of elasticity E_{min}
6.8	5800

MODIFICATION FACTOR

K_3, duration of loading is 1.0.

$$\frac{E_{min}}{\sigma_{c,||}} = \frac{5800}{6.8 \times 1.0} = 852.9 \text{ and } \lambda = 129.9$$

From *Table 6.7* by interpolation K_{12} is found to be 0.261.

$\dfrac{E_{min}}{\sigma_{c,||}}$	λ 120	129.9	140
800	0.280		0.217
852.9	0.293	0.261	0.228
900	0.304		0.237

AXIAL LOAD CAPACITY

Permissible compression stress parallel to grain is

$$\sigma_{c,adm,||} = \sigma_{c,g,||} K_3 K_{12} = 6.8 \times 1.0 \times 0.261 = 1.77 \text{ N mm}^{-2}$$

Hence axial long-term load capacity of column is

$$\sigma_{c,adm,||} A = 1.77 \times 10^4 \times 10^{-3} = 17.7 \text{ kN}$$

Example 6.6 Timber column resisting an axial load and moment

Check the adequacy of the column in *Example 6.5* to resist a long-term axial load of 10 kN and a bending moment of 350 kN mm.

SLENDERNESS RATIO

$$\lambda = L_e/i = 129.9 \not> 180 \quad (Example \ 6.5)$$

GRADE STRESSES AND MODULUS OF ELASTICITY

Values in N mm^{-2} for timber of strength class SC3 are as follows:

Bending parallel to grain $\sigma_{m,g,||}$	Compression parallel to grain $\sigma_{c,g,||}$	modulus of elasticity E_{min}
5.3	6.8	5800

MODIFICATION FACTORS

$$K_3 = 1.0$$

$$K_7 = \left(\frac{300}{h}\right)^{0.11} = \left(\frac{300}{100}\right)^{0.11} = 1.128$$

$$K_{12} = 0.261 \quad (Example \; 6.5)$$

COMPRESSION AND BENDING STRESSES

Permissible compression stress is

$$\sigma_{c,adm,||} = \sigma_{c,g,||} \, K_3 K_{12} = 6.8 \times 1.0 \times 0.261 = 1.77 \text{ N mm}^{-2}$$

Applied compression stress is

$$\sigma_{c,a,||} = \frac{\text{axial load}}{A} = \frac{10 \times 10^3}{10^4} = 1 \text{ N mm}^{-2}$$

Permissible bending stress is

$$\sigma_{m,adm,||} = \sigma_{m,g,||} \, K_3 K_7 = 5.3 \times 1.0 \times 1.128 = 5.98 \text{ N mm}^{-2}$$

Applied bending stress is

$$\sigma_{m,a,||} = \frac{M}{Z} = \frac{350 \times 10^3}{167 \times 10^3} = 2.10 \text{ N mm}^{-2}$$

Euler critical stress is

$$\sigma_e = \frac{\pi^2 E_{min}}{(L_e/i)^2} = \frac{\pi^2 \times 5800}{(129.9)^2} = 3.39$$

Since column is restrained at both ends, in position but not in direction, check that the column is so proportioned that

$$\frac{\sigma_{m,a,||}}{\sigma_{m,adm,||} \left(1 - \frac{1.5 \, \sigma_{c,a,||}}{\sigma_e} K_{12}\right)} + \frac{\sigma_{c,a,||}}{\sigma_{c,adm,||}} \leqslant 1$$

Substituting

$$\frac{2.10}{5.98 \left(1 - \frac{1.5 \times 1 \times 0.261}{3.39}\right)} + \frac{1}{1.77}$$

$$= 0.397 + 0.565 = 0.962 < 1$$

Therefore a 100×100 column is adequate to resist a long-term axial load of 10 kN and a bending moment of 350 kN mm.

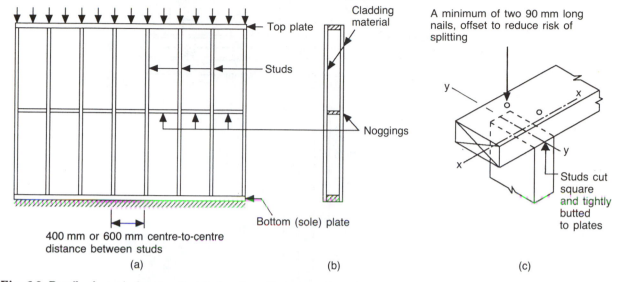

Fig. 6.9 *Details of a typical stud wall: (a) elevation; (b) section; (c) typical fixing of top and bottom plates to studs.*

6.9 Design of stud walls

In timber frame housing the loadbearing walls are normally constructed using stud walls (*Fig. 6.9*). These walls can be designed to resist not only the vertical loading but also loads normal to the wall due to wind, for example. Stud walls are normally designed in accordance with the requirements of Parts 6 and 6.1 of BS 5268. They basically consist of vertical timber members, commonly referred to as studs, which are held in position by nailing them to timber rails or plates, located along the top and bottom of the studs. The most common stud sizes are 100×50, 47, 38 mm and 75×50, 47, 38 mm. The studs are usually placed at 400 or 600 mm

centres depending upon preference, or on the loads they are required to transmit.

The frame is usually covered by a cladding material such as plasterboard which may be required for aesthetic reasons, but will also provide lateral restraint to the studs about the y–y axis. If the wall is not surfaced or only partially surfaced, the studs may be braced along their lengths by internal noggings. Bending about the x–x axis of the stud is assumed to be unaffected by the presence of the cladding material.

Since the centre to centre spacing of the stud is normally less than 610 mm, the load-sharing factor K_8 will apply to the design of stud walls. The design of stud walling is illustrated in the following example.

Example 6.7 Analysis of a stud wall

A stud wall panel has an overall height of 3.75 m including top and bottom rails and vertical studs at 600 mm centres with nogging pieces at mid-height. Assuming that the studs, rail framing and nogging pieces comprise 44×100 mm section of strength class SC4, calculate the maximum uniformly distributed long-term total load the panel is able to support.

SLENDERNESS RATIO

Effective height

$$L_{ex} = \text{coefficient} \times L = 1.0 \times 3750 = 3750 \text{ mm}$$
$$L_{ey} = \text{coefficient} \times L/2 = 1.0 \times 3750/2 = 1875 \text{ mm}$$

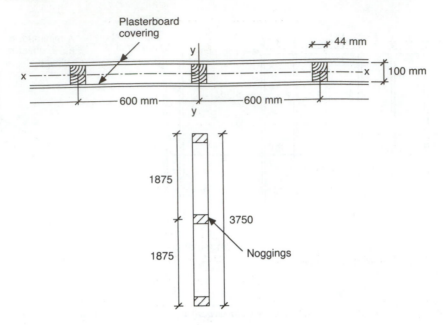

Radius of gyration

$$i_{xx} = \sqrt{\frac{I_{xx}}{A}} = \sqrt{\frac{(1/12) \times 44 \times 100^3}{44 \times 100}} = \frac{100}{\sqrt{12}}$$

$$i_{yy} = \sqrt{\frac{I_{yy}}{A}} = \sqrt{\frac{(1/12) \times 100 \times 44^3}{44 \times 100}} = \frac{44}{\sqrt{12}}$$

Slenderness ratio

$$\lambda_{xx} = \frac{L_{ex}}{i_{xx}} = \frac{3750}{100/\sqrt{12}} = 129.9 \ngtr 180$$

$$\lambda_{yy} = \frac{L_{ey}}{i_{yy}} = \frac{1875}{44/\sqrt{12}} = 147.6 \ngtr 180 \quad \text{(critical)}$$

Note that where two values of λ are possible the larger value must always be used to find $\sigma_{c,adm,||}$.

GRADE STRESSES AND MODULUS OF ELASTICITY
For timber of strength class SC4, values in N mm^{-2} are as follows:

Compression parallel to grain $\sigma_{c,g,||}$	Modulus of elasticity E_{min}
7.9	6600

MODIFICATION FACTORS

$$K_3 = 1.0 \qquad K_8 = 1.1$$

$$\frac{E_{min}}{\sigma_{c,||}} = \frac{6600}{7.9 \times 1.0} = 835.4 \qquad \lambda = 147.6$$

From *Table 6.7* $K_{12} = 0.206$ by interpolation.

	λ		
E_{min}	140	147.6	160
$\sigma_{c,\|\|}$			
800	0.217		0.172
835.4	0.224	0.206	0.178
900	0.237		0.188

AXIAL STRESSES

Permissible compression stress parallel to grain $\sigma_{c,adm,\|\|}$ is

$$\sigma_{c,adm,\|\|} = \sigma_{c,g,\|\|} K_3 K_8 K_{12} = 7.9 \times 1.0 \times 1.1 \times 0.206 = 1.79 \text{ N mm}^{-2}$$

Axial load capacity of stud is

$$\sigma_{c,adm,\|\|} A = 1.79 \times 44 \times 100 \times 10^{-3} = 7.87 \text{ kN}$$

Hence uniformly distributed load capacity of stud wall panel is

$$7.87/0.6 = 13.1 \text{ kN m}^{-1}$$

Note that the header spans 0.6 m and that this should also be checked as a beam in order to make sure that it is capable of supporting the above load.

6.10 Summary

This chapter has attempted to explain the concepts of stress grading and strength classes and the advantages that they offer to designers and contractors alike involved in specifying timber for structural purposes. The chapter has described the design of flexural and compression members and stud walling, to BS 5628: Part 2: *Structural Use of Timber*, which is based on permissible stress principles. In the design of flexural members (e.g. beams, rafters and joists), bending, shear and deflection are found to be the critical factors determining the design. With compression members (e.g. struts and columns), the slenderness ratio of the member influences its load-carrying capacity. Stud walls are normally designed on the assumption that the compression members act together to support a common load.

Questions

1. (a) Discuss the factors which influence the strength of timber and explain how the strength of timber is assessed in practice.
(b) A simply supported timber roof beam spanning 5 m supports a total uniformly distributed load of 11 kN. Determine a suitable section for the beam using timber of strength class SC3. Assume that the bearing length is 125 mm and that the compression edge is held in position.

2. (a) Give typical applications of timber in the construction industry and for each case discuss possible desirable properties.
(b) Redesign the timber joists in *Example 6.2* using timber of strength class SC4.

3. (a) Distinguish between softwood and hardwood and grade stress and permissible stress.
(b) Calculate the maximum long-term imposed load that a flat roof can support assuming the following construction details: roof joists are 50×225 mm of strength class SC3 at 600 mm centres; effective span is 4.2 m; unit weight of woodwool (50 mm thick) is 0.3 kN m^{-2}; unit weight of boarding, bitumen and roofing felt is 0.45 kN m^{-2}; unit weight of plasterboard and skim is 0.22 kN m^{-2}.

4. (a) Discuss the factors accounted for by the modification factor K_{12} in the

design of timber compression members.

(b) Design a timber column of effective length 2.8 m, capable of resisting the following loading: (i) medium-term axial load of 37.5 kN and (ii) long-term axial load of 30 kN and a bending moment of 300 kN mm.

5. (a) Explain with the use of sketches the connection details which will give rise to the following end conditions: (i) restrained in position and direction; (ii) restrained in position but not in direction and (iii) unrestrained in position and direction.

(b) Design a stud wall of length 4.2 m and height 3.8 m, using timber of strength class SC3 to support a long-term uniformly distributed load of 14 kN m^{-1}.

PART THREE

STRUCTURAL DESIGN TO THE EUROCODES

Part Two of this book has described the design of a number of structural elements in the four media: concrete, steel, masonry and timber to BS 8110, BS 5950, BS 5628 and BS 5268 respectively. The principal aim of this part of the book is to describe the salient features of the structural Eurocodes for these media, Eurocodes 2, 3, 6 and 5 respectively, and highlight the significant differences between the British Standard and the corresponding Eurocode.

The subject-matter has been divided into five chapters as follows:

1. *Chapter 7* introduces the Eurocodes and provides answers to some general questions regarding their nature, role, method of production and layout.

2. *Chapter 8* describes the contents of Part 1.1 of Eurocode 2 for the design of concrete buildings and illustrates the new procedures for designing beams, slabs, pad foundations and columns.

3. *Chapter 9* describes the contents of Part 1.1 of Eurocode 3 for the design of steel buildings and illustrates the new procedures for designing beams, columns and connections.

4. *Chapter 10* gives an overview of Eurocode 6 for the design of masonry structures, highlighting some of the difficulties which have delayed its progress.

5. *Chapter 11* describes the contents of Part 1.1 of Eurocode 5 for the design of timber structures and illustrates the new procedures for designing flexural and compression members.

Chapter 7

The structural Eurocodes: An introduction

This chapter describes the nature, objectives and mechanics of producing the new Eurocodes for structural design. The chapter highlights some of the difficulties associated with drafting these standards together with details of how these difficulties were resolved.

7.1 What are Eurocodes?

Eurocodes are the European standards for structural design. Table 7.1 shows the proposed range of Eurocodes currently under preparation. Like the present UK codes of practice, Eurocodes will come in a number of parts, covering a range of applications. These documents are at various stages of development. For instance, Parts 1.1 of Eurocodes 2 and 3 which are similar in scope to BS 8110 and BS 5950 respectively have already been finalized and issued as preliminary standards. However, much work remains to be done on many of the remaining parts of these codes.

Eurocodes will have the same legal standing as the national equivalent design standards or codes of practice do eventually. Eurocodes will be published first as preliminary standards (ENV) (*Prénorme Européenne*). This is equivalent to BSI's Draft for Development; ENVs are optional, and have a life of three years, with a possible extension of two years. They will then be revised and reissued as European Standards (EN) (*Norme Européenne*) which are mandatory in the sense that conflicting national standards must be withdrawn. All Eurocodes will be published in conjunction with a National Application Document (NAD), containing supplementary information specific to each member state. The NAD takes precedence over corresponding provisions in the Eurocode.

7.2 Why are Eurocodes necessary?

The establishment of international standards for structural design is not a new idea. Indeed, the first draft of Eurocode 2 for concrete structures was based on the CEB (Comité Européen du Béton) Model Code of 1978 drawn up by a number of experts from various European countries. Similarly Eurocode 3 was based on the 1977 *Recommendations for Design of Steel Structures* published by ECCS (European Convention for Constructional Steelwork) which was the work of several expert committees drawn from various countries in Europe and beyond. This process of drawing up of European design standards has been given a fresh impetus with the drive towards the political and economic unification of the EC.

There are several advantages to be gained from having design standards which are accepted by all member states. The first and foremost reason is that the provision of Eurocodes and the associated European Standards for construction products will help lower trade barriers between the member states. This will allow contractors and consultants from all member states to compete fairly for work within Europe. Hopefully this will lead to a pooling of resources and the sharing of expertise, thereby

Table 7.1 Structural Eurocodes currently under preparation

Eurocode	Subject
1	Basis of design and Actions on structure
2	Concrete
3	Steel
4	Composites
5	Timber
6	Masonry
7	Geotechnics
8	Seismic
9	Aluminium

lowering production costs. It is further believed that such standards will boost the international standing of European engineers which should help in increasing their chances of winning contracts abroad. A further benefit of having Eurocodes is that they will make it easier for engineers to practise within all EC countries.

7.3 Who produced the structural Eurocodes?

Generally, each Eurocode has been drafted by a small group of experts from various member states. These groups were formerly under contract to the EC Commission but are now under the direct control of CEN (Comité Européen de Normalisation), the European Standards Organization. In addition, a liaison engineer from each member state has been involved in evaluating the final document and discussing with the drafting group the acceptability of the Eurocode in relation to the national standard from the country which they represent.

7.4 Problems associated with the drafting of Eurocodes

At this stage very few Eurocodes are available, simply because they have yet to be drafted. It is difficult, therefore, to give a full picture of the problems encountered and the solutions developed by the drafting panels. At the time of writing, only Parts 1.1 of Eurocodes 2 and 3 have been issued as ENVs and the following discussion is based largely on the experiences of the people responsible for drafting these documents. Nevertheless, since there is a coordinating group linking the subcommittees producing the Eurocodes and the fact that these subcommittees and others are responsible to a main committee (CEN TC 250), it is likely that similar solutions will be adopted.

Specifically, the main problems faced by the drafting panels for Eurocodes 2 and 3 included agreeing a common terminology acceptable to all the member states, resolving differing opinions on technical issues, taking into account national differences in materials and design and construction practices, and regional differences in climatic conditions. In addition, it was also considered essential that all Eurocodes should be comprehensive but concise. The following subsections discuss how some of these issues were resolved without compromising the clarity or, indeed, simplicity of the codes.

P(2) In general, a minimum amount of shear reinforcement shall be provided, even where calculation shows that shear reinforcement is unnecessary. This minimum may be omitted in elements such as slabs, (solid, ribbed, hollow), having adequate provision for the transverse distribution of loads, where these are not subjected to significant tensile forces.
Minimum shear reinforcement may also be omitted in members of minor importance which do not contribute significantly to the overall strength and stability of the structure.

(3) Rules for minimum shear reinforcement are given in 5.4. An example of a member of minor importance would be a lintel of less than 2 m span.

Fig. 7.1 *Example of principles and application rules (Page 4-36, Oct. 1991 final text of EC2).*

7.4.1 TERMINOLOGY
At the outset of this work it was necessary to standardize the terminology used in the Eurocodes. Generally, this is similar to that already used in the equivalent UK documents. However, there are some minor differences; for example, loads are now called actions while dead and imposed loads are now termed permanent and variable actions respectively. Similarly, bending moments and axial loads are now called internal moments and internal forces respectively. These changes are so minor that they are unlikely to present any major problems to UK engineers.

7.4.2 PRINCIPLES AND APPLICATION RULES
In order to produce a document which is (a) concise, (b) describes the overall aims of design and (c) gives specific guidance as to how these aims can be achieved in practice, the material in the Eurocodes was divided into 'principles' and 'application rules'.

Principles are identified in Eurocode 2 by the letter P and are general statements, definitions, analytical methods, etc. for which no alternative is permitted (*Fig. 7.1*). The application rules are offset to the right of the page in Eurocode 2 and are generally recognized rules which follow the principles and satisfy their requirements (*Fig. 7.1*). In Eurocode 3 different typefaces are used for principles and application rules (*Chapter 9*). The latter approach is also expected to be followed in Eurocode 4.

Application rules each contain one suggested method for satisfying the corresponding principle. It is permissible to use alternative design rules provided that it can be shown that they satisfy the

		Exposure class, according to Table 4.1 of EC2								
		1	2a	2b	3	4a	4b	5a	5b	(3) 5c
(2) Minimum cover (mm)	Reinforcement	15	20	25	40	40	40	25	30	40
	Prestressing steel	25	30	35	50	50	50	35	40	50

Fig. 7.2 Minimum cover requirements (Table 4.2, Oct. 1991 final text of EC2)

relevant principles and do not negate the other aspects, e.g. serviceability, durability, of the structure.

7.4.3 BOXED VALUES

In Eurocode 2 a number of numerical values, e.g. partial safety factors, minimum concrete covers, coefficients in equations, etc. appear within boxes (*Fig. 7.2*). This signifies that these values are meant to be for guidance only and that other values may be used by individual member states, for the time being. In Eurocode 3 only safety elements, e.g. partial safety factors have been boxed. The actual values to be used in each country are given in the NAD.

This system of identifying certain parameters was introduced in order to account for national differences in material properties, design and construction practices, climatic conditions and so on. Unification of manufacturing and construction practices throughout the EC should see the gradual disappearance of most of these boxed values from the Eurocodes.

7.4.4 APPENDICES/ANNEXES

Some procedures which are not used in everyday design have been included in appendices in Eurocode 2 and annexes in Eurocode 3. Some of the annexes are labelled 'normative' while others are labelled 'informative'. The material which appears in the appendices and the 'normative' annexes has the same status as the rest of the Eurocode but appears there rather than in the body of the code in order to make the document as 'user-friendly' as possible. The material in the 'informative' annexes, however, does not have any status but has been included merely for information.

7.5 What are the differences between Eurocodes and British Standards?

Inevitably, there are many differences between Eurocodes and the national codes. Happily for UK engineers these changes are fairly minor, thanks largely to the work put in by the UK members of the various drafting panels. Consequently it is envisaged that it will not take very much time for engineers in the UK to become familiar with the contents of the Eurocodes and subsequently to use them. Of greater concern, perhaps, is the fact that many of the European standards which are needed to support the Eurocodes have yet to be drafted. Therefore, in the mean time, reference will have to be made to the appropriate national standards. This is bound to be frustrating and may cause a certain amount of confusion, especially for consultants bidding for work within EC countries. However, the strategy devised when the structural Eurocodes were initiated was to produce design standards first and allow these to generate the demand for relevant supporting standards.

Having discussed these more general aspects, the following chapters will describe the contents of Eurocodes 2, 3, 5 and 6 for concrete, steel, timber and masonry design respectively, and the significant differences between the Eurocode and the corresponding British Standard.

Chapter 8

Eurocode 2: Design of concrete structures

This chapter describes the contents of Part 1.1 of Eurocode 2, the new European standard for the design of buildings in concrete, which is expected to replace BS 8110 by about 1998. The chapter highlights the differences between Eurocode 2: Part 1.1 and BS 8110 and illustrates the new design procedures by means of a number of worked examples on beams, slabs, pad foundation and columns. To help comparison but primarily to ease understanding of the Eurocode the material here has been presented in a similar order to that in Chapter 3 of this book on BS 8110, rather than strictly adhering to the sequence of chapters and clauses adopted in the Eurocode.

8.1 Introduction

Eurocode 2 applies to the design of buildings and civil engineering works in plain, reinforced and prestressed concrete. It is based on limit state principles and comes in several parts as shown in Table 8.1.

Part 1.1 of Eurocode 2 gives a general basis for the design of buildings and civil engineering works in reinforced and prestressed concrete made with

Table 8.1 Overall scope of Eurocode 2

Part	Subject
1.1	Reinforced and prestressed concrete for ordinary buildings
1A	Plain concrete
1B	Precast concrete
1C	Lightweight aggregate concrete
1D	Unbonded and external tendons
1E	Fatigue
2	Bridges
3	Foundations
4	Liquid-retaining structures
5	Marine and maritime structures
6	Agriculture structures
7	Massive structures

normal weight aggregates. In addition, it gives some detailing rules which are mainly applicable to ordinary buildings. It is largely similar in scope to Part 1 of BS 8110 which it will replace by about 1998. Part 1.1 of Eurocode 2, hereafter referred to as EC2, was issued as a preliminary standard in 1992, ref. no. DD ENV 1992–1–1: 1992. Note that the letters DD signify that the document is a draft for development, the first 1992 is part of the document number and the second 1992 indicates the date it was issued.

The following subjects are covered in Part 1.1:

Chapter 1: Introduction
Chapter 2: Basis of design
Chapter 3: Material properties
Chapter 4: Section and member design
Chapter 5: Detailing provisions
Chapter 6: Construction and workmanship
Chapter 7: Quality control
Appendix 1: Time-dependent effects
Appendix 2: Non-linear analysis
Appendix 3: Additional design procedures for buckling
Appendix 4: Checking deflections by calculation.

The purpose of this chapter is to describe the contents of EC2 and to highlight the principal differences between it and BS 8110. A number of examples covering the design of beams, slabs, pad foundation and columns have also been included to illustrate the new design procedures.

8.2 Structure of EC2

Although the ultimate aim of EC2 and BS 8110 is the same, namely to give guidance on the design of reinforced and prestressed concrete structures, the organization of material in the two documents is rather different. For example, BS 8110 contains separate sections on the design of beams, slabs, columns, bases, etc. However, EC2 divides the material on the basis of structural action, i.e. bending, shear, deflection, torsion, which apply to any element. Furthermore, prestressed concrete is not dealt with separately in EC2 as in BS 8110, but each section on bending, shear, deflection, etc. contains rules relevant to the design of prestressed members.

There is a slight departure from this principle in Chapters 2 and 5 of EC2 which give guidance on the analysis and detailing respectively of specific member types.

8.3 Symbols

The following symbols which have largely been taken from EC2 have been used in this chapter.

GEOMETRIC PROPERTIES

b	width of section
d	effective depth of the tension reinforcement
h	overall depth of section
x	depth to neutral axis
z	lever arm
L_{eff}	effective span of beams and slabs
l_n	clear distance between the forces on the supports
a_1, a_2	distances between the faces of the support to the centre of the effective support at the two ends of the member
c	nominal cover to reinforcement
d'	depth to compression reinforcement

BENDING

g_k, G_k	characteristic permanent action
q_k, Q_k	characteristic variable action
w_k, W_k	characteristic wind load
F_k	characteristic action
F_d	design action
f_{ck}	characteristic compressive cylinder strength of concrete
f_{yk}	characteristic strength of reinforcement
f_{cd}	design concrete strength $= f_{ck} / 1.5$
f_{yd}	design steel strength $= f_{yk} / 1.5$
X_k	characteristic strength
X_d	design strength
γ_c	partial safety factor for concrete
γ_f	partial safety factor for actions
γ_s	partial safety factor for steel
γ_G	partial safety factor for permanent actions
γ_Q	partial safety factor for variable actions
γ_m	partial safety factor for material
α	reduction factor for sustained compression
C	concrete class
K_0	coefficient given by $M/f_{ck}bd^2$

K_0'	coefficient given by $M_u / f_{ck}bd^2 = 0.167$
M	design ultimate moment
M_u	design ultimate moment of resistance
A_{s1}	area of tension reinforcement
A_{s2}	area of compression reinforcement

SHEAR

S	steel class
τ_{Rd}	basic design shear strength
k	a constant relating to section depth and curtailment
ρ_1	reinforcement ratio corresponding to A_{s1}
σ_{cp}	average stress in concrete due to axial force
f_{ywd}	design yield strength of shear reinforcement
s	spacing of stirrups
ν	efficiency factor
V_{Sd}	design shear force due to ultimate loads
V_{Rd1}	design shear resistance of the concrete alone
V_{Rd2}	maximum design shear force that can be carried without crushing of the concrete
V_{Rd3}	design shear force of concrete and shear reinforcement
A_{sw}	cross-sectional area of the shear reinforcement
ρ_w	minimum shear reinforcement area

COMPRESSION

b	width of column
h	depth of section
l_{col}	clear height between centres of restraint
l_0	effective height
β	l_0/l_{col}
I_b	moment of inertia (gross section) of a beam
I_{col}	moment of inertia (gross section) of a column
k_A, k_B	coefficient describing the rigidity of restraint at column ends
α	factor taking into account the conditions of restraint of the beam at the opposite end
i	radius of gyration
λ_{crit}	critical slenderness ratio
υ	angle of inclination of the structure
υ_u	longitudinal force coefficient for an element
e_0	first-order eccentricity $= M_{sd1}/N_{sd}$
e_{01}, e_{02}	respectively the lower and higher values of the first-order eccentricity of the axial load at the ends of the member $|e_{01}| \leq |e_{02}|$
e_{min}	minimum eccentricity $= h/20$
e_e	equivalent eccentricity
e_a	eccentricity covering the effects of geometrical imperfections
e_2	second-order eccentricity
e_{tot}	total eccentricity
N_{Sd}	design axial load
N_{ud}	design ultimate capacity of column subject to axial load only

Table 8.2 Concrete strength classes and characteristic compressive cylinder strengths, f_{ck} (N mm^{-2}) (Table 3.1, EC2)

Strength class of concrete	C12/15	C16/20	C20/25	C25/30	C30/37	C35/45	C40/50	C45/55	C50/60
f_{ck}	12	16	20	25	30	35	40	45	50

N_{bal} design load capacity of balanced section
M_{Rd} minimum design resisting moment
M_{Sd1} first-order applied moment at end 1
M_{Sd2} first-order applied moment at end 2
A_c cross-section of the concrete
A_s area of longitudinal reinforcement

8.4 Material properties

8.4.1 CHARACTERISTIC STRENGTHS

8.4.1.1 Concrete (clause 3.1, EC2)

The design rules in EC2 are based on the characteristic 28-day strength of cylinders (f_{ck}). Equivalent cube strengths (f_{cu}) have also been included, but they are only regarded as an alternative method to prove compliance. *Table 8.2* shows the characteristic cylinder strengths of various classes of concrete recommended for use in reinforced and prestressed concrete design. Note that strength class C20/25, for example, refers to cylinder/cube strengths of 20 and 25 N mm^{-2} respectively.

8.4.1.2 Reinforcing steel (clause 3.2, EC2)

Unlike BS 8110, EC2 does not contain details of the reinforcement type to be used in design. For this information the designer must turn to the European standard, which is only available in draft form as provisional standard prEN10080 at this stage. Therefore, until such time that this standard comes into force, mild steel and high yield steel reinforcing bars of characteristic yield strength, f_{yk}, of 250 and 460 N mm^{-2} respectively will continue to be used in the UK. It should be noted that prEN10080 suggests that grade 460 steel will be replaced by a grade 500 steel and that grade 250 steel will not be included in the standard (see Table 5 of NAD for further information).

8.4.2 DESIGN STRENGTHS (CLAUSE 2.2.3.2, EC2)

The design strengths (X_d) are obtained by dividing the characteristic strengths (X_k) by the appropriate partial safety factor for materials (γ_m):

$$X_d = \frac{X_k}{\gamma_m} \qquad (8.1)$$

The partial safety factors for concrete and steel reinforcement are shown in *Table 8.3*. Note that in EC2 the partial safety factor for concrete has a single value of 1.5. This is different from BS 8110 where the partial safety factor varies depending upon the stress type under consideration (*Table 3.3*).

8.5 Actions (clause 2.2.2, EC2)

Actions is the Eurocode terminology for loads and imposed deformations. EC2 defines an action (F) as a force or load applied to a structure. Actions may be 'permanent' (G), e.g. self-weight of structure, fittings and fixed equipment, or 'variable' (Q), e.g. imposed, wind and snow loads.

8.5.1 CHARACTERISTIC ACTIONS (CLAUSE 2.2.2.2, EC2)

The characteristic values of actions (F_k), are specified in Eurocode 1: *Basis of Design and Actions on Buildings*, which is expected to be published as an ENV late in 1993. Therefore, in the mean time, the NAD requires the designer to refer to the relevant UK loading codes, e.g.

1. BS 648: *Schedule of Weights of Building Materials*;
2. BS 6399:Part 1:1984 *Code of Practice for Dead and Imposed Loads*;
3. CP 3:Chapter V:Part 2:1972 *Wind Loads*.

Clause 4 of the NAD notes certain modifications in using the above documents with EC2. However, these are not pertinent to the discussion.

Table 8.3 Partial safety factors for materials (Table 2.3, EC2)

Combination	Concrete γ_c	Steel reinforcement or prestressing tendons γ_s
Fundamental	1.5	1.15
Accidental (except earthquakes)	1.3	1.0

Table 8.4 Partial safety factors for actions in building structures for persistent and transient design situations (Table 2.2, EC2)

	Permanent actions (γ_G)	Variable actions (γ_Q)	
		One with its characteristic value	Others with their combination values
Favourable effect	1.0	0	0
Unfavourable effect	1.35	1.5	1.5

8.5.2 DESIGN ACTIONS (CLAUSE 2.2.2.4, EC2)

The design actions (F_d) are obtained by multiplying the characteristic actions (F_k) by the appropriate partial safety factor for actions (γ_F):

$$F_d = \gamma_F F_k \qquad (8.2)$$

For single beams, the partial safety factors for permanent, γ_G, and variable actions, γ_Q, will normally be 1.35 and 1.5 respectively (*Table 8.4*). The corresponding values in BS 8110 are 1.4 and 1.6 respectively (*Table 3.4*).

Note that where wind acts in combination with permanent and variable actions, all characteristic actions should be multiplied by 1.35. The corresponding value in BS 8110 is 1.2 (*Table 3.4*).

For continuous beams, clause 2.5.1.2 of EC2 recommends that the following load cases will generally be sufficient:

1. alternate spans carrying maximum loads with the others carrying minimum load
2. any two adjacent spans carrying maximum load with the remainder carrying minimum load.

These are different to the load cases suggested in BS 8110 (section 3.6.2). The methods of analysis permitted in EC2, however, are basically the same ones referred to in BS 8110.

8.6 Stress–strain diagrams

8.6.1 CONCRETE (CLAUSE 4.2.1.3.3, EC2)

Figure 8.1 shows the idealized and design stress – strain diagrams for concrete. The basic shape of the two curves, i.e. rectangular – parabolic, is similar to that adopted in BS 8110. Furthermore, the ultimate strain ($\varepsilon_{cu\ max}$) of the concrete in compression is taken to be 0.0035 in EC2 as in BS 8110.

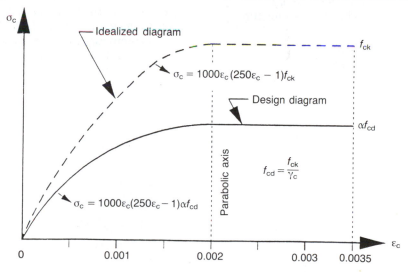

Fig. 8.1 *Parabolic–rectangular stress–strain diagram for concrete in compression (Fig. 4.2, EC2).*

The design concrete strength, f_{cd}, is obtained by dividing the characteristic strength, f_{ck}, of concrete by the partial safety factor for concrete, γ_c (*Table 8.3*):

$$f_{cd} = \frac{f_{ck}}{\gamma_c} \qquad (8.3)$$

However, the design concrete stress is obtained by applying a factor, α, to the design strength:

$$\text{Design stress} = \alpha\, f_{cd} \qquad (8.4)$$

where α is the reduction factor for sustained compression, generally taken as 0.85 (see Table 3 of NAD).

8.6.2 REINFORCING STEEL (CLAUSE 4.2.2.3.2, EC2)

The design steel stresses (f_{yd}) are derived from the idealized (characteristic) stresses (f_{yk}) by dividing by the partial safety factor for steel, γ_s:

$$f_{yd} = \frac{f_{yk}}{\gamma_s} \qquad (8.5)$$

Figure 8.2 shows the idealized and design stress – strain diagrams for reinforcing steel. However, it will not be possible to use this for design purposes until the Eurocode for reinforcement (EN10080) is published since there are no values of f_{tk} or ε_{uk}.

The stress – strain diagram may be modified with a flatter or horizontal top branch. If a horizontal top branch is assumed, the design stresses should not exceed f_{yk}/γ_s, although there is no limit to the steel strain. The design equations which have been developed later in this chapter assume that the stress – strain diagram for reinforcement has a horizontal top branch.

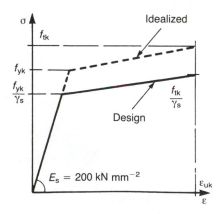

Fig. 8.2 *Design stress–strain diagram for reinforcing steel (Fig. 4.5, EC2).*

8.7 Fire resistance and durability

8.7.1 FIRE RESISTANCE

The European regulations regarding fire resistance of concrete structures are currently being drafted and will, once completed, be published either in an appendix to EC2 or separately as a further part of EC2. In the interim, the NAD requires UK engineers to continue to follow the recommendations contained in Part 2 of BS 8110: 1985.

8.7.2 DURABILITY

In EC2, like BS 8110, the durability of concrete structures is related to

1. environmental conditions
2. cover to reinforcement
3. concrete quality
4. maximum crack width.

With respect to the environment, EC2 identifies nine exposure classes to which a structure could be subject during its design life (*Table 8.5*). They correspond to the first four exposure conditions listed in BS 8110, namely mild, moderate, severe and very severe (*Table 3.5*). However, there is no class in EC2 which corresponds to the 'extreme' exposure condition of BS 8110.

Like BS 8110, EC2 allows the designer to reduce the thickness of the concrete cover to the reinforcing bars by specifying better-quality concrete. *Table 8.6* shows the nominal concrete covers to all reinforcement, including links, as a function of concrete quality and exposure class in accordance with EC2 and ENV 206: *Concrete – Performance, Production, Placing and Compliance Criteria: 1992.* Both BS 8110 and EC2 recommend that the maximum design crack width should not generally exceed 0.3 mm. This limiting crack width is achieved in practice by (a) providing a minimum amount of reinforcement and (b) limiting the maximum bar spacing or bar size. These requirements will be discussed individually for beams, slabs and columns in *sections 8.8.4, 8.9.2* and *8.11.6* respectively.

8.8 Design of singly and doubly reinforced rectangular beams

8.8.1 BENDING (CLAUSE 4.3.1, EC2)

In analysing a cross-section to determine its ultimate moment of resistance, clause 4.3.1.2 of EC2 recommends that the following assumptions can be made:

Table 8.5 Exposure classes related to environmental conditions (Table 4.1, EC2)

Exposure class		Examples of environmental conditions
1 Dry environment		Interior of buildings for normal habitation or offices
2 Humid environment	a Without frost	Interior of buildings where humidity is high (e.g. laundries) Exterior components Components in non-aggressive soil and/or water
	b With frost	Exterior components exposed to frost Components in non-aggressive soil and/or water and exposed to frost Interior components when the humidity is high and exposed to frost
3 Humid environment with frost and deicing salts		Interior and exterior components exposed to frost and deicing agents
4 Seawater environment	a Without frost	Components completely or partially submerged in seawater, or in the splash zone Components in saturated salt air (coastal area)
	b With frost	Components partially submerged in seawater or in the splash zone and exposed to frost Components in saturated salt air and exposed to frost
The following classes may occur alone or in combination with the above classes:		
5 Aggressive Chemical environment	a	Slightly aggressive chemical environment (gas, liquid or solid) Aggressive industrial atmosphere
	b	Moderately aggressive chemical environment (gas, liquid or solid)
	c	Highly aggressive chemical environment (gas, liquid or solid)

Table 8.6 Nominal cover to reinforcement and concrete quality for durability (based on Tables 6 and NA.1 of EC2 and ENV 206 respectively and the Concise Eurocode for the Design of Concrete Buildings)

Exposure class	Nominal cover (mm)				
1	20	20	20	20	20
2a	–	35	35	30	30
2b	–	–	35	30	30
3	–	–	40	35	35
4a	–	–	40	35	35
4b	–	–	40	35	35
5a	–	–	35	30	30
5b	–	–	–	30	30
5c	–	–	–	–	45
Max. free W/C ratio	0.65	0.60	0.55	0.50	0.45
Min. cement content (kg m^{-3})	260	280	300	300	300
Lowest concrete grade	C25/30	C30/37	C35/45	C40/50	C45/55

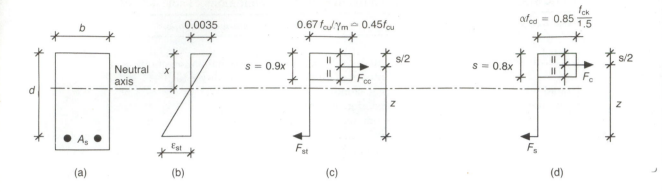

Fig. 8.3 *Singly reinforced section with rectangular stress block: (a) section; (b) strains; (c) stress block (BS 8110); (d) stress block (EC2).*

1. Plane sections remain plane.
2. The strain in bonded reinforcement is the same as that in the surrounding concrete.
3. The tensile strength of concrete is ignored.
4. The compressive stresses in the concrete may be derived from the design curve in *Fig. 8.1*. (Note that clause 4.2.1.3.3(10) of EC2 states that other idealized stress diagrams may be used, provided they are effectively equivalent to *Fig. 8.1*, e.g. the rectangular stress distribution shown in *Fig. 8.3(d)*.)
5. The stresses in the reinforcement may be derived from *Fig. 8.2*.

These assumptions are similar to those listed in clause 3.4.4.1 of BS 8110 except that in EC2 there is no limit on the lever arm depth.

Figure 8.3 shows the simplified stress blocks which are used in BS 8110 and EC2 to develop the design equations in bending. In EC2 the maximum concrete compressive stress is taken as $0.85f_{ck}/1.5$ (*Fig. 8.1*) which compares with $0.67f_{cu}/1.5$ in BS 8110 (*Fig. 3.6*). Furthermore, the depth of the compression block is taken as $0.8x$ in EC2 rather than $0.9x$ as in BS 8110, where x is the depth of the neutral axis.

The following subsections derive the equations relevant to the design of singly and doubly reinforced beams according to EC2. The corresponding equations in BS 8110 were derived in *section 3.9.1.1* to which the reader should refer for detailed explanations and the notation used.

8.8.1.1 Singly reinforced beams

Ultimate moment of resistance (M_u)

$$M_u = F_c z \qquad (8.6)$$

$$F_c = \frac{(0.85f_{ck})\,0.8 \times b}{1.5} \qquad (8.7)$$

$$z = d - 0.4x \qquad (8.8)$$

Clause 2.5.3.4.2 of EC2 limits the depth of the neutral axis (x) to $0.45d$ for concrete grades C12/15 to C35/45 (and $0.35d$ for concrete grades C40/50 and greater) in order to provide a ductile, i.e. under-reinforced, section. Thus

$$x = 0.45d \qquad (8.9)$$

Note that the corresponding value for x in BS 8110 is $0.5d$.

Combining equations (8.6)–(8.9) gives

$$M_u = 0.167f_{ck}bd^2 \qquad (8.10)$$

Compare $M_u = 0.156f_{cu}bd^2$ (BS 8110).

Area of tensile steel (A_{s1})

$$M = F_s z \qquad (8.11)$$

$$F_s = \frac{f_{yk}A_{s1}}{1.15} \qquad (8.12)$$

$$A_{s1} = \frac{M}{0.87f_{yk}z} \qquad (8.13)$$

Compare $A_s = M/0.87f_{yz}$ (BS 8110). Equation 8.13 can be used to calculate the area of tension reinforcement provided $M \leqslant M_u$.

Lever arm (z)

$$M = F_c z = \left(\frac{0.85f_{ck}}{1.5}\right)0.8bxz \quad \text{(from equation 8.7)}$$

$$= \frac{3.4}{3}f_{ck}b\,(d{-}z)\,z \quad \text{(from equation 8.8)}$$

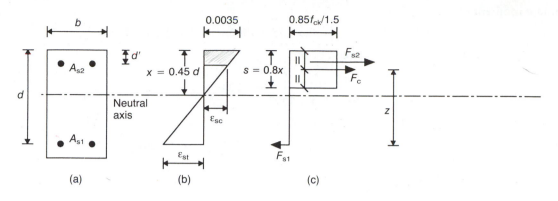

Fig. 8.4 *Doubly reinforced section: (a) section; (b) strains; (c) stress block (EC2).*

Solving for z gives

$$z = d[0.5 + \sqrt{(0.25 - 3K_0/3.4)}] \qquad (8.14)$$

where $K_0 = \dfrac{M}{f_{ck}bd^2}$. Compare

$$z = d[0.5 + \sqrt{(0.25 - K/0.9)}] \qquad \text{(BS 8110)}$$

where $K = \dfrac{M}{f_{cu}bd^2}$.

8.8.1.2 Doubly reinforced beams

If the design moment is greater than the ultimate moment of resistance, i.e. $M > M_u$, then compression reinforcement is required. Provided that

$d'/x \not> 0.43$ (i.e. compression steel has yielded)

where d' is the depth of the compression steel from the compression face and $x = (d - z)/0.4$, the area of compression reinforcement, A_{s2}, is given by

$$A_{s2} = \frac{M - M_u}{0.87 f_{yk}(d - d')} \qquad (8.15)$$

and the area of tension reinforcement, A_{s1}, is given by

$$A_{s1} = \frac{M_u}{0.87 f_{yk}z} + A_{s2} \qquad (8.16)$$

where $z = d[0.5 + \sqrt{(0.25 - 3K'_0/3.4)}]$
$\qquad K'_0 = 0.167$

Equations 8.15 and 8.16 have been derived using the stress block shown in *Fig. 8.4*. This is similar to that used to derive the equations for the design of singly reinforced beams (*Fig. 8.3(d)*) except for the additional compression force in the steel.

Example 8.1 Bending reinforcement for a singly reinforced beam

Determine the area of main steel, A_{s1}, required for the beam assuming the following material strengths: $f_{ck} = 25$ N mm^{-2} and $f_{yk} = 460$ N mm^{-2}.

Ultimate load, w, is

$$w = 1.35 g_k + 1.5 q_k = 1.35 \times 12 + 1.5 \times 8 = 28.2 \text{ kN m}^{-1}$$

Design moment, M, is

$$M = \frac{wl^2}{8} = \frac{28.2 \times 7^2}{8} = 172.7 \text{ kN m}$$

Ultimate moment of resistance, M_u, is

$$M_u = 0.167 f_{ck} b d^2 = 0.167 \times 25 \times 275 \times 450^2 \times 10^{-6} = 232.5 \text{ kN m}$$

Since $M_u > M$ design as a singly reinforced beam

$$K_0 = \frac{M}{f_{ck} b d^2} = \frac{172.7 \times 10^6}{25 \times 275 \times 450^2} = 0.124$$

$$z = d[0.5 + \sqrt{(0.25 - 3K_0/3.4)}]$$

$$= 450[0.5 + \sqrt{(0.25 - 3 \times 0.124/3.4)}] = 393.7 \text{ mm}$$

$$A_{s1} = \frac{M}{0.87 f_{yk} z} = \frac{172.7 \times 10^6}{0.87 \times 460 \times 393.7} = 1096 \text{mm}^2$$

Hence from *Table 3.7*, provide 4T20 (A_{s1} = 1260 mm^2).

Example 8.2 Bending reinforcement for a doubly reinforced beam

Design the bending reinforcement for the beam assuming the cover to the main steel is 40 mm, f_{ck} = 25 N mm^{-2} and f_{yk} = 460 N mm^{-2}.

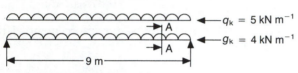

DESIGN MOMENT (M)
Total ultimate load, W, is

$$W = (1.35 g_k + 1.5 q_k) \text{ span} = (1.35 \times 4 + 1.5 \times 5)9 = 116.1 \text{ kN}$$

Design moment, M, is

$$M = \frac{Wl}{8} = \frac{116.1 \times 9}{8} = 130.6 \text{ kN m}$$

ULTIMATE MOMENT OF RESISTANCE (M_u)

Effective depth
Assume diameter of main bar (Φ) is 25 mm. Effective depth, d, is

$$d = h - c - \Phi/2 = 370 - 40 - 12.5 = 317 \text{ mm}$$

Ultimate moment
Ultimate moment of resistance (M_u) is

$$M_u = 0.167 f_{ck} b d^2 = 0.167 \times 25 \times 230 \times 317^2 \times 10^{-6} = 96.5 \text{ kN m}$$

Since $M_u < M$ design as a doubly reinforced beam.

COMPRESSION REINFORCEMENT (A_{s2})
Assume diameter of compression bars (ϕ) is 16 mm. Effective depth, d', is

$$d' = \text{cover} + \phi/2 = 40 + 16/2 = 48 \text{ mm}$$

$$\frac{d'}{x} = \frac{48}{0.45 d} = \frac{48}{0.45 \times 317} = 0.34 \ngtr 0.43$$

Hence

$$A_{s2} = \frac{M - M_u}{0.87 f_{yk} (d - d')} = \frac{(130.6 - 96.5) \, 10^6}{0.87 \times 460 \times (317 - 48)}$$

$$= 317 \text{ mm}^2$$

Provide 2T16 ($A_{s2} = 402 \text{ mm}^2$ from *Table 3.7*).

TENSION REINFORCEMENT (A_{s1})

$$z = d[0.5 + \sqrt{(0.25 - 3K'_o /3.4)}]$$

$$= d[0.5 + \sqrt{(0.25 - 3 \times 0.167/3.4)}] = 0.82d = 0.82 \times 317 = 260 \text{ mm}$$

$$A_{s1} = \frac{M_u}{0.87 f_{yk} \, z} + A_{s2} = \frac{96.5 \times 10^6}{0.87 \times 460 \times 260} + 317 = 1245 \text{ mm}^2$$

Provide 3T25 ($A_{s1} = 1470 \text{ mm}^2$ from *Table 3.7*).

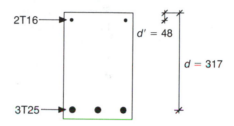

8.8.2 SHEAR (CLAUSE 4.3.2, EC2)

Unlike BS 8110, EC2 compares shear forces rather than shear stresses in order to assess whether shear reinforcement will be required. The notation used in EC2 tends, therefore, to be quite different from that in BS 8110. Furthermore, EC2 generally provides better guidance than BS 8110 regarding the spacing of shear reinforcement.

EC2 identifies four basic shear forces for design purposes, namely V_{Sd}, V_{Rd1}, V_{Rd2} and V_{Rd3} where V_{Sd} is the applied shear force and the remaining three parameters give an indication of the shear resistance of the member: V_{Rd1} is the actual shear resistance of the concrete alone, V_{Rd2} is the maximum shear resistance of the concrete and V_{Rd3} the sum of the shear resistances of the concrete and steel and is therefore a function of V_{Rd1}.

Here V_{Rd1} is given by

$$V_{Rd1} = [\tau_{Rd} k (1.2 + 40 \rho_1) + 0.15 \sigma_{cp}] \, b_w d \tag{8.17}$$

where τ_{Rd} is the basic shear strength (*Table 8.7*), $k = 1$ for members where > 50% of the bottom reinforcement is curtailed, otherwise put $k = 1.6 - d \not< 1$ (d in metres), $\rho_1 = A_{sl}/b_w d \not> 0.02$, A_{sl} is the (effective) area of tension reinforcement, b_w the width of section, $\sigma_{cp} = N_{Sd}/A_c$, N_{Sd} is the longitudinal force in section and A_c the cross-sectional area of concrete.

Here V_{Rd2} is given by

$$V_{Rd2} = \frac{1}{2} \nu f_{cd} b_w 0.9d \tag{8.18}$$

where

$$\nu = 0.7 - \frac{f_{ck}}{200} \not< 0.5 \quad (f_{ck} \text{ in N mm}^{-2}) \tag{8.19}$$

Table 8.7 Values of τ_{Rd} with $\gamma_c = 1.5$ for different concrete strengths (Table 4.8, EC2)

f_{ck}	12	16	20	25	30	35	40
τ_{Rd} (N mm^{-2})	0.18	0.22	0.26	0.30	0.34	0.37	0.41

Table 8.8 Minimum values for ρ_w (Table 5.5, EC2)

Concrete classes	Steel classes		
	S220	S400	S500
C12/15 and C20/25	0.0016	0.0009	0.0007
C25/30 to C35/45	0.0024	0.0013	0.0011
C40/50 to C50/60	0.0030	0.0016	0.0013

The total shear resistance, V_{Rd3}, is the sum of the shear resistance of the concrete (V_{cd}) and transverse reinforcement (V_{wd}). Here V_{Rd3} can be calculated using one of two methods, namely the **standard method** or the **variable strut inclination method**. Only the former method will be discussed here which is similar in principle to the method contained in BS 8110 (*section 3.9.1.3*). Thus

$$V_{Rd3} = V_{cd} + V_{wd} \qquad (8.20)$$

where $V_{cd} = V_{Rd1}$ and

V_{wd} = contribution of the shear reinforcement

$$= \frac{A_{sw}}{s} 0.9 f_{ywd} d \qquad (8.21)$$

where A_{sw} is the cross-sectional area of the shear reinforcement, s the spacing of shear reinforcement (*section 8.8.2.2*) and f_{ywd} the design yield strength of the shear reinforcement.

According to clause 4.3.2.2 of EC2, if $V_{Sd} < V_{Rd1}$, minimum shear reinforcement must be provided (*section 8.8.2.1*) except in members of minor importance where such reinforcement may be omitted. An example of a member of minor importance would be a lintel of less than 2 m span. If $V_{Rd1} < V_{Sd} < V_{Rd2}$,

Table 8.9 Spacing of shear reinforcement

If	$V_{Sd} \leq 1/5\ V_{Rd2}$	$s_{max} = 0.8d \not> 300$ mm
If $1/5\ V_{Rd2} \leq$	$V_{Sd} \leq 2/3\ V_{Rd2}$	$s_{max} = 0.6d \not> 300$ mm
If	$V_{Sd} > 2/3\ V_{Rd2}$	$s_{max} = 0.3d \not> 200$ mm

design shear reinforcement must be provided according to equation 8.20, with $V_{Rd3} = V_{Sd}$.

8.8.2.1 Minimum shear reinforcement areas (ρ_w) (clause 5.4.2.2, EC2)

Minimum values of the shear ratio (ρ_w are shown in *Table 8.8*, ρ_w being defined as

$$\rho_w = A_{sw}/sb_w \sin \alpha \qquad (8.22)$$

where A_{sw} is the area of shear reinforcement within length s, s the spacing of the shear reinforcement, b_w the breadth of the member and $\alpha = 90°$ for vertical stirrups, i.e. $\sin \alpha = 1$. Since the above steel classes do not correspond with those used in the UK, appropriate values of the shear ratio can be obtained by interpolation.

8.8.2.2 Diameter and spacing of shear reinforcement (clause 5.4.2.2, EC2)

EC2 recommends that the diameter of the shear reinforcement should not generally exceed 12 mm. In addition, EC2 recommends that the maximum spacing of shear reinforcement (s_{max}) should lie in the range $0.3d$–$0.8d$ (*Table 8.9*); the actual value will depend upon the ratio of the design shear force to the maximum shear resistance of the concrete (i.e. V_{Sd}/V_{Rd2}). In BS 8110, the only guidance given is that the spacing of the links should not exceed $0.75d$.

Example 8.3 Design of shear reinforcement for a beam

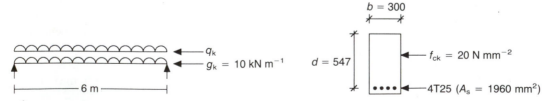

Design the shear reinforcement for the beam using mild steel ($f_{ywd} = 250$ N mm^{-2}) links for the following load cases: (i) $q_k = 0$, (ii) $q_k = 10$ kN m^{-1}, (iii) $q_k = 27$ kN m^{-1}, (iv) $q_k = 40$ kN m^{-1}.

Shear resistance of concrete alone (V_{Rd1})

$$f_{ck} = 20 \text{ N mm}^{-2} \quad \tau_{Rd} = 0.26 \text{ N mm}^{-2}$$
$$k = 1.6 - d = 1.6 - 0.547 = 1.053$$

$$\rho_1 = \frac{A_{sl}}{b_w d} = \frac{1960}{300 \times 547} = 0.012 \qquad \sigma_{cp} = 0$$

150N/m²

$$V_{Rd1} = [\tau_{Rd}\, k\, (1.2 + 40\rho_1) + 0.15\sigma_{cp}]\, b_w d$$

$$= [0.26 \times 1.053(1.2 + 40 \times 0.012)]300 \times 547$$

$$= 75\,478 \text{ N}$$

Maximum shear resistance of concrete (V_{Rd2})

$$\nu = 0.7 - \frac{f_{ck}}{200} = 0.7 - \frac{20}{200} = 0.6$$

$$V_{Rd2} = \frac{1}{2}\nu f_{cd} b_w 0.9d$$

$$= \frac{1}{2} \times 0.6 \times \frac{20}{1.5} \times 300 \times 0.9 \times 547 = 590760 \text{ N}$$

Minimum shear reinforcement (ρ_w)

$$\rho_w = 0.0015 \quad \text{(by interpolation between the values in *Table 8.8*)}$$

Substituting this into $\rho_w = A_{sw}/sb_w \sin \alpha$ gives

$$0.0015 = A_{sw}/s \times 300 \times 1$$

$$\frac{A_{sw}}{s} = 0.45$$

$q_k = 0$

Shear force (V_{Sd})

Total ultimate load, W, is

$$W = (1.35g_k + 1.5q_k)\, \text{span} = (1.35 \times 10 \times 10^3)\, 6 = 81\,000 \text{ N}$$

Since beam is symmetrical, maximum shear force (V_{Sd}) = $R_A = R_B = \dfrac{W}{2} = 40\,500$ N

Diameter and spacing of links

Although $V_{Sd} < V_{Rd1}$, member is > 2 m long and cannot therefore be considered to be of minor structural importance (clause 4.3.2.1(2)). Hence provide minimum shear reinforcement, i.e.

$$\frac{A_{sw}}{s} = 0.45 \qquad \frac{V_{Sd}}{V_{Rd2}} = \frac{40\,500}{590\,760} = 0.07$$

Hence from *Table 8.9*, maximum spacing of links, s_{max}, is given by

$$s_{max} = 0.8d = 0.8 \times 547 = 437 \text{ mm} \not> 300 \text{ mm}$$

Therefore, provide 30-R8 links at 200 mm centres, $A_{sw}/s = 0.503$, for whole length of beam (*Table 3.10*).

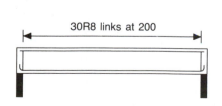

30R8 links at 200

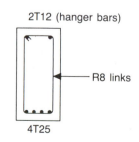

2T12 (hanger bars)

R8 links

4T25

$q_k = 10$ kN m^{-1}

Shear force (V_{Sd})

Total ultimate load, W, is

$$W = (1.35g_k + 1.5q_k) \text{ span} = (1.35 \times 10 + 1.5 \times 10)6 = 171 \text{ kN}$$

Since beam is symmetrical, maximum shear force, V_{Sd}, is

$$V_{Sd} = R_A = R_B = \frac{W}{2} = 85\,500 \text{ N}$$

Diameter and spacing of links

Since $V_{Rd1} < V_{Sd} < V_{Rd2}$ provide shear reinforcement according to

$$V_{Rd3} = V_{cd} + V_{wd}$$

where $V_{cd} = V_{Rd1} = 75\,478$ N, $V_{Rd3} = V_{Sd} = 85\,500$ N and

$$V_{wd} = \frac{A_{sw}}{s} 0.9f_{ywd}d = \frac{A_{sw}}{s} 0.9 \times \frac{250}{1.15} \times 547 = 107\,021\frac{A_{sw}}{s}$$

Hence

$$85\,500 = 75\,478 + 107\,021\frac{A_{sw}}{s}$$

$$\frac{A_{sw}}{s} = \frac{85\,500-75\,478}{107\,021} = 0.094 \not< \text{min req} (= 0.45)$$

$$\frac{V_{Sd}}{V_{Rd2}} = \frac{85\,500}{590\,760} = 0.14$$

Hence from *Table 8.9*, maximum spacing of links, s_{max}, is given by

$$s_{max} = 0.8d = 0.8 \times 547 = 437 \text{ mm} \not> 300 \text{ mm}$$

Therefore, provide 30-R8 links at 200 mm centres, $A_{sw}/s = 0.503$, for whole length of beam as for case (i), $q_k = 0$, above.

$q_k = 27$ kN m^{-1}

Shear force (V_{Sd})

Total ultimate load, W, is

$$W = (1.35g_k + 1.5q_k) \text{ span} = (1.35 \times 10 + 1.5 \times 27)6 = 324 \text{ kN}$$

Since beam is symmetrical, maximum shear force, V_{Sd}, is

$$V_{Sd} = R_A = R_B = \frac{W}{2} = 162\,000 \text{ N}$$

Diameter and spacing of links

Minimum shear reinforcement required where shear force in beam, V_c, is

$$V_c \leq V_{cd} + V_{wd}$$

where $V_{cd} = V_{Rd1} = 75\,478$ N and

$$V_{wd} = \frac{A_{sw}}{s} 0.9f_{ywd}d = \frac{0.45 \times 0.9 \times 250 \times 547}{1.15} = 48\,160 \text{ N}$$

Hence

$$V_c = 75\,478 + 48\,160 = 123\,638 \text{ N}$$

$$V_c/V_{Rd2} = 123\ 638/590\ 760$$
$$= 0.21 \Rightarrow s_{max} \not> 300 \text{ mm} \quad (Table\ 8.9)$$

Hence from *Table 3.12*, provide R8 links at 200 mm centres ($A_{sw}/s = 0.503$), where $V \leqslant 123\ 638$ N, i.e. 2290 mm either side of the mid-span of beam.

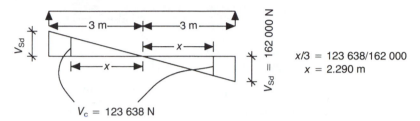

Where shear force in beam $> V_c$, provide shear reinforcement according to

$$V_{Sd} = V_{cd} + V_{wd}$$
$$162\ 000 = 75\ 478 + 107\ 021 A_{sw}/s$$

$$\frac{A_{sw}}{s} = \frac{162\ 000 - 75\ 478}{107\ 021} = 0.81$$

$$\frac{V_{Sd}}{V_{Rd2}} = \frac{162\ 000}{590\ 760} = 0.27$$

$$\Rightarrow s_{max} = 0.6d = 0.6 \times 547 = 328 \text{ mm} \not> 300 \text{ mm}$$

Hence from *Table 3.10*, provide R8 links at 100 mm centres ($A_{sw}/s = 1.006$) where shear force in beam exceeds 123 638 N, i.e. 710 mm in from both supports.

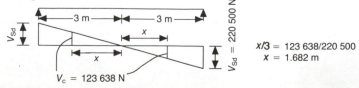

$q_k = 40 \text{ kN m}^{-1}$

Shear force (V_{Sd})
Total ultimate load, W, is

$$W = (1.35g_k + 1.5q_k) \text{ span} = (1.35 \times 10 + 1.5 \times 40)\ 6 = 441 \text{ kN}$$

Since beam is symmetrical, maximum shear force V_{Sd}, is

$$V_{Sd} = R_A = R_B = W/2 = 220\ 500 \text{ N}$$

Diameter and spacing of links
As above, minimum shear reinforcement required where shear force in beam, $V_c \leqslant 123\ 638$ N. From *Table 3.12*, provide R10 links at 300 mm centres ($A_{sw}/s = 0.523$) where $V \leqslant 123\ 638$ N, i.e. 1682 mm either side of the mid-span of beam.

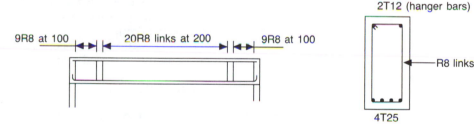

Where shear force in beam > V_c, provide shear reinforcement according to

$$V_{Sd} = V_{cd} + V_{wd}$$

$$220\ 500 = 75\ 478 + 107\ 021\ \frac{A_{sw}}{s}$$

$$\frac{A_{sw}}{s} = \frac{220\ 500 - 75\ 478}{107\ 021} = 1.36$$

$$\frac{V_{Sd}}{V_{Rd2}} = \frac{220\ 500}{590\ 760} = 0.37$$

$$\Rightarrow s_{max} = 0.6d = 0.6 \times 547 = 328\ \text{mm} \not> 300\ \text{mm}$$

Hence from *Table 3.10*, provide R10 links at 100 mm centres ($A_{sw}/s = 1.57$) where shear force in beam exceeds 123 638 N, i.e. 1318 mm in from both supports.

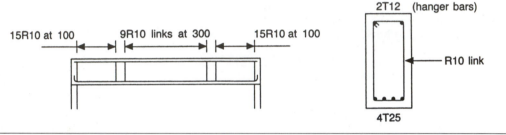

8.8.3 DEFLECTION (CLAUSE 4.4.3, EC2)

The approach used to check deflection in EC2 is basically the same as in BS 8110 but the presentation has been simplified considerably.

As in BS 8110, EC2 does not require the actual deflections to be calculated explicitly, but recommends that the span/depth ratios shown in Table 8.10 should not be exceeded. Compliance with these values for the span/depth ratios will normally ensure

that the deflection limits in clause 4.4.3.1 of EC2 are satisfied, namely:

1. The calculated sag of a beam, slab or cantilever subjected to the quasi-permanent loads should not exceed span/250.
2. The deflection occurring after construction of the element should not exceed span/500.

The values in Table 8.10 assume the steel stress at

Table 8.10 Basic ratios of span/effective depth for reinforced concrete members without axial compression (Table 4.14 of EC2 as amended by Table 7 of NAD)

Structural system	Concrete highly stressed	Concrete lightly stressed	Concrete nominally reinforced[a]
1. Simply supported beam, one or two-way spanning simply supported slab	18	25	34
2. End span of continuous beam or one-way continuous slab or two-way spanning slab continuous over one long side	23	32	44
3. Interior span of beam or one- or two-way spanning slab	25	35	48
4. Slab supported on columns without beams (flat slab) (based on longer span)	21	30	41
5. Cantilever	7	10	14

[a] Apply to all stress levels.

the critical section, σ_s, is 250 N mm^{-2}, corresponding roughly to $f_{yk} = 400$ N mm^{-2}. Where other stress levels are used, the values in the table can be multiplied by $250/\sigma_s$. However, the final span/depth ratio should not exceed the basic span/depth ratios for nominally reinforced concrete members given in the table. It will normally be conservative to assume that

$$\sigma_s = \frac{250 f_{yk}\, A_{s,req}}{400\, A_{s,prov}} \qquad (8.23)$$

where $A_{s,req}$ is the area of steel required and $A_{s,prov}$ the area of steel provided. Furthermore it should be noted that in *Table 8.10* the actual span/depth ratios for each member type are related to the reinforcement ratio ρ, where $\rho = A_s/bd$. Nominally reinforced values correspond to $\rho = 0.15\%$; lightly stressed members correspond to $\rho = 0.5\%$; highly stressed members correspond to $\rho = 1.5\%$. Values between these cases may be obtained by interpolation.

8.8.4 REINFORCEMENT DETAILS FOR BEAMS

This section outlines EC2 requirements regarding the detailing of beams with respect to:

1. reinforcement percentages
2. spacing of reinforcement
3. anchorage lengths
4. curtailment of reinforcement
5. lap lengths.

8.8.4.1 Reinforcement percentages (clause 5.4.2.1.1, EC2)

The cross-sectional area of the longitudinal tensile reinforcement, A_{s1}, in beams should be not less than the following:

$$A_{s1} \geqslant \frac{0.6bd}{f_{yk}} \not< 0.0015bd$$

where b is the breadth of section, d the effective depth and f_{yk} the characteristic yield stress of reinforcement (N mm^{-2}). The area of the tension, A_{s1}, and of the compression reinforcement, A_{s2}, should not be greater than the following other than at laps:

$$A_{s1},\, A_{s2} \leqslant 0.04A_c$$

where A_c is the cross-sectional area of concrete.

8.8.4.2 Spacing of reinforcement (clauses 5.2.1.1 and 4.4.2.3, EC2)

The clear horizontal or vertical distance between reinforcing bars should not be less than the following:

1. maximum bar diameter

Table 8.11 Maximum bar spacing for high bond bars (Table 4.12, EC2)

Steel stress (MPa)	Maximum bar spacing (mm)		
	Pure flexure	*Pure tension*	*Prestressed sections (bending)*
160	300	200	200
200	250	150	150
240	200	125	100
280	150	75	50
320	100	–	–
360	50	–	–

2. 20 mm
3. $d_g + 5$ mm, if d_g, the maximum aggregate size, exceeds 32 mm.

The maximum spacing between bars is based on the need to ensure that the maximum crack width does not exceed 0.3 mm which may generally be achieved by limiting the bar spacings to the values shown in *Table 8.11*. (clause 4.4.2.3 of EC2 also states that cracks greater than 0.3 mm can be avoided by limiting bar sizes, but this aspect is not discussed here.) The steel stresses in the table can conservatively be estimated using equation 8.23.

8.8.4.3 Anchorage (clause 5.2.2.3, EC2)

EC2 distinguishes between basic and required anchorage lengths. The basic anchorage length, l_b, is the length of bar required to resist the maximum force in the reinforcement, $A_s f_{yd}$, assuming constant bond stress equal to f_{bd} and is given by

$$l_b = (\phi/4)(f_{yd}/f_{bd}) \qquad (8.24)$$

where ϕ is the diameter of bar to be anchored, f_{yd} the design strength of reinforcement and f_{bd} the ultimate bond stress (*Table 8.12*). The ultimate bond stress depends upon the quality of the bond between the concrete and steel. EC2 states that bond conditions are considered to be good for:

1. all bars in members with an overall depth (h) of less than or equal to 250 mm (*Fig. 8.5(a)*);
2. all bars in the lower half of members with an overall depth between 250 mm and 600 mm (*Fig. 8.5(b)*);
3. all bars located at a depth greater than or equal to 300 mm in members with an overall depth of 600 mm or greater (*Fig. 8.5(c)*).

All other conditions will give rise to poor bond conditions.

Table 8.12 Ultimate bond stress, f_{bd} (N mm^{-2}), assuming good bond conditions (Table 5.3, EC2)

f_{ck}	12	16	20	25	30	35	40	45	50
Plain bars	0.9	1.0	1.1	1.2	1.3	1.4	1.5	1.6	1.7
High bond bars where $\phi \leqslant 32$ mm or welded mesh fabrics made of ribbed wires	1.6	2.0	2.3	2.7	3.0	3.4	3.7	4.0	4.3

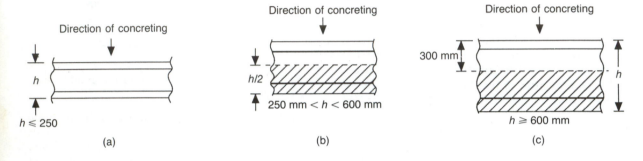

Fig. 8.5 *Definition of bond conditions (Fig. 5.1, EC2).*

In conditions of good bond, the design value for the ultimate bond stress, f_{bd}, is given in *Table 8.12*. In all other cases, the values in *Table 8.12* should be multiplied by a coefficient 0.7.

The required anchorage length, $l_{b,net}$, is used to determine the curtailment length of bars in beams as discussed below and is given by

$$l_{b,net} = \alpha_a \, l_b \, \frac{A_{s,req}}{A_{s,prov}} \not< l_{b,min} \qquad (8.25)$$

where $\alpha_a = 1$ for straight bars and 0.7 for curved bars in tension if the cover perpendicular to the plane of curvature is at least 3ϕ. Also l_b is the basic anchorage length, $A_{s,req}$ the area of reinforcement required by design, $A_{s,prov}$ the area of reinforcement actually provided and $l_{b,min}$ the minimum anchorage length given by

$$l_{b,min} = 0.3 l_b \not< 10\phi \text{ or } \not< 100 \text{ mm} \qquad (8.26)$$

(for anchorages in tension)

$$l_{b,min} = 0.6 l_b \not< 10\phi \text{ or } \not< 100 \text{ mm} \qquad (8.27)$$

(for anchorages in compression)

where ϕ is the diameter of bar to be anchored.

Table 8.13 shows the anchorage lengths for straight and curved bars as multiples of bar size for plain ($f_{yk} = 250$ N mm^{-2}) and high yield bars ($f_{yk} =$ 460 N mm^{-2}) embedded in a range of concrete strengths. Note that the values in the table apply to good bond conditions and to bar sizes less than or equal to 32 mm. Where the bond conditions are poor the values in the table should be divided by 0.7 and where the bar diameter exceeds 32 mm the values should be divided by $[(132 - \phi)/100]$. The minimum radii to which reinforcement may be bent is shown in *Table 8.14* while the anchorage lengths for straights, hooks, bends and loops are shown in *Fig. 8.6*.

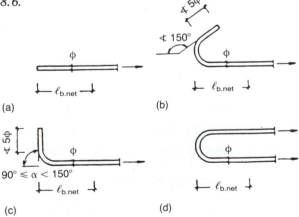

Fig. 8.6 *Required anchorage length (Fig. 5.2, EC2): (a) straight anchor; (b) hook; (c) bend; (d) loop.*

Example 8.4 Calculation of anchorage lengths

Calculate the anchorage lengths for straight and curved bars in tension as multiples of bar size assuming:

1. The bars are high yield of diameter ≤ 32 mm.
2. Concrete strength class is C20/25.
3. Bond conditions are good.

STRAIGHT BARS

High yield bars, f_{yk}	$= 460 \text{ N mm}^{-2}$
Concrete strength, f_{ck}	$= 20 \text{ N mm}^{-2}$
Ultimate bond stress, f_{bd}	$= 2.3$ (*Table 8.12*)
Design strength of bars, f_{yd}	$= f_{yk}/\gamma_s = 460/1.15$
Coefficient α	$= 1$ (for straight bar)

The basic anchorage length, l_b, is

$$l_b = (\phi/4) \, (f_{yd}/f_{bd})$$

$$= (\phi/4) \, (460/1.15) \div 2.3 \approx 44 \, \phi$$

Hence the anchorage length, $l_{b,net}$, is

$$l_{b,net} = \alpha_a l_b = 1 \times 44\phi = 44\phi \quad (\textit{Table 8.13})$$

CURVED BARS

The calculation is essentially the same for this case except that $\alpha_a = 0.7$ for curved bars and therefore, $l_{b,net}$, is

$$l_{b,net} = \alpha_a l_b = 0.7 \times 44\phi \approx 31\phi$$

Table 8.13 Anchorage lengths, $l_{b,net}$, as multiples of bar size

Steel grade		Concrete class				
		C20/25	C25/30	C30/37	C35/45	C40/45
Plain $f_{yk} = 250 \text{ N mm}^{-2}$	Straight bars, compression tension	50	46	42	39	37
	Curved bars, tension[a]	35	32	30	28	26
Deformed bars type 2 $f_{yk} = 460 \text{ N mm}^{-2}$	Straight bars, compression tension	44	37	34	30	27
	Curved bars, tension[a]	31	26	24	21	19

[a] In the anchorage region, cover perpendicular to the plane of curvature should be at least 3ϕ.

Table 8.14 Minimum diameter of hooks, bends and loops (based on Table 5.1 of EC2 as amended by Table 8 of NAD)

	Bar diameter	
	$\phi < 20$ mm	$\phi \geq 20$ mm
Plain bars S 250	4ϕ	4ϕ
High bond bars S460	6ϕ	8ϕ

8.8.4.4 Curtailment of bars (clause 5.4.2.1.3, EC2)

The curtailment length of bars in beams is obtained from *Fig. 8.7*. The theoretical cut-off point is based on the design bending moment curve which has been horizontally displaced in the direction of decreasing moment by an amount a_l. The physical (actual) cut-off point occurs an anchorage length, $l_{b,net}$, beyond the 'theoretical' cut-off point.

If the shear resistance is calculated according to the 'standard method' a_l is given by

$$a_l = \frac{z}{2}(1 - \cot \alpha) \not< 0 \qquad (8.28)$$

where α is the angle of the shear reinforcement with the longitudinal axis, and z can normally be taken as $0.9d$. With vertical links, for instance, a_l will be equal to

$$a_l = \frac{0.9d}{2}(1 - \cot 90°) = 0.45d$$

The *Concise Eurocode for the Design of Concrete Buildings* recommends the following simplified rules for the curtailment of reinforcement in beams subjected to predominantly uniformly distributed loads. These are based on the present rules in BS 8110.

Near internal supports in continuous beams of approximately equal spans where $Q_k < G_k$. For curtailment of top reinforcement, at least 25% of the reinforcement required at the supports for the ultimate limit state should be effectively continuous through the spans. All of the reinforcement needed at the supports should extend into the span for a distance from the face of the support of $0.1l + l_{b,net} + 0.45d$. At least 50% of the reinforcement at the support should extend into the span for a distance from the face of the support of $0.25l + l_{b,net} + 0.45d$ (*Fig. 8.8*).

For curtailment of bottom reinforcement, at least 30% of the reinforcement required at mid-span

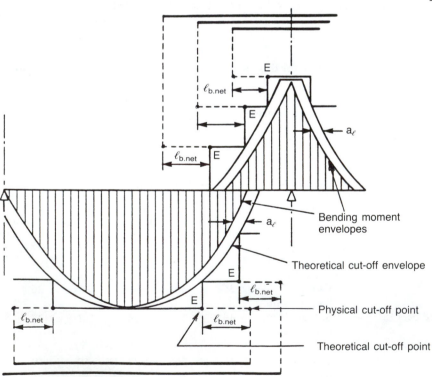

Fig. 8.7 Envelope line for curtailment of reinforcement in flexural members (Fig. 5.11, EC2).

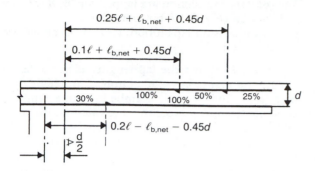

ℓ – effective span

Fig. 8.8 *Curtailment of reinforcement at internal supports of continuous beams.*

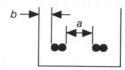

Fig. 8.10 *Evaluation of α_1 (Fig. 5.6, EC2).*

should extend to the support. The remainder should extend to within $0.2l - l_{b,net} - 0.45d$ of the centre line of the support.

Bottom reinforcement near end supports. At least 50% of the reinforcement provided at mid-span should be taken into the support and be anchored as shown in *Fig. 8.9*. The remaining reinforcement should extend to within $0.15l - l_{b,net} - 0.45d$ of the centre line of the support.

Table 8.15   Lap lengths as multiples of bar size

Steel grade		Concrete class				
		C20/25	C25/30	C30/37	C35/45	C40/45
Plain $f_{yk} = 250$ N mm^{-2}	Laps – compression and tension[a]	50	46	42	39	37
	Laps – tension [b]	70	64	60	56	52
	Laps – tension [c]	100	92	84	78	74
Deformed bars type 2 $f_{yk} = 460$ N mm^{-2}	Laps – compression and tension [a]	44	37	34	30	27
	Laps – tension[b]	62	52	48	42	38
	Laps – tension[c]	88	74	68	60	54

[a] $\alpha_1 = 1.$
[b] $\alpha_1 = 1.4.$
[c] $\alpha_1 = 2.$

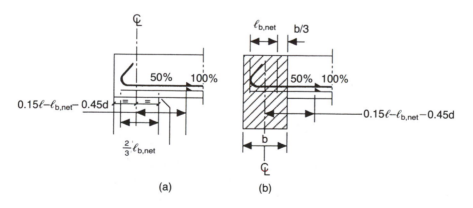

(a)　　　　(b)

Fig. 8.9 *Curtailments and anchorages of bottom reinforcement at end supports: (a) direct support; (b) indirect support (based on Fig. 5.12, EC2).*

8.8.4.5 Lap lengths (clause 5.2.4.13, EC2)

Lap length, l_s, is given by

$$l_s = l_{b,net}\alpha_1 \not< l_{s,min} \qquad (8.29)$$

where $l_{b,net}$ is the anchorage length, $l_{s,min}$ the minimum lap length which should be not less than 15ϕ or 200 mm and α_1 a coefficient which takes the following values:

$\alpha_1 = 1$ for compression laps

$\alpha_1 = 1$ for tension laps where less than 30% of the bars in the section are lapped and where $a \geq 6\phi$ and $b \geq 2\phi$ (Fig. 8.10)

$\alpha_1 = 1.4$ for tension laps where either (i) 30% or more of bars at a section are lapped, or (ii) $a < 6\phi$ or $b < 2\phi$ (Fig. 8.10) but not both

$\alpha_1 = 2$ for tension laps if both (i) and (ii) above are satisfied

Table 8.15 shows the lap lengths as multiples of bar size for plain ($f_{yk} = 250$ N mm^{-2}) and high yield bars ($f_{yk} = 460$ N mm^{-2}) embedded in a range of concrete strengths. Note that the values in the table apply to good bond conditions and to bar sizes ≤ 32 mm. Where the bond conditions are poor the values in the table should be divided by 0.7 and where the bar diameter exceeds 32 mm the values should be divided by $[(132 - \phi)/100]$.

Example 8.5 Design of a simply supported beam

Design the main steel and shear reinforcement for the beam shown below assuming the following material strengths: $f_{ck} = 25$ N mm^{-2}, $f_{yk} = 460$ N mm^{-2} and $f_{ywd} = 250$ N mm^{-2}. The environmental conditions fall within exposure class 1.

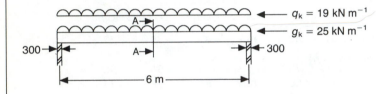

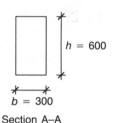

Section A–A

DESIGN MOMENT (M)

Loading

Permanent Self-weight of beam $= 0.6 \times 0.3 \times 24 = 4.32$ kN m^{-1}
Total permanent load (g_k) $= 25 + 4.32 = 29.32$ kN m^{-1}

Variable Total variable load (q_k) $= 19$ kN m^{-1}

Ultimate load Total ultimate load (W) $= (1.35g_k + 1.5q_k)$ span
$$= (1.35 \times 29.32 + 1.5 \times 19)\ 6$$
$$= 408.5 \text{ kN}$$

Design moment

Maximum design moment (M) $= \dfrac{Wl}{8} = \dfrac{408.5 \times 6}{8} = 306.4$ kN m

ULTIMATE MOMENT OF RESISTANCE (M_u)

Effective depth

Assume diameter of main bar (ϕ) $= 25$ mm
Assume diameter of links (ϕ') $= 10$ mm
Cover for exposure class 1 (c) $= 20$ mm

Effective depth, d, is

$$d = h - \phi/2 - \phi' - c = 600 - 12.5 - 10 - 20 = 557 \text{ mm}$$

Ultimate moment

Ultimate moment of resistance, M_u is

$$M_u = 0.167 f_{ck} b d^2 = 0.167 \times 25 \times 300 \times 557^2 \times 10^{-6} = 388.6 \text{ kN m}$$

Since $M_u > M$ design as a singly reinforced beam.

MAIN STEEL (A_{s1})

$$K_0 = \frac{M}{f_{ck} b d^2} = \frac{306.4 \times 10^6}{25 \times 300 \times 557^2} = 0.132$$

$$z = d[0.5 + \sqrt{(0.5 - 3K_0/3.4)}]$$

$$= 557[0.5 + \sqrt{(0.25 - 3 \times 0.132/3.4)}]$$

$$= 482.3 \text{ mm}$$

$$A_{s1} = \frac{M}{0.87 f_{yk} z} = \frac{306.4 \times 10^6}{0.87 \times 460 \times 482.3} = 1588 \text{ mm}^2$$

Therefore, from *Table 3.7*, provide 4T25 ($A_{s1} = 1960 \text{ mm}^2$).

SHEAR REINFORCEMENT

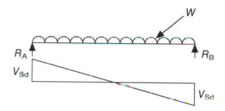

Ultimate load (W) = 408.5 kN

Shear force (V_{Sd})

$$V_{Sd} = \frac{W}{2} = \frac{408\,500}{2} = 204\,250 \text{ N}$$

Shear resistance of concrete alone, V_{Rd1}

$$f_{ck} = 25 \text{ N mm}^{-2} \quad \tau_{Rd} = 0.30 \text{ N mm}^{-2} \quad (\textit{Table 8.7})$$
$$k = 1.6 - d = 1.6 - 0.557 = 1.043$$

$$\rho_1 = \frac{A_{s1}}{b_w d} = \frac{1960}{300 \times 557} = 0.01173$$

$$\sigma_{cp} = 0$$
$$V_{Rd1} = [\tau_{Rd}\, k(1.2 + 40\rho_1) + 0.15\sigma_{cp}]\, b_w d$$
$$V_{Rd1} = [0.30 \times 1.043(1.2 + 40 \times 0.011\,73)]\, 300 \times 557$$
$$= 87\,275 \text{ N}$$

Maximum shear resistance of concrete (V_{Rd2})

$$\nu = 0.7 - \frac{f_{ck}}{200} = 0.7 - \frac{25}{200} = 0.575$$
$$V_{Rd2} = \tfrac{1}{2}\nu f_{cd} b_w 0.9d$$

$$= \frac{1}{2} \times 0.575 \times \frac{25}{1.5} \times 300 \times 0.9 \times 557 = 720\ 619 \text{ N} > V_{\text{Sd}}$$

Minimum shear reinforcement

$\rho_w = 0.0022$ by interpolation between the values in *Table 8.8*, hence

$$\rho_w = A_{\text{sw}}/sb_w \sin \alpha$$
$$0.0022 = A_{\text{sw}}/s \times 300 \times 1$$
$$\frac{A_{\text{sw}}}{s} = 0.66$$

Diameter and spacing of links

Minimum shear reinforcement required where shear force in beam, V_c, is

$$V_c \leqslant V_{\text{cd}} + V_{\text{wd}}$$

where $V_{\text{cd}} = V_{\text{Rd1}} = 872\ 75$ N and

$$V_{\text{wd}} = \frac{A_{\text{sw}}}{s}\ 0.9 f_{\text{ywd}} d = \frac{0.66 \times 0.9 \times 250 \times 557}{1.15}$$
$$= 0.66 \times 108\ 978 = 71\ 925 \text{ N}$$

Hence

$$V_c = 87\ 275 + 71\ 925 = 159\ 200 \text{ N}$$
$$V_c/V_{\text{Rd2}} = 159\ 200/720\ 619$$
$$= 0.22 \Rightarrow s_{\text{max}} \not> 300 \text{ mm} \quad (\textit{Table 8.9})$$

Hence, from *Table 3.10*, provide R10 links at 200 mm centres ($A_{\text{sw}}/s = 0.785$), where $V \leqslant 159\ 200$ N, i.e. 2338 mm either side of the mid-span of beam.

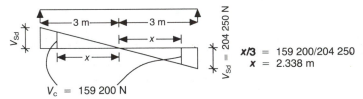

Where shear force in beam $> V_c$, provide shear reinforcement according to

$$V_{\text{Sd}} = V_{\text{cd}} + V_{\text{wd}}$$
$$204\ 250 = 87\ 275 + 108\ 978 A_{\text{sw}}/s$$
$$\frac{A_{\text{sw}}}{s} = \frac{204\ 250 - 87\ 275}{108\ 978} = 1.07$$
$$\frac{V_{\text{Sd}}}{V_{\text{Rd2}}} = \frac{204\ 250}{720\ 619} = 0.28$$
$$\Rightarrow s_{\text{max}} = 0.6 d = 0.6 \times 557 = 334 \text{ mm} \not> 300 \text{ mm}$$

Hence from *Table 3.10*, provide R10 links at 125 mm centres ($A_{\text{sw}}/s = 1.256$) where shear force in beam exceeds 159 200 N, i.e. 662 mm in from both supports.

DEFLECTION

Steel ratio, $\rho = A_s/bd = 1960/557 \times 300 = 0.01\ 173 = 1.173\%$. From *Table 8.10*, basic span/depth ratios for simply supported beams corresponding to $\rho = 0.5$ and 1.5% are 25 and 18 respectively. Interpolating between these values gives a span/depth ratio of 20.3 for $\rho = 1.173\%$.

$$\text{Design service stress, } \sigma_s = \frac{5 f_{\text{yk}} A_{\text{req}}}{8 A_{\text{prov}}} = \frac{5 \times 460 \times 1588}{8 \times 1960}$$
$$= 233 \text{ N mm}^{-2}$$

$$\text{Modification factor} = \frac{250}{\sigma_s} = \frac{250}{233} = 1.07$$

Hence modified span/depth ratio corresponding to $\rho = 1.173\%$ is $20.3 \times 1.07 = 21.7 >$ actual span/depth ratio ($= 6000/557 = 10.7$).

REINFORCEMENT DETAILS

The sketches below show the main reinforcement requirements for the beam. For reasons of buildability the actual reinforcement details may well be slightly different.

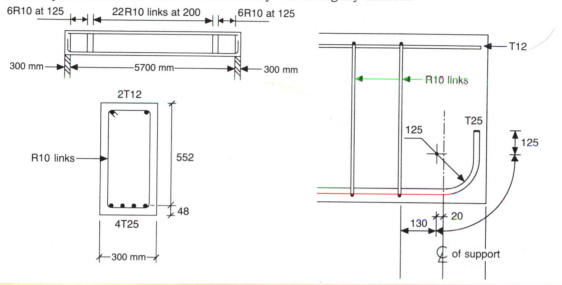

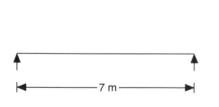

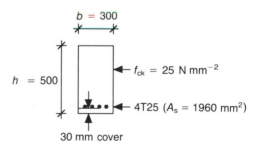

Example 8.6 Analysis of a singly reinforced beam

Calculate the maximum variable load that the beam shown below can carry assuming that the load is (i) uniformly distributed and (ii) occurs as a point load at mid-span.

MOMENT CAPACITY OF SECTION

Effective depth, d, is

$$d = h - \text{cover} - \phi/2$$
$$= 500 - 30 - 25/2 = 457 \text{ mm}$$
$$K_o = \frac{M}{f_{ck}bd^2} = \frac{M}{25 \times 300 \times 457^2}$$
$$z = d[0.5 + \sqrt{(0.25 - 3K_o/3.4)}] \tag{1}$$
$$= 457[0.5 + \sqrt{(0.25 - 3M/3.4 \times 25 \times 300 \times 457^2)}]$$

$$A_{s1} = 1960 \text{ mm}^2 = \frac{M}{0.87 f_{yk} z} = \frac{M}{0.87 \times 460 z} \qquad (2)$$

Solving equations (1) and (2) simultaneously gives

$$M = 286.1 \text{ kN m} \qquad z = 364.7 \text{ mm}$$

Maximum uniformly distributed load, q_k

Permanent load, g_k $\quad = 0.5 \times 0.3 \times 24 = 3.6 \text{ kN m}^{-1}$

Total ultimate load (W) $\quad = (1.35 g_k + 1.5 q_k) \text{ span}$

$\qquad\qquad\qquad\qquad\quad = (1.35 \times 3.6 + 1.5 q_k) \, 7$

Substituting into

$$M = \frac{Wl}{8}$$

$$286.1 = \frac{(1.35 \times 3.6 + 1.5 q_k) \, 7^2}{8}$$

Hence

$$q_k = 27.9 \text{ kN m}^{-1}$$

Maximum point load, Q_k

Factored permanent load $(W_D) = (1.35 g_k) \text{ span}$

$\qquad\qquad\qquad\qquad\qquad\quad = 1.35 \times 3.6 \times 7 = 34.02 \text{ kN}$

Factored variable load (W_I) $\quad = 1.5 Q_k$

Substituting into

$$M = \frac{W_D l}{8} + \frac{W_I l}{4}$$

$$286.1 = \frac{34.02 \times 7}{8} + \frac{1.5 Q_k 7}{4}$$

Hence

$$Q_k = 97.6 \text{ kN}$$

8.9 Design of one-way spanning solid slabs

8.9.1 DEPTH, BENDING, SHEAR

EC2 requires a slightly different approach to BS 8110 for designing one-way spanning solid slabs. To calculate the depth of the slab, the designer must first estimate the percentage of steel required in the slab for bending. Generally, most slabs will be lightly reinforced, i.e. $\rho < 0.5\%$. The percentage of reinforcement can be used, together with the support conditions, to select an appropriate span/depth ratio from *Table 8.10*. The effective depth of the slab can then be determined by multiplying the span/depth ratio by the span of the slab (*Example 8.7*). The actual area of steel required in the slab for bending can be calculated using the equations developed in *section 8.8.1*. Provided this agrees with the assumed value, the calculated depth of the slab is acceptable.

The designer will also need to check that the slab will not fail in shear. The shear resistance of the slab can be calculated as for beams (*section*

8.8.2). According to clause 4.3.2.1(2), where the design shear force (V_{Sd}) is less than the design shear resistance of the concrete alone (V_{Rd1}) no shear reinforcement need be provided. The same requirement appears in BS 8110. Where $V_{Sd} > V_{Rd1}$, shear reinforcement should be provided such that $V_{Sd} \leqslant V_{Rd3}$ where V_{Rd3} is the design shear force due to the concrete and shear reinforcement.

The actual area of shear reinforcement in slabs can be calculated as for beams and should not be less than the values in *Table 8.8* for beams (see clause 5.4.3.3(2) as amended by NAD). The maximum longitudinal spacing of successive series of links can be determined using the equations in *Table 8.9*. However, the limits in millimetres should be ignored and the maximum spacing of links should not exceed $0.75d$. Clause 5.4.3.3 also notes that shear reinforcement should not be provided in slabs less than 200 mm deep.

8.9.2 REINFORCEMENT DETAILS FOR SOLID SLABS

This section outlines EC2 requirements regarding the detailing of slabs with respect to:

1. reinforcement percentages
2. spacing of reinforcement
3. anchorage and curtailment of reinforcement
4. crack control.

8.9.2.1 Reinforcement areas (clause 5.4.2.1.1, EC2)

In EC2, the maximum and minimum percentages of longitudinal steel permitted in beams and slabs are the same, namely

$$\frac{0.6\,bd}{f_{yk}} \not< 0.0015\,bd \leqslant A_s \leqslant 0.04A_c \quad \text{(other than at laps)}$$

where b is the breadth of section, d the effective depth, A_c the cross-sectional area of concrete (bh) and f_{yk} the characteristic yield stress of reinforcement. The area of transverse or secondary reinforcement should not generally be less than 20% of the principal reinforcement, i.e.

$$A_s(\text{trans}) \not< 0.2.A_s(\text{main})$$

8.9.2.2 Spacing of reinforcement

The clear distance between reinforcing bars should not be less than the following:

1. maximum bar diameter
2. 20 mm

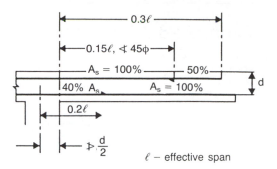

Fig. 8.11 *Curtailment of reinforcement in continuous slabs.*

3. $d_g + 5$ mm if d_g (maximum aggregate size) > 32 mm.

The maximum bar spacing in slabs, s_{max}, for the main and secondary reinforcement should not exceed the following:

$$s_{max} < 3h \not> 500 \text{ mm}$$

where h denotes the overall depth of the slab.

8.9.2.3 Anchorage and curtailment

For detailing the main reinforcement in slabs, the same provision outlined earlier for beams will apply with $a_l = d$. Clause 5.4.3.2.2 of EC2 requires that in slabs near end supports, half the calculated span reinforcement should continue up to the support and be anchored in accordance with *Fig. 8.9*.

For continuous slabs, the *Concise Eurocode for the Design of Concrete Buildings* recommends that the simplified curtailment rules for continuous slabs, given in BS 8110, may still be suitable (Fig. 8.11).

8.9.2.4 Crack widths (clause 4.4.2.3, EC2)

According to EC2, where the overall depth of the slab does not exceed 200 mm and the code provisions with regard to reinforcement areas, spacing of reinforcement, anchorage and curtailment of bars, etc. discussed above have been applied, no further measures specifically to control cracking are necessary. For slab depths greater than 200 mm, where at least the minimum reinforcement area has been provided, the limitation of crack width to less than 0.3 mm will generally be achieved by limiting bar spacing (or bar size) in accordance with the provisions in sizes *Table 8.11*.

Example 8.7 Design of a one-way spanning floor

Design the floor shown below for an imposed load of 3.5 kN m^{-2}, assuming the following material strengths: $f_{ck} = 30$ N mm^{-2}, $f_{yk} = 460$ N mm^{-2}. The environmental conditions fall within exposure class 1.

DETERMINE EFFECTIVE DEPTH OF SLAB AND AREA OF MAIN STEEL

Estimate overall depth of slab

Assume reinforcement ratio, ρ, is 0.35%. From *Table 8.10*, basic span/depth ratios for lightly and nominally reinforced, simply supported slabs are 25 and 34 respectively. Hence by interpolation, basic span/depth ratio for slab with $\rho = 0.35\%$ is approximately 28.8.

$$\text{Minimum effective depth } (d) = \frac{\text{span}}{\text{basic ratio}}$$

$$= \frac{4500}{28.8} \approx 156 \text{ mm}$$

Take $d = 160$ mm and assume diameter of main steel (Φ) = 10 mm and cover to reinforcement (c) = 20 mm. Overall depth of slab, h, is

$$h = d + \Phi/2 + c = 160 + 10/2 + 20 = 185 \text{ mm}$$

Loading

Permanent Self-weight of slab (g_k) = 0.185 × 24 (kN m^{-3}) = 4.44 kN m^{-2}

Variable Total variable load (q_k) = 3.5 kN m^{-2}

Ultimate load For 1 m width of slab, total ultimate load is

$$(1.35g_k + 1.5q_k) \text{ span} = (1.35 \times 4.44 + 1.5 \times 3.5) \, 4.5$$

$$= 50.6 \text{ kN}$$

Design moment

Maximum design moment $(M) = \dfrac{Wl}{8} = \dfrac{50.6 \times 4.5}{8} = 28.5$ kN m

Ultimate moment

Ultimate moment of resistance, M_u, is

$$M_u = 0.167 f_{ck} b d^2$$
$$= 0.167 \times 30 \times 1000 \times 160^2 \times 10^{-6} = 128 \text{ kN m}$$

Since $M_u > M$ no compression reinforcement is required.

Main reinforcement (A_{s1})

$$K_o = \frac{M}{f_{ck}bd^2} = \frac{28.5 \times 10^6}{30 \times 1000 \times 160^2} = 0.0371$$

$$\begin{aligned}
z &= d[0.5 + \sqrt{(0.25 - 3K_o/3.4)}] \\
&= 160[0.5 + \sqrt{(0.25 - 3 \times 0.0371/3.4)}] \\
&= 160 \times 0.966 = 154 \text{ mm}
\end{aligned}$$

$$A_{s1} = \frac{M}{0.87f_{yk}z} = \frac{28.5 \times 10^6}{0.87 \times 460 \times 154} = 462 \text{ mm}^2$$

Hence from *Table 3.17*, T10 at 150 mm centres ($A_{s1} = 523$ mm^2 m^{-1}) would be suitable.

$$\text{Maximum bar spacing} < 3h = 3 \times 185 = 555 \text{ mm} \not> 500 \text{ mm} \quad \text{OK}$$
$$\text{Minimum reinforcement area} = 0.0015bd = 0.0015 \times 10^3 \times 160$$
$$= 240 \text{ mm}^2 \text{ m}^{-1} \quad \text{OK}$$

The reinforcement ratio, ρ, in this case is

$$\rho = \frac{A_s}{bd} = \frac{523}{1000 \times 160} = 0.00327 = 0.327\%$$

Check estimated effective depth of slab

Design service stress, $\sigma_s = \dfrac{5f_{yk}A_{req}}{8A_{prov}} = \dfrac{5 \times 460 \times 462}{8 \times 523}$
$$= 254 \text{ N mm}^{-2}$$

Modification factor $\quad = \dfrac{250}{\sigma_s} = \dfrac{250}{254} = 0.984$

Hence modified span/depth ratios for $\rho = 0.5$ and 0.15% are 24.6 ($= 25 \times 0.984$) and 33.46 ($= 34 \times 0.984$) respectively. Interpolating between these values gives a span/depth ratio of 28.9 for $\rho = 0.327\%$. Hence minimum effective depth, d is

$$d = \frac{\text{span}}{\text{span / depth ratio}}$$

$$= \frac{4500}{28.9} = 156 \text{ mm} < \text{assumed} \quad (= 160 \text{ mm})$$

Therefore, use a slab with an effective depth = 160 mm, overall depth = 185 mm and main steel = T10 at 150 mm centres.

SECONDARY REINFORCEMENT
Provide T8 at 300 mm centres ($A_{s \text{ (trans)}} = 168$ mm^2 m^{-1}):
$$A_{s \text{ (trans)}} \not< 0.2A_{s \text{ (main)}} = 0.2 \times 523 = 105 \text{ mm}^2 \text{ m}^{-1} \quad \text{OK}$$
Maximum spacing $< 3h = 3 \times 185 = 555$ mm $\not> 500$ mm OK

SHEAR REINFORCEMENT

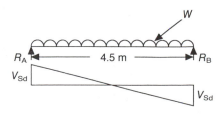

Ultimate load $(W) = 50.6$ kN

Shear force (V_{Sd})

$$V_{Sd} = \frac{W}{2} = \frac{50\ 600}{2} = 25\ 300 \text{ N}$$

Shear resistance of concrete alone, V_{Rd1}

$f_{ck} = 30$ N mm^{-2} $\tau_{Rd} = 0.34$ N mm^{-2}

$k = 1.6 - d = 1.6 - 0.160 = 1.440$

Assuming that 50% of midspan steel does not extend to the supports $A_{s1} = 260$ mm^2/m.

$$\rho_1 = \frac{A_{s1}}{b_w d} = \frac{260}{1000 \times 160} = 0.0016 \qquad \sigma_{cp} = 0$$

$$\begin{aligned}
V_{Rd1} &= [\tau_{Rd}\ k(1.2 + 40\rho_1) + 0.15\sigma_{cp}]\ b_w d \\
&= [0.34 \times 1.440(1.2 + 40 \times 0.0016)]1000 \times 160 \\
&= 99\ 017 \text{ N}
\end{aligned}$$

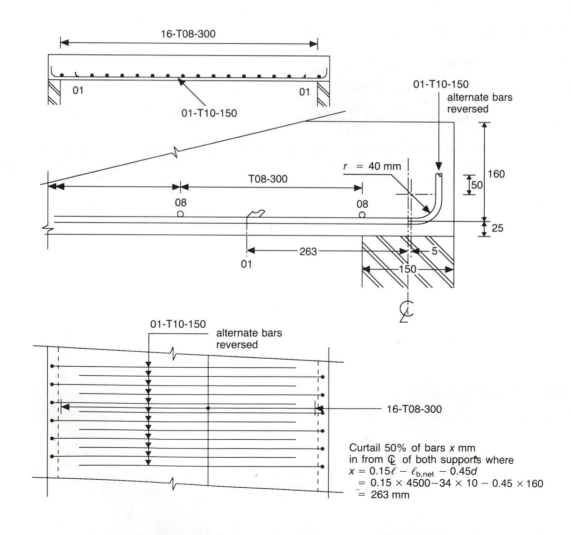

Curtail 50% of bars x mm
in from $\mathbb{C}$ of both supports where
$x = 0.15\ell - \ell_{b,net} - 0.45d$
$\quad = 0.15 \times 4500 - 34 \times 10 - 0.45 \times 160$
$\quad = 263$ mm

Since $V_{Rd1} > V_{Sd}$, no shear reinforcement is required.

REINFORCEMENT DETAILS
The sketches on page 258 show the main reinforcement requirements for the slab.

Reinforcement areas and bar spacing
Minimum reinforcement areas and bar spacing rules were checked above and found to be satisfactory.

Crack width
Since the overall depth of the slab does not exceed 200 mm and the rest of the code provisions have been met, no further measures specifically to control cracking are necessary.

Example 8.8 Analysis of one-way spanning floor
Calculate the maximum uniformly distributed variable load that the floor shown below can carry.

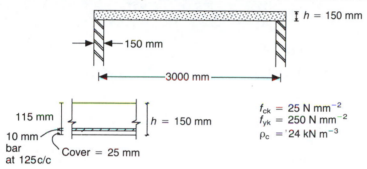

$h = 150$ mm
150 mm
3000 mm

115 mm
10 mm bar at 125 c/c
Cover = 25 mm
$h = 150$ mm

$f_{ck} = 25$ N mm^{-2}
$f_{yk} = 250$ N mm^{-2}
$\rho_c = 24$ kN m^{-3}

EFFECTIVE SPAN
From clause 2.5.2.2.2 of EC2, effective span of slab, l_{eff}, is

$$l_{eff} = l_n + a_1 + a_2 = 2850 + \frac{150}{3} + \frac{150}{3} = 2950 \text{ mm}$$

MOMENT CAPACITY
Effective depth of slab $(d) = h$ - cover - $\Phi/2$
$$= 150 - 25 - 10/2 = 120 \text{ mm}$$
Assume $z = 0.96d = 0.96 \times 120 = 115.2$ mm:

$$A_{s1} = 628 \text{ mm}^2 \text{ m}^{-1} = \frac{M}{0.87 f_{yk} z} = \frac{M}{0.87 \times 250 \times 115.2}$$

Hence moment capacity $M = 157\,350\,00$ N mm $= 15.735$ kN m.

LEVER ARM (z)
Check the assumed value of z:

$$K_o = \frac{M}{f_{ck} b d^2} = \frac{15.735 \times 10^6}{25 \times 10^3 \times 120^2} = 0.0437$$
$$z = d[0.5 + \sqrt{(0.25 - 3K_o/3.4)}]$$
$$= d[0.5 + \sqrt{(0.25 - 3 \times 0.0437/3.4)}]$$
$$= 0.96d \text{ (as assumed)}$$

MAXIMUM UNIFORMLY DISTRIBUTED VARIABLE LOAD (q_k)

Loading

Permanent

$$\text{Self-weight of slab } (g_k) = 0.15 \times 24 \text{ kN m}^{-3} = 3.6 \text{ kN m}^{-2}$$

Ultimate load

$$\text{Total ultimate load } (W) = (1.35g_k + 1.5q_k) \text{ span}$$

$$= (1.35 \times 3.6 + 1.5q_k) \, 2.95$$

Variable load
Maximum design moment M, is

$$M = 15.7 = \frac{Wl}{8} = (4.86 + 1.5q_k) \frac{2.95^2}{8}$$

$$q_k = 6.4 \text{ kN m}^{-2}$$

Hence the maximum uniformly distributed variable load the slab can support is 6.4 kN m^{-2}.

8.10 Design of pad foundations

The design of pad foundations was discussed in *Chapter 3* and found to be similar to that for slabs in respect of bending. However, the designer must also check that the pad will not fail due to face, transverse or punching shear. EC2 requirements in respect of these three modes of failure are discussed next.

8.10.1 FACE SHEAR
According to clause 6.4(d) of the NAD, the shear stress at the perimeter of the column should not exceed the following:

$$v_{Rd2} \ngtr 0.9\sqrt{f_{ck}} \tag{8.30}$$

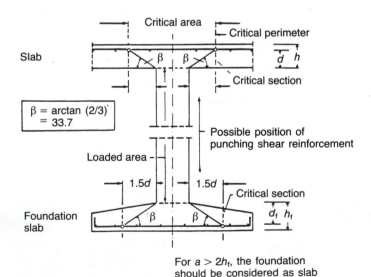

Fig. 8.12 *Design model for punching shear at ULS (Fig. 4.16, EC2).*

8.10.2 TRANSVERSE SHEAR (CLAUSE 4.3.4.5, EC2)

The critical section for transverse shear failure normally occurs a distance d from the face of the column. No shear reinforcement will be required provided $v_{Sd} < v_{Rd1}$, where v_{Sd} is the applied shear per unit length and is given by

$$v_{Sd} = \frac{\text{total design force } (Fig.\ 3.58)}{\text{length of critical section}} \quad (8.31)$$

and where v_{Rd1} is the shear resistance per unit length and is given by

$$v_{Rd1} = \tau_{Rd}k\,(1.2 + 40\rho_1)d \quad (8.32)$$

where τ_{Rd} is given in *Table 8.7*,

$$k = (1.6 - d) \geqslant 1.0 \quad (d \text{ in metres})$$
$$\rho_1 = \sqrt{(\rho_{1x}\rho_{1y})} \not> 0.015$$

where ρ_{1x} and ρ_{1y} relate to the tension steel in the x and y directions respectively.

8.10.3 PUNCHING SHEAR (CLAUSE 4.3.4, EC2)

No shear reinforcement is required if $v_{Sd} < v_{Rd1}$, where v_{Rd1} is the design shear resistance per unit length of the critical perimeter (see above) and v_{Sd} is the applied shear per unit length and is given by

$$v_{Sd} = \frac{V_{Sd}\beta}{u} \quad (8.33)$$

where V_{Sd} total design shear force. For a foundation this is calculated along the perimeter of the base of the truncated punching shear cone, assumed to form

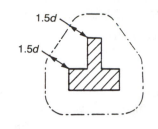
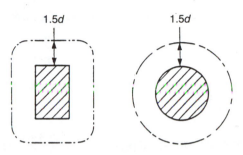

Fig. 8.13 *Critical perimeters (Fig. 4.18, EC2).*

at 33.7°, provided this falls within the foundation (*Fig. 8.12*). Here u is the perimeter of the critical section (*Fig. 8.13*) and β is a coefficient which takes account of the effects of eccentricity of loading. In cases where no eccentricity of loading. In cases where no eccentricity of loading is possible, β may be taken as 1.0. If v_{Sd} exceeds v_{Rd1}, shear reinforcement will need to be provided such that $v_{Sd} \leqslant v_{Rd3}$, where v_{Rd3} is the design shear resistance per unit length of the critical perimeter, for a slab with shear reinforcement calculated in accordance with clause 4.3.4.5.2 of EC2 and clause 6.4(d) of the NAD.

Example 8.9 Analysis of a pad foundation

The pad footing shown below supports a column which is subject to axial characteristic permanent and variable actions of 900 and 300 kN respectively. Assuming the following material strengths and the cover to the main steel is 40 mm, check the suitability of the design in shear:

$$f_{ck} = 30 \text{ N mm}^{-2} \qquad f_{yk} = 460 \text{ N mm}^{-2}$$

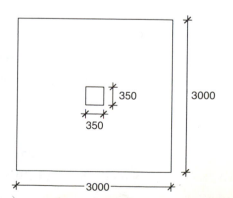

$G_k = 900$ kN
$Q_k = 300$ kN

T20 at 250c/c each way

$h = 600$

350

350

3000

3000

Shear failure could arise:

1. at the face of the column;
2. at a distance d from the face of the column;
3. punching failure of the slab.

FACE SHEAR

Load on footing due to column is

$$1.35 \times 900 + 1.5 \times 300 = 1665 \text{ kN}$$

Design shear stress at perimeter of column, v_{Sd}, is

$$v_{Sd} = \frac{V_{Sd}}{\text{perimeter} \times d} = \frac{1665 \times 10^3}{4 \times 350 \times 540} = 2.2 \text{ N mm}^{-2}$$

where $d = h$ - cover - diameter of bar = 600 - 40 - 20 = 540 mm. Maximum shear stress at perimeter of column, v_{Rd2}, is

$$v_{Rd2} = 0.9\sqrt{f_{ck}} = 0.9\sqrt{30} = 4.93 \text{ N mm}^{-2} > v_{Sd} \quad \text{OK}$$

TRANSVERSE SHEAR

Design shear force on footing (V_{Sd}) is

$$V_{Sd} = \text{load from column}$$
$$= (1.35 \times 900 + 1.5 \times 300) = 1665 \text{ kN}$$

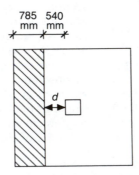

785   540
mm   mm

d

$$\text{Earth pressure } (p_E) = \frac{\text{ultimate load}}{\text{area}} = \frac{1665}{9}$$
$$= 185 \text{ kN m}^{-2}$$

Ultimate load on shaded area is

$$p_E \times \text{area} = 185(0.785 \times 3) = 435.7 \text{ kN}$$

Design shear force per unit length, v_{Sd}, is

$$v_{Sd} = 435.7/3 = 145.2 \text{ kN m}^{-1}$$

Shear resistance of concrete alone, v_{Rd1}, is

$$v_{Rd1} = \tau_{Rd}k(1.2 + 40\rho_1)d$$

$$= [0.34 \times 1.06(1.2 + 40 \times 0.0023)]540$$

$$= 251 \text{ N mm}^{-1} = 251.7 \text{ kN m}^{-1} > v_{Sd} \ (= 145.2 \text{ kN m}^{-1}) \quad \text{OK}$$

where $\tau_{Rd} = 0.34 \text{ N mm}^{-2}$ since $f_{ck} = 30 \text{ N mm}^{-2}$ from *Table 8.7* and where

$$k = 1.6 - d = 1.6 - 0.54 = 1.06$$

$$\rho_1 = \rho_{1y} = \rho_{1x} = \frac{A_{s1}}{b_w d} = \frac{1260}{1000 \times 540} = 0.0023 \qquad \sigma_{cp} = 0$$

PUNCHING SHEAR

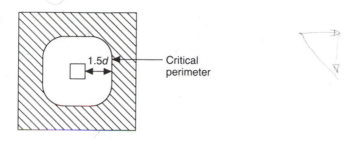

Perimeter of critical section, u, is

$$u = \text{column perimeter} + 2\pi(1.5d) = 4 \times 350 + 2\pi(1.5 \times 540) = 6489 \text{ mm}$$

Area within critical perimeter is

$$4 \times 350(1.5 \times 540) + 350^2 + \pi(1.5 \times 540)^2 = 3.32 \times 10^6 \text{ mm}^2$$

Design shear force, V_{Sd}, is

$$V_{Sd} = p_E A = 185(9 - 3.32) = 1050.8 \text{ kN}$$

Applied shear per unit length, v_{Sd}, is

$$v_{Sd} = \frac{V_{Sd}\beta}{u} = \frac{1050.8 \times 1}{6.489} = 161.9 \text{ kN m}^{-1} < v_{Rd1} = 251.7 \text{ kN m}^{-1} \quad \text{OK}$$

Hence no shear reinforcement is required.

8.11 Design of columns

Column design is covered in EC2 mostly within the chapter on buckling. The design procedure is slightly more complex than that used in BS 8110, although the final result is similar in both codes. This is partly due to the style of presentation and partly the differences in design approach in the two codes, e.g. in EC2 the slenderness ratio above which columns are considered as slender (λ_{min}) has to be evaluated while in BS 8110 it is taken as 15 for braced columns and 10 for unbraced columns. In addition, EC2 does not contain any formulae equivalent to equations 38 and 39 in BS 8110 (*section 3.13.5*) which relate to the design of short-braced columns and are so useful for initial sizing of members in compression.

In this chapter only the design of the most common types of columns found in building structures, i.e. those in non-sway structures subject to an axial load and bending will be described. Their design involves consideration of the following aspects which are discussed individually below:

1. braced columns
2. sway structures
3. slenderness ratio
4. slender columns
5. eccentricities.

8.11.1 BRACED COLUMNS (CLAUSE 4.3.5.3.2, EC2)

A column may be considered to be braced in a given plane if the bracing element or system (e.g. core or shear walls) is sufficiently stiff to resist at least 90% of all lateral forces in that plane. Otherwise it should be considered as unbraced. Note that this condition is more realistically stated in EC2 than in BS 8110 which requires the bracing element/structure should be capable of resisting all lateral forces on the structure.

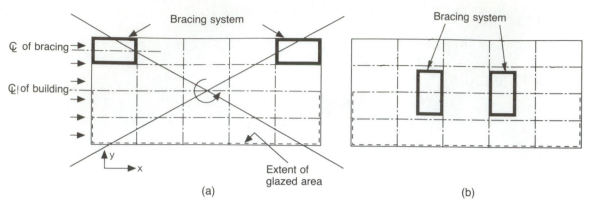

Fig. 8.14 *Braced sway and braced, non-sway structures.*

8.11.2 SWAY STRUCTURES (CLAUSE 4.3.5.3.3, EC2)

Unlike BS 8110, EC2 recognizes the fact that some structures which are braced may still undergo significant displacements at the connections due to the lateral loading. These displacements will increase the design moment in the member and must therefore be taken into account. For example, the columns shown in both the floor plans in *Fig. 8.14* may well be braced. However, because the shear centre of the bracing system in plan (a) is not concentric with the axis of the building, the structure is liable to rotate when the direction of the wind loads are parallel to the x – x axis, thereby generating additional forces in the columns. Such a layout may be necessary because of architectural/client requirements for an open area which is glazed as shown in *Fig. 8.14*, for instance. For engineering purposes it would be better to adopt the arrangement of shear walls shown in plan (b).

Where there is doubt, Appendix 3 of EC2 should be used which gives a criterion for classifying braced structures as non-sway, based on the height of the structure, number of storeys, flexural stiffnesses of the bracing elements and service loads.

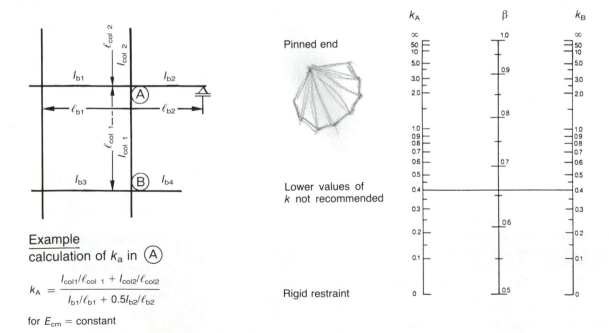

Example
calculation of k_a in Ⓐ

$$k_A = \frac{I_{col1}/\ell_{col\ 1} + I_{col2}/\ell_{col2}}{I_{b1}/\ell_{b1} + 0.5I_{b2}/\ell_{b2}}$$

for $E_{cm} = $ constant

Fig. 8.15 *Nomograph for evaluating effective lengths of columns: (a) non-sway frame; (b) sway frame (Fig. 4.27, EC2).*

8.11.3 SLENDERNESS RATIO (CLAUSE 4.3.5.3.5, EC2)

In EC2 slenderness ratio (λ) is given by

$$\lambda = \frac{l_0}{i} \qquad (8.34)$$

where l_0 is the effective height of the column and i the radius of gyration. Note that in BS 8110 the slenderness ratio is based on the lateral dimensions of the column (b or h) and not the radius of gyration.

8.11.3.1 Effective height

The effective height of a column (l_0) is given by

$$l_0 = \beta \, l_{col} \qquad (8.35)$$

where l_{col} is the height of column measured between centres of restraint and β is a coefficient. Values of β can be determined from the nomograph shown in *Fig. 8.15* via the coefficients k_A and k_B which denote the rigidity of restraint at the column ends and are given by

$$k_A \text{ (or } k_B) = \frac{\Sigma E_{cm} \, I_{col}/l_{col}}{\Sigma E_{cm} \, \alpha I_b/l_{eff}} \not< 0.4 \qquad (8.36)$$

where E_{cm} is the modulus of elasticity of the concrete (*Table 8.16*), I_{col}, and I_b are the second moments of area of the column and beam respectively, l_{col} the height of column between centres of restraint, l_{eff} the effective span of beam and α the factor taking into account the condition of restraint of the beam at the opposite end to the joint:

$\alpha = 1.0$ opposite end elastically or rigidly restrained, i.e. a continuous end

$\alpha = 0.5$ opposite end free to rotate, i.e. a simply supported end

$\alpha = 0$ for a cantilever beam

8.11.3.2 Radius of gyration

The radius of gyration (i) is given by

$$i = \sqrt{(I/A)} \qquad (8.37)$$

where I is the second moment of area of the column and A the gross cross-sectional area of the column.

8.11.4 SLENDER AND NON-SLENDER COLUMNS (CLAUSE 4.3.5.5.3, EC2)

If the slenderness ratio of an isolated column, λ,

exceeds a critical value, λ_{min}, then it may be assumed to be slender. λ_{min} is taken as the greater of

$$\lambda_{min} = 25 \text{ and } \frac{15}{\sqrt{\upsilon_u}} \qquad (8.38)$$

where υ_u is the longitudinal force coefficient for an element and is given by

$$\upsilon_u = \frac{N_{Sd}}{A_c f_{cd}} \qquad (8.39)$$

where N_{Sd} is the design axial load, A_c the cross-sectional area of concrete and f_{cd} the design concrete strength $= \dfrac{f_{ck}}{1.5}$.

EC2 only gives simplified rules for the design of columns in non-sway structures. Since sway structures are fairly uncommon in practice, however, this is unlikely to present any major drawbacks to designers. The following discusses EC2 rules for the design of columns in non-sway structures.

8.11.5 DESIGN OF COLUMNS IN NON-SWAY STRUCTURES

A column in a non-sway structure is classified as slender if the slenderness ratio exceeds the greater of 25 or $\dfrac{15}{\sqrt{\upsilon_u}}$.

According to Flow chart 3 in Appendix A of EC2, where the column is non-slender, i.e. $\lambda < \lambda_{min}$, it should be designed simply for first order internal moments and forces, using the design charts shown in *Fig. 8.16*.

Where the column is slender, however, the design procedure depends upon whether the slenderness ratio lies above or below λ_{crit}, given by,

$$\lambda_{crit} = 25[2 - (e_{01}/e_{02})] \qquad (8.40)$$

where e_{01} and e_{02} the first-order eccentricities are defined as

$$e_{01} = M_{Sd1}/N_{Sd} \qquad e_{02} = M_{Sd2}/N_{Sd} \quad (8.41)$$

where M_{Sd1} and M_{Sd2} are the first-order design moments and N_{Sd} the design axial load. It is also assumed that $|e_{01}| \leq |e_{02}|$. The following subsections describe the design of slender columns where (a) $\lambda_{min} < \lambda < \lambda_{crit}$ and (b) $\lambda > \lambda_{crit}$.

Table 8.16 Values of the secant modulus of elasticity E_{cm} (kN mm^{-2}) (Table 3.2, EC2)

Strength class C	C12/15	C16/20	C20/25	C25/30	C30/37	C35/45	C40/50	C45/55	C50/60
E_{cm}	26	27.5	29	30.5	32	33.5	35	36	37

8.11.5.1 Slender columns where $\lambda_{min} < \lambda < \lambda_{crit}$ (clause 4.3.5.5.3, EC2)

If $\lambda_{min} < \lambda < \lambda_{crit}$ the column is designed for the design axial load (N_{Sd}) and the first-order design moment (M_{Sd1}) obtained by analysis. The latter is compared with the minimum design moment (M_{Rd}) and the larger value taken. Here M_{Rd} is given by

$$M_{Rd} = N_{Sd}\frac{h}{20} \qquad (8.42)$$

Note that this is similar to BS 8110.

Once N_{Sd} and M_{Sd1} have been determined, the area of longitudinal steel can be calculated by strain compatibility using an iterative procedure. However, this is not practical for everyday design and therefore the British Cement Association have produced a series of design charts, similar to those which appear in BS 8110:Part 3, which can be used to evaluate the required steel areas. Typical charts for the design of columns to EC2 are shown in *Fig. 8.16*. Examples 8.10 and 8.13. show the procedure involved.

8.11.5.2 Slender columns where $\lambda > \lambda_{crit}$ (clause 4.3.5.6, EC2)

If the slenderness of a column exceeds λ_{crit} allowance has to be made for the additional moments caused by the deformations. This is achieved by firstly evaluating the various eccentricities and secondly calculating the resulting design moments. Critical conditions may occur at the top, middle or bottom of the column. The total design eccentricity (e_{tot}) at the **top** and **bottom** of the column will be given by

$$e_{tot} = e_0 + e_a \qquad (8.43)$$

where e_0 is the first-order eccentricity (equation 8.41) and e_a the allowance for imperfections given by:

$$e_a = \upsilon\frac{l_0}{2} \qquad (8.44)$$

where l_0 is the effective height of the column (*section 8.11.3*) and υ is the angle of inclination of the structure and is equal to the greater of

$$\upsilon = \frac{1}{100\sqrt{l}} \text{ and } \frac{1}{200} \qquad (8.45)$$

where l is the total height of the structure in metres. This eccentricity will give rise to a design bending moment, M_{Sd}, given by

$$M_{Sd} = N_{Sd1}e_{tot} \qquad (8.46)$$

The total design eccentricity at the **middle** of the column will depend on the relative positions of the first-order eccentricities. EC2 considers two cases: (i) $e_{01} = e_{02}$ (*Fig. 8.17(a)*) and (ii) $e_{01} \neq e_{02}$ (*Fig. 8.17(b), (c)*). In case (i)

$$e_{tot} = e_0 + e_a + e_2 \qquad (8.47)$$

where e_0 is the first-order eccentricity (equation 8.41), e_a the allowance for imperfections (equation 8.44) and e_2 the second-order eccentricity given by

$$e_2 = K_1\frac{l_0^2}{10}(1/r) \qquad (8.48)$$

where

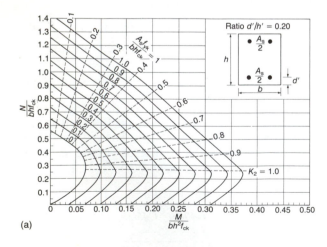

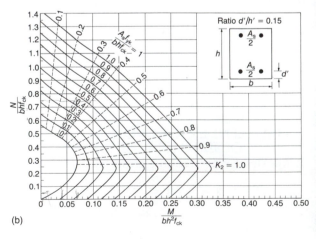

Fig. 8.16 *Typical column design charts for use with EC2 (Concise Eurocode for the Design of Concrete Buildings)*

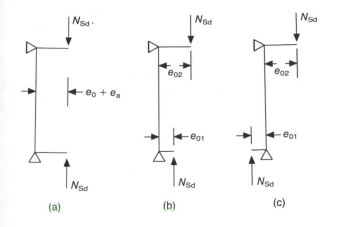

Fig. 8.17 *Design model for the calculation of total eccentricity (Fig. 4.29, EC2).*

$$K_1 = \frac{\lambda}{20} - 0.75 \quad \text{for } 15 \leqslant \lambda \leqslant 35 \qquad (8.49)$$

and where $K_1 = 1$ for $\lambda > 35$ and l_0 the effective length of the column (*section 8.11.3*).

In cases where great accuracy is not required, the curvature $(1/r)$ may be derived from

$$1/r = 2K_2 \frac{\varepsilon_{yd}}{0.9d} \qquad (8.50)$$

where ε_{yd} is the design yield strain of reinforcement:

$$\varepsilon_{yd} = f_{yd}/E_s \qquad (8.51)$$

and where d is the effective depth of the section and K_2 the curvature factor:

$$K_2 = \frac{N_{ud} - N_{Sd}}{N_{ud} - N_{bal}} \leqslant 1 \qquad (8.52)$$

where N_{ud} is the design ultimate capacity of the section:

$$N_{ud} = 0.85f_{cd}A_c + f_{yd}A_s \qquad (8.53)$$

and where N_{Sd} is the actual design axial force and N_{bal} the design load capacity of the balanced section:

$$N_{bal} = 0.4f_{cd}A_c \qquad (8.54)$$

Note that it is always conservative to assume $K_2 = 1$.

For case (ii), i.e. $e_{01} \neq e_{02}$, the total design eccentricity is given by

$$e_{tot} = e_e + e_a + e_2 \qquad (8.55)$$

where e_a and e_2 are as previously defined in equations 8.44 and 8.48 respectively and e_e is the equivalent eccentricity taken as the greater of

$$0.6e_{02} + 0.4e_{01} \qquad (8.56)$$

and

$$0.4e_{02} \qquad (8.57)$$

Here e_{01} and e_{02} are the first-order eccentricities at the two ends, and $|e_{02}| > |e_{01}|$ *Fig. 8.17(b),(c)*. The design moment can then be evaluated using

$$M_{Sd} = N_{Sd1}e_{tot} \qquad (8.58)$$

Summarizing, the design axial load (N_{Sd1}) is determined by analysing the loads acting on the structure. The design moment (M_{Sd}) is determined from equation 8.58 and e_{tot} is taken as the greater of equations 8.43 and 8.47 if $e_{01} = e_{02}$ or equations 8.43 and 8.55 if $e_{01} \neq e_{02}$. Once N_{Sd1} and M_{Sd} are known, the area of longitudinal steel can be evaluated using design charts similar to those in *Fig. 8.16* (*Example 8.13*).

8.11.6 REINFORCEMENT DETAILS FOR COLUMNS

8.11.6.1 Longitudinal reinforcement (clause 5.4.1.2.1, EC2)

Number of bars. Columns with rectangular cross-sections should be reinforced with a minimum of four longitudinal bars; columns with circular cross-sections should be reinforced with a minimum of six. Each bars should have a diameter of not less than 12 mm.

Reinforcement percentages. The area of longitudinal reinforcement, A_s, should lie with the following limits:

$$\frac{0.15N_{Sd}}{f_{yd}} \not< 0.003A_c \leqslant A_s \leqslant 0.08\,A_c$$

where f_{yd} is the design yield strength of the reinforcement, N_{Sd} the design axial compression force and A_c the cross-section of the concrete. Note that the upper limit should not be exceeded even where lapped joints occur.

8.11.6.2 Transverse reinforcement (link) (clause 5.4.1.2.2, EC2)

Size and spacing of links. The diameter of the links should not be less than 6 mm or one-quarter of the maximum diameter of the longitudinal bar, whichever is the greater. However, as noted in *Chapter 3*, 6mm bars may not be freely available and a minimum bar size of 8 mm is preferable. The spacing of links along the column should not exceed the smallest of the following three dimensions:

1. 12 times the minimum diameter of the longitudinal bars;
2. the smallest lateral dimension of the column;
3. 300 mm.

Arrangement of links. With regard to the arrangement of links around the longitudinal reinforcement, EC2 stipulates that (a) every longitudinal bar placed in a corner should be supported by a link passing around the bar and (b) a maximum of five bars in or close to each corner can be secured against buckling by any set of links.

Example 8.10 Design of a slender column

A slender column for a non-sway structure is required to resist an ultimate axial load (N_{Sd}) of 2000 kN and bending moment (M_{Sd1}) of 60 kN m. Design the column using class C30/37 concrete and grade 460 reinforcement assuming that the slenderness ratio is < λ_{crit}.

CROSS-SECTION
Since the design bending moment is relatively small, equation 8.52 may be used to size the column:

$$N_{ud} = 0.85 f_{cd} A_c + f_{yd} A_{sc}$$

Clause 5.4.1.2.1 of EC2 stipulates that the percentage of longitudinal reinforcement, A_s, should generally lie within the following limits:

$$0.3\% \, A_c < A_s < 8\% \, A_c$$

Assuming that the percentage of reinforcement is equal to 3% (say) gives

$$A_s = 0.03 A_c$$

Substituting this into the above equation gives

$$2 \times 10^6 = \frac{0.85 \times 30 \times A_c}{1.5} + \frac{460 \times 0.03 A_c}{1.15}$$

therefore $A_c = 68\,966$ mm^2. For a square column $b = h = \sqrt{68\,966} = 263$ mm. Therefore a 300 mm square column is suitable.

LONGITUDINAL STEEL

Check minimum moment
Minimum eccentricity,

$$e_{min} = \frac{h}{20} = \frac{300}{20} = 15 \text{ mm}$$

Minimum design moment, M_{Rd}, is

$$M_{Rd} = N_{Sd} e_{min} = 2 \times 10^3 \times 15 \times 10^{-3} = 30 \text{ kN m} < M_{Sd1} = 60 \text{ kN m}$$

Therefore design moment is 60 kN m.

Design chart
Cover to links for exposure class 1	= 20 mm
Assume diameter of longitudinal bars (Φ)	= 32 mm
Assume diameter of links	= 8 mm

Therefore

$$d' = 20 + 32/2 + 8 = 44 \text{ mm}$$
$$\frac{d'}{h} = \frac{44}{300} = 0.147$$

Round up to 0.15 and use chart no. 1 (*Fig. 8.16*).

Longitudinal steel area

$$\frac{N}{bhf_{ck}} = \frac{2 \times 10^6}{300 \times 300 \times 30} = 0.74$$

$$\frac{M}{bh^2f_{ck}} = \frac{60 \times 10^6}{300^3 \times 30} = 0.074$$

$$\frac{A_s f_{yk}}{bhf_{ck}} = \frac{A_s \times 460}{300^2 \times 30} = 0.43 \quad (Fig.\ 8.16)$$

$$A_s = 2524 \text{ mm}^2$$

Use 4T32 (3220 mm^2).

$$\%A_s/A_c = 3220/300 \times 300 = 3.5\% \quad \text{(acceptable)}$$

LINKS

Diameter of links is the greater of (i) 8 mm and (ii) $\frac{1}{4}\Phi = \frac{1}{4} \times 32 = 8$ mm. Spacing of links should not exceed the least of (i) $12\Phi = 12 \times 32 = 384$ mm, (ii) least dimension of column = 300 mm and (iii) 300 mm. Therefore, provide R8 links at 300 mm centres.

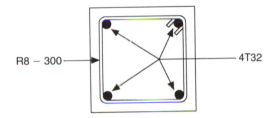

R8 – 300   4T32

Example 8.11 Classification of a column section

Determine if column GH shown in *Fig. 8.18(a)* is slender assuming that it is designed to resist the design loads and moments in *Fig. 8.18(b)*. Assume that the structure is non-sway and $f_{ck} = 25$ N/mm^2.

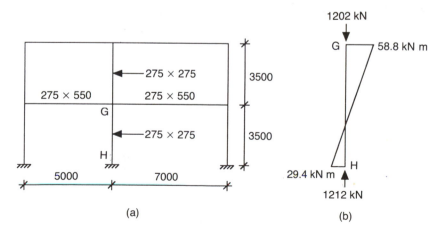

Fig. 8.18

SLENDERNESS RATIO OF COLUMN GH

Effective height
Moment of inertia of column, I_{col}, is given by

$$I_{col} = \frac{bd^3}{12} = \frac{275 \times 275^3 \times 10^{-12}}{12} = 4.766 \times 10^{-4} \text{ m}^4$$

Moment of inertia of beam, I_{beam}, is given by

$$I_{beam} = \frac{bd^3}{12} = \frac{275 \times 550^3 \times 10^{-12}}{12} = 3.813 \times 10^{-3} \text{ m}^4$$

From equation 8.36

$$k_G = \frac{E_{cm}\,(I_{col\ upper}\ /\ l_{col\ upper} + I_{col\ lower}\ /\ l_{col\ lower})}{E_{cm}(\alpha I_{beam\ 1}\ /\ l_{eff\ 1} + \alpha I_{beam\ 2}\ /\ l_{eff\ 2})} \not< 0.04$$

$$= \frac{(4.766 \times 10^{-4}\ /\ 3.5 + 4.766 \times 10^{-4}\ /\ 3.5)}{(1.0 \times 3.813 \times 10^{-3}\ /\ 5 + 1.0 \times 3.813 \times 10^{-3}\ /\ 7)}$$

$$= 0.208 \not< 0.4$$

To calculate the slenderness ratio it is conservative to assume that the column is pinned at H. Hence $\alpha = 0$ and $k_H = \infty$.
From nomograph in *Fig. 8.15*, $\beta = 0.8$. Effective height of column GH, l_0, is given by

$$l_0 = \beta\, l_{col} = 0.8 \times 3500 = 2800 \text{ mm}$$

Radius of gyration
Radius of gyration, i, is given by

$$i = \sqrt{(I/A)} = \sqrt{[(bd^3/12)/bd]} = d/\sqrt{12} = 275/\sqrt{12} = 79.4 \text{ mm}$$

Slenderness ratio
Slenderness ratio, λ, is given by

$$\lambda = \frac{l_0}{i} = \frac{2800}{79.4} = 35.3$$

Classification of column
Minimum slenderness ratio, λ_{min}, is

$$\lambda_{min} = 25 \text{ or } \frac{15}{\sqrt{\upsilon_u}} \text{ whichever is the greater}$$

$$\upsilon_u = \frac{N_{Sd}}{A_c f_{cd}} = \frac{1202 \times 10^3}{275^2 \times 25/1.5} = 0.954$$

$$\frac{15}{\sqrt{\upsilon_u}} = \frac{15}{\sqrt{0.954}} = 15.3$$

Hence $\lambda_{min} = 25$
Since $\lambda > \lambda_{min}$ column GH is slender

Example 8.12 Classification of a column

Determine if column PQ is slender assuming that it is designed to resist the design loads and moment shown in *Fig. 8.19(b)*. Assume that the structure is non-sway and $f_{ck} = 25$ N mm^{-2}.

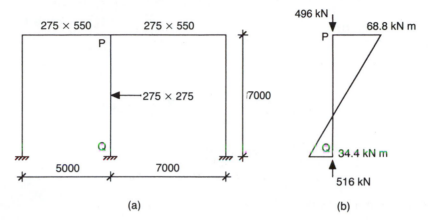

Fig. 8.19

(a)　　　　　　　　　(b)

SLENDERNESS RATIO OF COLUMN PQ

Effective height

Moment of inertia of column, I_{col}, is given by

$$I_{col} = \frac{bd^3}{12} = \frac{275 \times 275^3 \times 10^{-12}}{12} = 4.766 \times 10^{-4} \text{ m}^4$$

Moment of inertia of beam, I_{beam}, is given by

$$I_{beam} = \frac{bd^3}{12} = \frac{275 \times 550^3 \times 10^{-12}}{12} = 3.813 \times 10^{-3} \text{ m}^4$$

From equation 8.36

$$k_P = \frac{E_{cm}(I_{col\ upper}/l_{col\ upper} + I_{col\ lower}/l_{col\ lower})}{E_{cm}(\alpha I_{beam\ 1}/l_{eff\ 1} + \alpha I_{beam\ 2}/l_{eff\ 2})} \not< 0.4$$

$$= \frac{(4.766 \times 10^{-4}/7)}{E_{cm}(1.0 \times 3.813 \times 10^{-3}/5 + 1.0 \times 3.813 \times 10^{-3}/7)}$$

$$= 0.052 \not< 0.4$$

To calculate the slenderness ratio it is conservative to assume that column is pinned at Q. Hence $\alpha = 0$ and $k_Q = \infty$.

From nomograph in *Fig. 8.15*, $\beta = 0.8$. Effective height of column PQ, l_0, is given by

$$l_0 = \beta\ l_{col} = 0.8 \times 7000 = 5600 \text{mm}$$

Radius of gyration

Radius of gyration, i, is given by
$$i = \sqrt{(I/A)} = \sqrt{[(bd^3/12)/bd]} = d/\sqrt{12} = 275/\sqrt{12} = 79.4 \text{ mm}$$

Slenderness ratio

Slenderness ratio, λ, is given by

$$\lambda = \frac{l_0}{i} = \frac{5600}{79.4} = 70.5$$

Classification of column

Critical slenderness ratio, λ_{min}, is

λ_{min} is the greater of 25 or $\dfrac{15}{\sqrt{\upsilon_u}}$

$$\upsilon_u = \frac{N_{Sd}}{A_c f_{cd}} = \frac{496 \times 10^3}{275^2 \times 25/1.5} = 0.394$$

$$\frac{15}{\sqrt{\upsilon_u}} = 23.9$$

$$\therefore \lambda_{min} = 25$$

Since $\lambda > \lambda_{min}$ column is slender

Example 8.13 Column design: (i) $\lambda_{min} < \lambda < \lambda_{crit}$; (ii) $\lambda < \lambda_{crit}$

Design the columns in *Examples 8.11* and *8.12*.

COLUMN GH

CRITICAL SLENDERNESS RATIO, λ_{crit}

Eccentricity at end G, $e_{02} = \dfrac{M_{Sd2}}{N_{Sd}} = \dfrac{58.8 \times 10^6}{1202 \times 10^3} = 48.9$ mm

Eccentricity at end H, $e_{01} = \dfrac{M_{Sd1}}{N_{Sd}} = \dfrac{-29.4 \times 10^6}{1212 \times 10^3} = -24.5$ mm

Critical slenderness ratio, λ_{crit}, is given by

$$\lambda_{crit} = 25(2 - e_{01}/e_{02}) = 25(2 - -24.5/48.9) = 62.5$$

Since $\lambda = 35.3$ (*Example 8.11*), and $\lambda < \lambda_{crit}$ column GH does not need to be designed for second order effects.

Longitudinal steel

$$N_{Sd} = 1202 \times 10^3 \text{ N}$$

$$M_{Sd} = 58.8 \times 10^6 \text{ N mm} > M_{Rd} = N_{Sd} \frac{h}{20} = (1202 \times 10^3) \frac{275}{20} = 16.5 \times 10^6 \text{ N mm OK.}$$

Assume:

> Diameter of longitudinal steel (Φ) = 32 mm
> Diameter of links (Φ') = 8 mm
> Cover to all reinforcement (c) = 30 mm

Therefore

$$d' = \Phi/2 + \Phi' + c = 32/2 + 8 + 30 = 54 \text{ mm}$$

$$\frac{d'}{h} = \frac{54}{275} = 0.196$$

Use graph with $d'/h = 0.20$:

$$\frac{N}{bhf_{ck}} = \frac{1202 \times 10^3}{275^2 \times 25} = 0.635$$

$$\frac{M}{bh^2 f_{ck}} = \frac{58.8 \times 10^6}{275^3 \times 25} = 0.113$$

$$\frac{A_s f_{yk}}{bhf_{ck}} = \frac{A_s \times 460}{275^2 \times 25} = 0.5 \quad (Fig.\ 8.16)$$

$$A_s = 2055 \text{ mm}^2$$

Provide 4T32 ($A_s = 3220$ mm^2).

Links

Diameter of links is the greater of (i) 8 mm and (ii) $\frac{1}{4}\Phi = \frac{1}{4} \times 32 = 8$ mm. Spacing of links should not exceed the least of (i) $12\Phi = 12 \times 32 = 384$ mm, (ii) least dimension of column = 275 mm and (iii) 300 mm. Therefore, provide R8 at 275 mm centres.

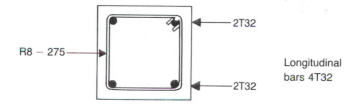

2T32

R8 – 275

2T32

Longitudinal bars 4T32

COLUMN PQ

Critical slenderness ratio, λ_{crit}

Eccentricity at end P, $e_{02} = \dfrac{M_{Sd2}}{N_{Sd}} = \dfrac{68.8 \times 10^6}{496 \times 10^3} = 139$ mm

Eccentricity at end Q, $e_{01} = \dfrac{M_{Sd1}}{N_{Sd}} = \dfrac{-34.4 \times 10^6}{516 \times 10^3} = -66.7$ mm

Critical slenderness ratio, λ_{crit}, is given by

$$\lambda_{crit} = 25(2 - e_{01}/e_{02}) = 25(2 - -66.7/139) = 62$$

Since $\lambda = 70.5$ (*Example 8.12*) and $\lambda > \lambda_{crit}$ column PQ needs to be designed for second order effects.

Eccentricities

Calculate e_e

e_e is the greater of

$$0.6e_{02} + 0.4e_{01} = 0.6 \times 139 + 0.4 \times -66.7 = 57 \text{ mm}$$

$$0.4e_{02} = 0.4 \times 139 = 56 \text{ mm}$$

Therefore take $e_e = 56$ mm.

Calculate e_a

$$l = 7000 \text{ mm}$$

$$\upsilon = \frac{1}{100\sqrt{l}} = \frac{1}{100\sqrt{7}} = \frac{1}{264.5} \not< \frac{1}{200}$$

$$l_0 = 5600 \text{ mm} \quad (\textit{Example 8.12})$$

Hence

$$e_a = \frac{\upsilon l_0}{2} = \frac{(1/200)\,5600}{2} = 14 \text{ mm}$$

Calculate e_2. Assume:

Diameter of longitudinal steel (Φ) = 20 mm
Diameter of links (Φ')   = 8 mm
Cover to all reinforcement (c)   = 35 mm

Thus

$$d = h - (\Phi/2 + \Phi' + c)$$
$$= 275 - (20/2 + 8 + 35) = 222 \text{ mm}$$
$$K_1 = 1 \quad (\text{since } \lambda > 35)$$

Assume $K_2 = 1$:

$$\varepsilon_{yd} = \frac{f_{yk}}{E_s} = \frac{460}{200 \times 10^3} = 2.3 \times 10^{-3}$$

$$1/r = \frac{2K_2\varepsilon_{yd}}{0.9d} = \frac{2 \times 1 \times 2.3 \times 10^{-3}}{0.9 \times 222} = 2.3 \times 10^{-5}$$

$$e_2 = \frac{K_1 l_0^2 (1/r)}{10} = \frac{1 \times 5600^2 \times (2.3 \times 10^{-5})}{10} = 72 \text{ mm}$$

Calculate e_{tot}. Here e_{tot} is taken as the greater of:

$$e_{tot} = e_0 + e_a = 139 + 14 = 153 \text{ mm}$$

and

$$e_{tot} = e_e + e_a + e_2 = 57 + 14 + 72 = 143 \text{ mm}$$

Hence $e_{tot} = 153$ mm.

Design moment

$$M_{Sd} = N_{Sd}e_{tot} = 496 \times 0.153 = 75.9 \text{ kN m}$$

Longitudinal steel area

Assuming that the diameter of longitudinal steel (Φ) is 20 mm, the diameter of links (Φ') 8 mm and the cover to all reinforcement (c) 35 mm

$$d' = \Phi/2 + \Phi' + c = 20/2 + 8 + 35 = 53 \text{ mm}$$

$$\frac{d'}{h} = \frac{53}{275} = 0.193$$

Use graph with $d'/h = 0.20$:

$$\frac{N}{bhf_{ck}} = \frac{496 \times 10^3}{275^2 \times 25} = 0.262$$

$$\frac{M}{bh^2 f_{ck}} = \frac{75.9 \times 10^6}{275^3 \times 25} = 0.146$$

$$\frac{A_s f_{yk}}{bh f_{ck}} = \frac{A_s \times 460}{275^2 \times 25} = 0.3$$

$$A_s = 1230 \text{ mm}^2$$

Provide 4T20 ($A_s = 1260$ mm^2).

Check assumed value of K_2.

$$N_{ud} = 0.85 f_{cd} A_c + f_{yd} A_s$$

$$= \frac{0.85 \times 25 \times 275^2}{1.5} + \frac{460 \times 1260}{1.15} = 1575 \times 10^3 \text{ N}$$

$$N_{Sd} = 496 \times 10^3 \text{ N}$$

$$N_{bal} = 0.4 f_{cd} A_c = \frac{0.4 \times 25 \times 275^2}{1.5} = 504 \times 10^3 \text{ N}$$

$$K_2 = \frac{N_{ud} - N_{Sd}}{N_{ud} - N_{bal}} = \frac{(1575 - 496)\,10^3}{(1575 - 504)\,10^3} = 1.0$$

Therefore assumed value of K_2 is correct.

Links
Diameter of links is the greater of (i) 8 mm and (ii) ¼Φ = ¼ × 20 = 5 mm. Spacing of links should not exceed the least of (i) 12Φ = 12 × 20 = 240 mm, (ii) least dimension of column = 275 mm and (iii) 300 mm. Therefore, provide R8 links at 240 mm centres.

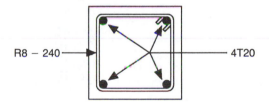

R8 − 240   4T20

Chapter 9

Eurocode 3: Design of steel structures

This chapter describes the contents of Part 1.1 of Eurocode 3, the new European standard for the design of buildings in steel, which is expected to replace BS 5950 by about 1998. The chapter highlights the differences between Eurocode 3: Part 1.1 and BS 5950 and illustrates the new design procedures by means of a number of worked examples on beams, columns and connections. To help comparison but primarily to ease understanding of the Eurocode, the material has here been presented in a similar order to that in Chapter 4 of this book on BS 5950, rather than strictly adhering to the sequence of chapters and clauses adopted in the Eurocode.

9.1 Introduction

Eurocode 3 applies to the design of buildings and civil engineering works in steel. It is based on limit state principles and comes in several parts as shown in *Table 9.1*.

Part 1.1 of Eurocode 3, which is the subject of this discussion, gives a general basis for the design of buildings and civil engineering works in steel. It is largely similar in scope to Part 1 of BS 5950, which was discussed in *Chapter 4*. Part 1.1 of Eurocode 3, hereafter referred to as EC3, was published in draft

form in 1984 and then a European pre-standard, reference no. DD ENV 1993–1–1: 1992, in September 1992. It is due to be issued as a European standard some time after 1998.

EC3 is published in two volumes and deals with the following subjects:

Volume 1
Chapter 1: Introduction (covers layout of code, important conventions and assumptions, symbols)
Chapter 2: Basis of design
Chapter 3: Materials
Chapter 4: Serviceability limit states
Chapter 5: Ultimate limit states
Chapter 6: Connections subject to static loading
Chapter 7: Fabrication and erection
Chapter 8: Design assisted by testing
Chapter 9: Fatigue

Volume 2
Annex B: Reference standards
Annex C: Design against brittle fracture
Annex E: Buckling length of a compression member
Annex F: Lateral – torsional buckling
Annex J: Beam-to-column connections
Annex K: Hollow section lattice girder connections
Annex L: Column bases
Annex M: Alternative method for fillet welds
Annex Y: Guidelines for loading tests

The purpose of the following discussion is to describe the contents of EC3 and to highlight the main differences between it and Part 1 of BS 5950. A number of design examples on beams, columns and connections are also included in *sections 9.11, 9.12* and *9.13* respectively to illustrate the new design procedures.

Table 9.1 Overall scope of Eurocode 3

Part	Subject
1.1	General rules and rules for buildings
1.2	Fire resistance
1.3	Cold formed thin gauge members and sheeting
2	Bridges and plated structures
3	Towers, masts and chimneys
4	Tanks, silos and pipelines
5	Piling
6	Crane structures
7	Marine and maritime structures
8	Agriculture

9.2 Structure of EC3

As can be appreciated from the above contents list, the organization of material in EC3 is quite different from that in BS 5950. BS 5950 contains separate sections on the design of individual elements, e.g. beams, columns and connections. EC3, however, divides the material on the basis of design criteria, e.g. deflection, tension, compression, bending, shear and buckling, which may apply to any element.

9.3 Principles and application rules (clause 1.2, EC3)

For the reasons discussed in *Chapter 7*, the clauses in EC3 have been divided into principle and application rules. Principles encompass statements, definitions, requirements and analytical methods for which there is no permitted alternative. Application rules are generally recognized rules which follow the principles, but for which EC3 permits the use of alternative techniques. The principles are printed in roman type. The application rules, on the other hand, are printed in italics (*Fig. 9.1*).

9.4 Boxed values (clause 1.3, EC3)

As pointed out in *Chapter 7*, a number of the safety elements given in EC3 appear in a box as shown below. This signifies that these values are meant to be for guidance only and that other values may be used by individual member states, for the time being.

$$\boxed{1.5}$$

6.5.2.2.   Design shear rupture resistance

(1) 'Block shear' failure at a group of fastener holes near the end of a beam web or bracket, see figure 6.5.5. shall be prevented by using appropriate hole spacing. This mode of failure generally consists of tensile rupture along the line of fastener holes on the tension face of the hole group, accompanied by gross section yielding in shear at the row of fastener holes along the shear face of the hole group, see figure 6.5.5.

(2) *The design value of the effective resistance to block shear $V_{eff.Rd}$ should be determined from*

$$V_{eff.Rd} = (f_y/\sqrt{3})\, A_{v.eff}/\gamma_{MO} \qquad (6.1)$$

where $A_{v.eff}$ is the effective shear area

Fig. 9.1 *Example of principle and application rules (page 145, EC3).*

The actual values to be used in the UK can be found in the UK National Application Document (NAD) for EC3.

9.5 Symbols (clause 1.6, EC3)

The symbols used in EC3 are extremely schematic but a little cumbersome. For example, $M_{pl.y.Rd}$ denotes the design plastic moment of resistance about the y – y (major) axis. Symbols such as this makes many expressions and formulae seem much longer than they actually are, but on the other hand the traditional list of definitions of symbols is no longer required. A number of symbols which are used to identify particular dimensions of universal sections have changed in EC3 as discussed in section 9.6. Furthermore, the elastic and plastic section moduli are denoted in EC3 by the symbols W_{el} and W_{pl} respectively.

9.6 Member axes (clause 1.6.7, EC3)

Member axes for commonly used steel members are shown in *Fig. 9.2*. This system is different from that used in BS 5950. Thus, x–x is the axis along the member, while axis y–y is what is termed in BS 5950 as axis x–x, the major axis. The minor axis of bending is taken as axis z–z in EC3. Considering steel design in isolation, this change need not have happened. It was insisted on for consistency with the structural Eurocodes for other materials, e.g. concrete and timber.

9.7 Basis of design

Like BS 5950, EC3 uses limit state principles and for design purposes considers the two principal categories of limit states: ultimate and serviceability. The terms 'ultimate limit state' (ULS) and 'serviceability limit state' (SLS) apply in the same way as we understand them in BS 5950. Thus, ultimate limit states are 'those associated with collapse, or with other forms of structural failure which may endanger the safety of people' while serviceability limit states 'correspond to states beyond which specified service criteria are no longer met'.

The serviceability limit states requirements for steel structures are discussed in Chapters 2 and 4 of EC3. The principal serviceability limit states in EC3 are listed below. These are broadly similar to those specified in BS 5950 (*Table 4.1*):

1. deformations or deflections which affect the appearance or effective use of the structure;

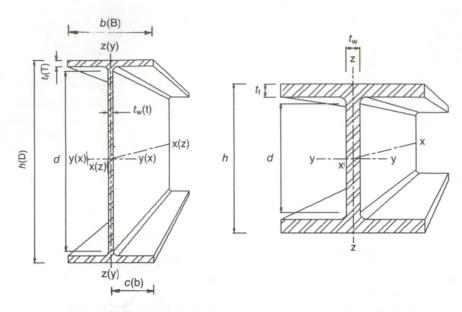

Fig. 9.2 *Member definitions and axes used in EC3 (N.B. symbols in parentheses denote symbols used in BS 5950).*

2. vibration, oscillation or sway (i.e. dynamic effects) which causes discomfort to the occupants of a building or damage to its contents;
3. damage to finishes or non-structural elements due to deformations, deflections or dynamic effect.

According to clause 2.3.2 of EC3, the ultimate limit states for steel structures and components are:

1. static equilibrium of the structure;
2. rupture or excessive deformation of a member;
3. transformation of the structure into a mechanism;
4. instability induced by second-order effects, e.g. lack of fit, thermal effects, sway;
5. fatigue;
6. accidental damage (may include fire resistance).

In addition, there is a separate (third) limit state of durability. As can be appreciated, these are roughly similar to the limit states used in BS 5950 (*Table 4.1*) although there are slight changes in the terminology.

In the context of elemental design, the designer is principally concerned with the ultimate limit states which affect the strength of the member, e.g. yield, buckling and rupture and the serviceability limit state of deflection. EC3 requirements in respect of these limit states will be discussed later for individual element types.

In a similar way to BS 5950, clause 5.2 of EC3 gives three possible sets of design assumptions for analysing structural frames:

1. Simple framing – this assumes that all joints are pinned and the structure is fully braced.
2. Continuous framing – this assumes that all joints are rigid (for elastic design) or full strength (for plastic analyses) and can transmit all moments and forces.
3. Semi-continuous framing – these methods take the real stiffness and strength of the joints into account.

9.8 Actions (clause 2.2.2, EC3)

'Actions' is Eurocode terminology for loads and imposed deformations. Dead and imposed loads are generally referred to in EC3 as permanent and variable actions respectively. The characteristic values of permanent and variable actions will be specified in Eurocode 1: *Basis of Design and Actions on Structures*, which is due to be published as an ENV late in 1993. In the mean time, the UK NAD requires designers to continue using BS 648, BS 6399: Part 1 and CP 3: Chapter 5: Part 2 for characteristic values of actions. Indeed, the NAD requires these UK standards to be used throughout the life of the ENV, even after Eurocode 1 has been published.

The design values of actions (F_d) are obtained by multiplying the characteristic actions (F_k) by the appropriate partial safety factor for actions (γ_F) obtained from *Table 9.2*:

Table 9.2 Partial safety factors for actions on building structures for persistent and transient design situations (Table 2.2, EC3)

	Permanent actions (γ_G)	Variable actions (γ_Q)	
		One with its characteristic value	Others with their combination values
Favourable effect $\gamma_{F,inf}$	1.0	0	0
Unfavourable effect $\gamma_{F,sup}$	1.35	1.5	1.5

$$F_d = \gamma_F \times F_k \qquad (9.1)$$

For buildings, the partial safety factor for permanent actions (γ_G) is 1.35 and variable actions (γ_Q) is 1.5. The corresponding values in BS 5950 are 1.4 and 1.6 respectively.

9.9 Materials

9.9.1 DESIGN STRENGTHS (CLAUSE 2.2.3.2, EC3)

EC3 reflects the model Eurocode clause that design strengths, X_d, are obtained by dividing the characteristic strengths, X_k, by the partial safety factor for materials, γ_m, i.e.

$$X_d = X_k/\gamma_m \qquad (9.2)$$

But this is not in fact used in EC3. Instead, EC3 divides the cross-section resistance, R_k, by a partial safety factor to give the design resistance, R_d for the member, i.e.

$$R_d = R_k/\gamma_M \qquad (9.3)$$

where γ_M is the partial safety factor for the resistance. Thus, γ_M in EC3 is not the same as γ_m in BS 5950. Here γ_m is applied to strength; γ_M is applied to structures.

9.9.2 CHARACTERISTIC STRENGTHS (CLAUSE 3.2.2, EC3)

Table 9.3 shows the characteristic yield and ultimate strength of structural steelwork recommended for use by EC3. Note that the equivalents of grade 43 and 50 steel, which are referred to in BS 5950 are given, but not grade 55. Instead a lower-strength steel (Fe 360) is given which is the normal steel grade used on the Continent. This lower-grade steel has a higher fracture toughness, which enables the use of thicker ply in colder temperatures. Grade 55 steel can still be specified, however, since a grade Fe E 460 steel ($f_y = 460$ N mm^{-2}) similar to grade 55 is included in Annex D in Part 1A of Eurocode 3, due out in 1993.

9.9.3 PARTIAL SAFETY FACTORS (CLAUSE 5.1.1, EC3)

The partial safety factor for materials is taken as 1 in BS 5950 and consequently there is no difference between characteristic and design strength. However, as noted above, in EC3 the partial safety factor, γ_M, is applied to structures and components, rather than strengths and it is therefore not easy to compare the changes in the two standards.

According to clauses 5.1.1 and 3 of EC3 and the UK NAD respectively, the values for partial safety factors should be those given below:

Resistance of class 1, 2 or 3 cross-section γ_{M0} = 1.05

Resistance of class 4 cross-section γ_{M1} = 1.05

Resistance of member to buckling γ_{M1} = 1.05

Resistance of net section at bolt holes γ_{M2} = 1.25

Table 9.3 Nominal values of yield strength, f_y, and ultimate tensile strength, f_u, for structural steel to EN 10025 (Table 3.1, EC3)

Nominal steel grade	Thickness t (mm)			
	$t \leq 40$ mm		40 mm $< t \leq 100$ mm	
	f_y (N mm^{-2})	f_u (N mm^{-2})	f_y (N mm^{-2})	f_u (N mm^{-2})
Fe 360	235	360	215	340
Fe 430	275	430	255	410
Fe 510	355	510	335	490

Note. t is the nominal thickness of the element.

9.9.4 MATERIAL COEFFICIENTS (CLAUSE 3.2.5, EC3)

The following coefficients are specified in EC3:

Modulus of elasticity E = 210 000 N mm^{-2}

Shear modulus G = $E/2\,(1+v)$ N mm^{-2} (9.4)

Poisson's ratio v = 0.3

Coefficient of linear thermal expansion α = 12×10^6 per deg C

Density ρ = 7850 kg m^{-3}

Note that the value of E is slightly higher than that specified in BS 5950, being equal to 205 kN mm^{-2}.

9.10 Classification of cross-sections (clause 5.3, EC3)

Classification has the same purpose as in BS 5950, and the four classifications are identical:

Class 1 cross-sections: 'plastic' in BS 5950
Class 2 cross-sections: 'compact' in BS 5950
Class 3 cross-sections: 'semi-compact' in BS 5950
Class 4 cross-sections: 'slender' in BS 5950

Classification of a cross-section depends upon the proportions of each of its compression elements. The highest (least favourable) class number is generally quoted for a particular section.

Appropriate sections of EC3's classification are given in *Table 9.4*. Briefly, for rolled sections, d/t_w (d/t in BS 5950) ratios are slightly more onerous for webs, and c/t_f (b/T in BS 5950) ratios are slightly less onerous for outstand flanges. It should be noted that the factor ε is given by

Table 9.4 Maximum width-to-thickness ratios for compression elements (Table 5.3.1, EC3)

Type of element	Class of element		
	Class 1	Class 2	Class 3
Outstand flange for rolled section	$\dfrac{c}{t_f} \leq 10\,\varepsilon$	$\dfrac{c}{t_f} \leq 11\varepsilon$	$\dfrac{c}{t_f} \leq 15\,\varepsilon$
Web with neutral axis at mid depth, rolled section	$\dfrac{d}{t_w} \leq 72\,\varepsilon$	$\dfrac{d}{t_w} \leq 83\,\varepsilon$	$\dfrac{d}{t_w} \leq 124\,\varepsilon$
Web subject to compression, rolled sections	$\dfrac{d}{t_w} \leq 33\,\varepsilon$	$\dfrac{d}{t_w} \leq 38\,\varepsilon$	$\dfrac{d}{t_w} \leq 42\,\varepsilon$
f_y	235	275	355
ε	1	0.92	0.81

$$\varepsilon = (235/f_y)^{0.5} \qquad (9.5)$$

and not $(275/p_y)^{0.5}$ as in BS 5950. For class 4 sections effective cross-sectional properties can be calculated using effective widths of plate which in some cases put the member back into class 3.

9.11 Design of beams

The procedure for the design of beams is given in clause 5.1.5 of EC3. However, to ease understanding of this part of the code, as in *Chapter 4* of this book, we will consider the design of fully laterally restrained and unrestrained beams separately. Thus, the following section (*9.11.1*) will consider only the design of beams which are fully laterally restrained. *Section 9.11.2* will then look at EC3's rules for designing beams which are not laterally torsionally restrained.

9.11.1 FULLY LATERALLY RESTRAINED BEAMS

Generally, such members should be checked for:

1. resistance of cross-section to bending ULS;
2. resistance to shear buckling ULS;
3. resistance to flange-induced buckling ULS;
4. resistance of the web to transverse forces ULS;
5. deflection SLS.

9.11.1.1 Resistance of cross-sections – bending moment (clause 5.4.5, EC3)

When shear force is absent or of a low value, the design value of the bending moment, M_{Sd}, should at no point exceed the moment of resistance of the section, $M_{c.Rd}$, i.e.

$$M_{Sd} \leq M_{c.Rd} \qquad (9.6)$$

$M_{c.Rd}$ may be taken as follows:

1. The design plastic resistance moment of the gross section

$$M_{pl.Rd} = W_{pl} f_y / \gamma_{M0} \qquad (9.7)$$

where W_{pl} is the plastic section modulus, for class 1 and 2 sections only.

2. The design elastic resistance moment of the gross section

$$M_{el.Rd} = W_{el} f_y / \gamma_{M0} \qquad (9.8)$$

where W_{el} is the elastic section modulus for class 3 sections.

3. The design local buckling resistance moment of the gross section

$$M_{o.Rd} = W_{eff} f_y / \gamma_{M1} \qquad (9.9)$$

where W_{eff} is the effective section modulus, for class 4 cross-sections only.

4. The design ultimate resistance moment of the net section at bolt holes $M_{u.Rd}$, if this is less than the appropriate values above. In calculating this value, fastener holes in the compression zone do not need to be considered unless they are oversize or slotted. In the tension zone holes do not need to be considered provided that

$$0.9(A_{f.net}/A_f) \geqslant (f_y/f_u)(\gamma_{M2}/\gamma_{M0}) \quad (9.10)$$

When this is not the case a reduced flange area may be assumed which satisfies the above. Consideration of fastener holes in bending is not clearly covered in BS 5950.

9.11.1.2 Resistance of cross-sections – shear
(clause 5.4.6, EC3)

The design value of the shear force, V_{Sd}, at each cross-section should satisfy

$$V_{Sd} \leqslant V_{pl.Rd} \quad (9.11)$$

where $V_{pl.Rd}$ is the design plastic shear resistance, given by

$$V_{pl.Rd} = A_v(f_y/\sqrt{3})/\gamma_{M0} \quad (9.12)$$

where A_v is the shear area, which for rolled I- and H-sections, loaded by gravity is

$$A_v = A - 2bt_f + (t_w + 2r)\,t_f \quad (9.13)$$

where A is the cross-sectional area, b the overall breadth, r the root radius, t_f the flange thickness and t_w the web thickness.

As this will generally give a slightly higher shear area than in BS 5950, for simplicity Av may be taken as

$$A_v = 1.04ht_w \quad (9.14)$$

where h is the overall depth of the section.

The shear resistance calculated as above will give almost identical answers to that calculated by BS 5950, if it is assumed that γ_{M0} is cancelled out by the lower loading partial safety factors γ_F.

Fastener holes in the web do not have to be considered provided that

$$A_{v.net}/A_v \geqslant f_y/f_u \quad (9.15)$$

When this is not the case an effective shear area of $A_{v.net}f_u/f_y$ may be used. Explicit consideration of fastener holes in shear is a departure from BS 5950. The shear buckling resistance of unstiffened webs must additionally be considered when

$$d/t_w > 69\varepsilon \quad (9.16)$$

When the changed definition of ε is taken into account, this value has an almost identical effect to BS 5950 in this respect.

For a stiffened web shear buckling resistance will need to be considered when

$$d/t_w > 30\varepsilon\sqrt{k_\tau} \quad (9.17)$$

where k_τ is the buckling factor for shear and is given by

$k_\tau = 5.34$ (for webs with transverse stiffeners at supports only)

$k_\tau = 4 + \dfrac{5.34}{(a/d)^2}$ (for webs with transverse stiffeners at supports and intermediate transverse stiffeners with $a/d < 1$)

$k_\tau = 5.34 + \dfrac{4}{(a/d)^2}$ (for webs with transverse stiffeners at supports and intermediate transverse stiffeners with $a/d \geqslant 1$)

9.11.1.3 Resistance of cross-sections – bending and shear (clause 5.4.7, EC3)

The plastic resistance moment of the section is reduced by the presence of shear. When the design value of the shear force exceed 50% of the design plastic shear resistance, the design resistance moment of the section should be reduced to $M_{V.Rd}$, obtained as follows for equal flanged sections:

$$M_{V.Rd} = f_y(W_{pl} - \rho\,A_v^2/4t_w)/\gamma_{M0} \leqslant M_{c.Rd} \quad (9.18)$$

where $\rho = (2V_{Sd}/V_{pl.Rd} - 1)^2$. In other cases the design strength should be reduced to $(1-\rho)f_y$ for the shear area only, when calculating the design resistance moment.

9.11.1.4 Shear buckling resistance (clause 5.6, EC3)

The shear buckling resistance of a web has to be checked when $d/t_w > 69\varepsilon$. For standard rolled beams and columns this check is rarely necessary since d/t_w is always less than 69ε, except for just one universal beam (UB) section, which has a d/t_w ratio which is marginally over this limit. See also *section 9.11.1.2*.

9.11.1.5 Flange-induced buckling (clause 5.7.7, EC3)

To prevent the possibility of the compression flange buckling in the plane of the web, EC3 requires that the ratio d/t_w of the web should satisfy the following criterion:

$$d/t_w \leqslant k(E/f_{yf})\,[A_w/A_{fc}]^{0.5} \quad (9.19)$$

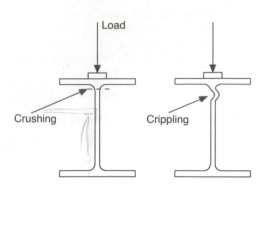

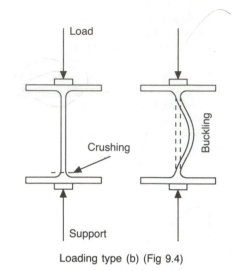

Loading type (a) (Fig 9.4)   Loading type (b) (Fig 9.4)

Fig. 9.3 *Web failure.*

where A_w is the area of the web, A_{fc} the area of the compression flange and f_{yf} is the yield strength of the compression flange. The factor k assumes the following values: class 1 flanges: 0.3; class 2 flanges: 0.4; class 3 or 4 flanges: 0.55.

9.11.1.6 Resistance of the web to transverse forces (clause 5.7, EC3)

EC3 identifies not two, as in BS 5950, but three possible modes of failure due to loads applied to the web through a flange (*Fig. 9.3*):

1. crushing of the web close to the flange, accompanied by plastic deformation of the flange;
2. crippling of the web in the form of localized buckling and crushing of the web close to the flange, accompanied by plastic deformation of the flange;
3. buckling of the web over most of the depth of the member.

A distinction is also made between two types of load application: (a) forces applied through a flange and resisted by shear in the web (*Fig. 9.4.(a)*); (b) forces applied through one flange and transferred through the web directly to the other flange (*Fig. 9.4.(b)*). For loading type (a) the web resistance should be taken as the smaller of (1) and (2) above, i.e. (1) the crushing resistance and (2) the crippling resistance. For loading type (b) the web resistance should be taken as the smaller of (1) and (3), i.e. (1) the crushing resistance and (3) the buckling resistance.

Crushing resistance. For an I- or H-section, the design crushing resistance is

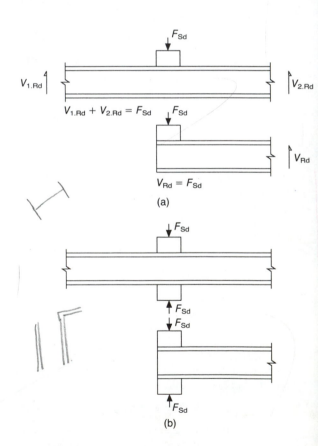

Fig. 9.4 *Forces applied through a flange: (a) forces resisted by shear in the web; (b) forces transmitted directly through the web (Fig. 5.7.1, EC3).*

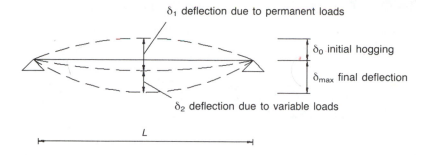

Fig. 9.5 *Vertical deflections.*

$$R_{y.Rd} = (s_s + s_y)t_w f_{yw}/\gamma_{M1} \qquad (9.20)$$

where

$$s_y = 2t_f (b_f/t_w)^{1/2} (f_{yf}/f_{yw})^{1/2} (1 - (\sigma_{f.Ed}/f_{yf})^2)^{1/2} \qquad (9.21)$$

in which b_f should not be more than $25t_f$, s_s is the length of stiff bearing and $\sigma_{f.Ed}$ is the longitudinal stress in the flange.

For a rolled I-, H- or U-section, s_y may alternatively be obtained from

$$s_y = \frac{2.5 (h - d) (1 - (\sigma_{f.Ed}/f_{yf})^2)^{1/2}}{(1 + 0.8 s_s / (h - d))} \qquad (9.22)$$

At the end of a member s_y should be halved.

Crippling resistance. For an I- or H-section, the design crippling resistance is

$$R_{a.Rd} = 0.5 t_w^2 (E f_{yw})^{1/2} ((t_f/t_w)^{1/2} + 3 (t_w/t_f) (s_s/d))/\gamma_{M1} \qquad (9.23)$$

in which s_s/d should not be more than 0.2. Where the member is also subject to bending moments the following relationship should be satisfied:

$$\frac{F_{Sd}}{R_{a.Rd}} + \frac{M_{Sd}}{M_{c.Rd}} \leqslant 1.5 \qquad (9.24)$$

Buckling resistance. For the web of I- or H-sections the design buckling resistance $R_{b.Rd}$ should be obtained by considering the web as a virtual compression member with an effective breadth

$$b_{eff} = (h^2 + s_s^2)^{1/2} \qquad (9.25)$$

although b_{eff} must be reduced at the ends of a member. The buckling resistance can then be obtained as discussed in *section 9.12.1.2*, using buckling curve c and $\beta_A = 1$.

Table 9.5 Recommended limiting values for vertical deflections (Table 4.1, EC3)

Conditions	Limits (see Fig 9.5)	
	δ_{max}	δ_2
Roofs generally	$L/200$	$L/250$
Roofs frequently carrying personnel other than for maintenance	$L/250$	$L/300$
Floors generally	$L/250$	$L/300$
Floors and roofs supporting plaster or other brittle finish or non-flexible partitions	$L/250$	$L/350$
Floors supporting columns (unless the deflection has been included in the global analysis for ultimate limit state)	$L/400$	$L/500$
Where δ_{max} can impair the appearance of the building	$L/250$	–

9.11.1.7 Deflections (clause 4.2, EC3)

Unlike BS 5950, EC3 recommends not one but two limiting values for vertical deflections, δ_2 and δ_{max}, as shown in *Table 9.5* and *Fig. 9.5*. The deflection in BS 5950 is comparable to the deflection δ_2, due to unfactored variable loads, in EC3. However, there are slight variations between the limiting values in the two codes, e.g. the recommended limiting deflection for beams carrying plaster or other brittle finishes is $L/360$ in BS 5950 and $L/350$ in EC3. The other deflection, δ_{max}, is not mentioned in BS 5950 and is the deflection due to the total permanent and variable loading. (*Note:* For cantilever beams the length L should be considered as twice the projecting length of the cantilever.)

Example 9.1 Analysis of a laterally restrained beam

Check the suitability of $356 \times 171 \times 51$ kg m^{-1} UB section in grade Fe 430 steel loaded by uniformly distributed loading $g_k = 8$ kN m^{-1} and $q_k = 6$ kN m^{-1} as shown below. Assume that the beam is fully laterally restrained and that the beam sits on 100 mm bearings at each end. Ignore self-weight of beam.

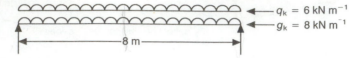

$q_k = 6$ kN m^{-1}
$g_k = 8$ kN m^{-1}
8 m

DESIGN BENDING MOMENT

Design action $(F_d) = (\gamma_G g_k + \gamma_Q q_k) \times$ span
$$= (1.35 \times 8 + 1.5 \times 6)\, 8 = 158.4 \text{ kN}$$

Design bending moment $(M_{Sd}) = \dfrac{F_d l}{8} = \dfrac{158.4 \times 8}{8}$
$$= 158.4 \text{ kN m}$$

STRENGTH CLASSIFICATION

Flange thickness $(t_f) = 11.5$ mm, steel grade Fe 430. Hence from *Table 9.3*, $f_y = 275$ N mm^{-2}.

SECTION CLASSIFICATION

$\varepsilon = (235/f_y)^{0.5} = (235/275)^{0.5} = 0.924$

$\dfrac{c}{t_f} = 7.46 < 10\varepsilon = 10 \times 0.924 = 9.24$

$\dfrac{d}{t_w} = 42.8 < 72\varepsilon = 72 \times 0.924 = 66.5$

Hence from Table 9.4, $356 \times 171 \times 51$ UB section belongs to class 1.

RESISTANCE OF CROSS-SECTION

Bending moment

Since the beam section belongs to class 1, design moment of resistance is equal to the plastic moment of resistance, $M_{pl.Rd}$, which is given by

$$M_{pl.Rd} = \frac{W_{pl} f_y}{\gamma_{M0}} = \frac{895 \times 10^3 \times 275}{1.05}$$

$$= 234.4 \times 10^6 \text{ N mm}$$

$$= 234.4 \text{ kN m} > M_{Sd} \,(= 158.4 \text{ kN m}) \quad \text{OK}$$

Shear

Design shear force, V_{Sd}, is

$$V_{Sd} = \frac{F_d}{2} = \frac{158.4}{2} = 79.2 \text{ kN}$$

For class 1 section design plastic shear resistance, $V_{pl.Rd}$, is given by

$$V_{pl.Rd} = A_v \, (f_y/\sqrt{3})/\gamma_{M0}$$

where

$$A_v = 1.04 h t_w = 1.04 \times 355.6 \times 7.3 = 2699.7 \text{ mm}^2$$

Hence

$$V_{pl.Rd} = 2699.7(275/\sqrt{3})/1.05$$
$$= 408\,224 \text{ N} = 408 \text{ kN} > V_{Sd} \quad \text{OK}$$

Bending and shear

According to EC3 the theoretical plastic resistance moment of the section, i.e. $M_{pl.Rd}$, is reduced if

$$V_{Sd} > 0.5V_{pl.Rd}$$

But in this case

$$V_{Sd} = 79.2 \text{ kN} \not> 0.5V_{pl.Rd} = 0.5 \times 408 = 204 \text{ kN}$$

Hence no check is required.

SHEAR BUCKLING RESISTANCE

As $d/t_w = 42.8 < 69\varepsilon = 69 \times 0.924 = 64$, no check on shear buckling is required.

FLANGE-INDUCED BUCKLING

Area of web is

$$A_w = (h - 2t_f)t_w = (355.6 - 2 \times 11.5)\,7.3 = 2428 \text{ mm}^2$$

Area of compression flange is

$$A_{fc} = bt_f = 171.5 \times 11.5 = 1972.2 \text{ mm}^2$$

For class 1 section, $k = 0.3$

$$k(E/f_{yf})[A_w/A_{fc}]^{1/2} = 0.3(210 \times 10^3/275)[2428/1972.2]^{1/2}$$
$$= 254.2 > d/t_w = 42.8 \quad \text{OK}$$

RESISTANCE OF THE WEB TO TRANSVERSE FORCES

According to EC3 when loads are resisted by shear in the web, the web should be checked for (a) crushing resistance and (b) crippling resistance.

Crushing resistance

For a rolled I-section

$$s_y = \frac{2.5\,(h - d)\,(1 - (\sigma_{f.Ed}/f_{yf})^2)^{1/2}}{(1 + 0.8s_s/(h - d))}$$

$$= \frac{2.5\,(355.6 - 312.3)\,(1 - (0)^2)^{1/2}}{(1 + 0.8 \times 100\,/\,(355.6 - 312.3))} = 38 \text{ mm}$$

Clause 5.7.3(3) of EC3 recommends that s_y should be halved at the ends of members. Hence, the design crushing resistance of web, $R_{y.Rd}$, is given by

$$R_{y.Rd} = (s_s + s_y)t_w\,f_{yw}/\gamma_{M1}$$
$$= (100 + 19)7.3 \times 275/1.05$$
$$= 227\,516 \text{ N} = 227.5 \text{ kN} > V_{Sd} = 79.2 \text{ kN} \quad \text{OK}$$

Crippling resistance

Design crippling resistance of web, $R_{a.Rd}$, is given by

$$R_{a.Rd} = 0.5t_w^2\,(Ef_{yw})^{1/2}\,[(t_f/t_w)^{1/2} + 3\,(t_w/t_f)\,(s_s/d)]\,/\gamma_{M1}$$

$$= 0.5 \times 7.3^2(210 \times 10^3 \times 275)^{1/2}[(11.5/7.3)^{1/2}$$
$$+ 3(7.3/11.5)\,(100/312.3)]/1.05$$

$$= 359\ 633\ \text{N} = 360\ \text{kN} > V_{Sd}\ (= 79.2\ \text{kN}) \quad \text{OK}$$

Hence section satisfies web crippling requirements.

DEFLECTION

The maximum bending moment due to working load is

$$M_{max} = 1/8\ (g_k + q_k)L^2$$

$$= 1/8\ (8 + 6)8^2 = 112\ \text{kN m}$$

Elastic resistance is

$$(M_{c.Rd})_{el} = \frac{W_{el}f_y}{\gamma_{M0}} = \frac{796 \times 10^3 \times 275}{1.05} = 208 \times 10^6\ \text{N mm}$$

$$= 208\ \text{kN m} > M_{max}$$

Hence deflection can be calculated elastically.
Deflection due to permanent and variable loading (w), i.e.

$$w = g_k + q_k = 8 + 6 = 14\ \text{kN m}^{-1} = 14\ \text{N mm}^{-1}$$

is

$$\delta_{max} = \frac{5}{384}\frac{wL^4}{EI} = \frac{5}{384} \times \frac{14 \times (8 \times 10^3)^4}{210 \times 10^3 \times 14200 \times 10^4}$$

$$= 25\ \text{mm} < \frac{\text{span}}{250} = \frac{8 \times 10^3}{250} = 32\ \text{mm} \quad \text{OK}$$

Deflection due to variable loading (q_k) = 6 kN m^{-1} = 6 N mm^{-1}.
Hence

$$\delta_2 = \frac{5wL^4}{384EI} = \frac{5 \times 6 \times (8 \times 10^3)^4}{384 \times 210 \times 10^3 \times 14200 \times 10^4}$$

$$= 11\ \text{mm} < \frac{\text{span}}{350} = \frac{8 \times 10^3}{350} = 22\ \text{mm} \quad \text{OK}$$

Example 9.2 Design of a laterally restrained beam

Select and check a suitable beam section using grade Fe 360 steel to support the loads shown below. Assume beam is fully laterally restrained and that it sits on 125 mm bearings at each end. Ignore self-weight of beam.

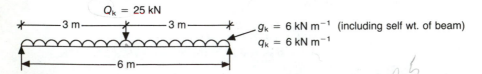

DESIGN BENDING MOMENT
Design action (F_d) = $(F_d)_{udl} + (F_d)_{pl}$

$$= (\gamma_G g_k + \gamma_Q q_k) \times \text{span} + \gamma_Q Q_k$$
$$= (1.35 \times 6 + 1.5 \times 6)6 + 1.5 \times 25$$
$$= 102.6 + 37.5 = 140.1\ \text{kN}$$

Design bending moment, M_{Sd}, is

$$M_{Sd} = \frac{(F_d)_{udl}\,l}{8} + \frac{(F_d)_{pl}\,l}{4} = \frac{102.6 \times 6}{8} + \frac{37.5 \times 6}{4}$$

$$= 133.2 \text{ kN m}$$

SECTION SELECTION

(Refer to 'Resistance of cross-sections – bending moments'). Assume suitable section belongs to class 1. Hence the minimum required plastic moment of resistance, W_{pl}, is given by

$$W_{pl} = \frac{M_{pl.Rd}\gamma_{M0}}{f_y}$$

Putting $M_{pl.Rd} = M_{Sd}$ and $f_y = 235 \text{ N mm}^{-2}$ (*Table 9.3*) gives

$$W_{pl} \geqslant \frac{133.2 \times 10^6 \times 1.05}{235} = 595.1 \times 10^3 \text{ mm}^3 = 595.1 \text{ cm}^3$$

From the steel table, try $356 \times 171 \times 45 \text{ kg m}^{-1}$ UB ($W_{pl} = 774 \text{ cm}^3$).

CHECK STRENGTH CLASSIFICATION

Flange thickness (t_f) = 9.7 mm. Hence from *Table 9.3*, $f_y = 235 \text{ N mm}^{-2}$ as assumed.

CHECK SECTION CLASSIFICATION

$$\varepsilon = (235/f_y)^{0.5} = (235/235)^{0.5} = 1$$

$$\frac{c}{t_f} = 8.81 < 10\varepsilon = 10 \quad \text{and}$$

$$\frac{d}{t_w} = 45.3 < 72\varepsilon = 72$$

Hence from *Table 9.4*, section belongs to class 1 as assumed.

RESISTANCE OF CROSS-SECTION

Bending moment

Plastic moment of resistance of $356 \times 171 \times 45$ UB is given by

$$M_{pl.Rd} = \frac{W_{pl}f_y}{\gamma_{M0}} = \frac{774 \times 10^3 \times 235}{1.05}$$

$$= 173 \times 10^6 \text{ N mm} = 173 \text{ kN m} > M_{Sd} \,(= 133.2 \text{ kN m}) \quad \text{OK}$$

Shear

Design shear force, $V_{Sd} = \dfrac{F_d}{2} = \dfrac{140.1}{2} = 70 \text{ kN}$

For class 1 section, design plastic shear resistance, $V_{pl.Rd}$, is given by

$$V_{pl.Rd} = A_v(f_y/\sqrt{3})/\gamma_{M0}$$

where

$$A_v = 1.04ht_w = 1.04 \times 352 \times 6.9 = 2526 \text{ mm}^2$$

Hence,

$$V_{pl.Rd} = 2526(235/\sqrt{3})/1.05$$

$$= 326.4 \times 10^3 \text{ N} = 326 \text{ kN} > V_{Sd} (= 70 \text{ kN}) \quad \text{OK}$$

Bending and shear

$$V_{Sd} (= 70 \text{ kN}) \not> 0.5V_{pl.Rd} = 0.5 \times 326 = 163 \text{ kN}$$

Hence no check is required.

SHEAR BUCKLING RESISTANCE

As $d/t_w = 45.3 < 69\varepsilon = 69$, no check on shear buckling is required.

FLANGE-INDUCED BUCKLING

Area of web is

$$A_w = (h - 2t_f)t_w = (352 - 2 \times 9.7)6.9 = 2294.9 \text{ mm}^2$$

Area of compression flange is

$$A_{fc} = bt_f = 171 \times 9.7 = 1658.7 \text{ mm}^2$$

$$k(E/f_{yf})[A_w/A_{fc}]^{1/2} = 0.3(210 \times 10^3/235)[2294.9/1658.7]^{1/2}$$

$$= 315.3 > d/t_w = 45.3 \quad \text{OK}$$

RESISTANCE OF THE WEB TO TRANSVERSE FORCES

Since loads are resisted by shear in the web, check web for (a) crushing resistance and (b) crippling resistance.

Crushing resistance

For a rolled I-section

$$s_y = \frac{2.5 (h - d) (1 - (\sigma_{f.Ed}/f_{yf})^2)^{1/2}}{(1 + 0.8s_s/(h - d))}$$

$$= \frac{2.5 (352 - 312.3) (1 - 0)^{1/2}}{1 + 0.8 \times 125/(352 - 312.3)} = 28 \text{ mm}$$

Clause 5.7.3(3) of EC3 recommends that s_y should be halved at the ends of members. Hence, design crushing resistance of web, $R_{y.Rd}$, is given by

$$R_{y.Rd} = (s_s + s_y) \ t_w f_{yw}/\gamma_{M1} = (125 + 14)6.9 \times 235/1.05$$

$$= 214.6 \times 10^3 \text{ N} = 214.6 \text{ kN} > V_{Sd} (= 70 \text{ kN}) \quad \text{OK}$$

Crippling resistance

Design crippling resistance of web, $R_{a.Rd}$, is given by

$$R_{a.Rd} = 0.5t_w^2(Ef_{yw})^{1/2}[(t_f/t_w)^{1/2} + 3(t_w/t_f) \ (s_s/d)]/\gamma_{M1}$$

$$= 0.5 \times 6.9^2(210 \times 10^3 \times 235)^{1/2}[(9.7/6.9)^{1/2}$$

$$+ 3(6.9/9.7) \ (125/312.3)]/1.05$$

$$= 324 \ 873 \text{ N} = 324.8 \text{ kN} > V_{Sd} (= 70 \text{ kN}) \quad \text{OK}$$

Hence section satisfies web crippling requirements.

DEFLECTION

Deflection due to variable uniformly distributed loading $q_k = 4$ kN m^{-1} = 4 N mm^{-1} and variable point load, $Q_k = 25\,000$ N is given by

$$\delta_2 = \frac{5q_kL^4}{384EI} + \frac{Q_kL^3}{48EI}$$

Hence

$$\delta_2 = \frac{5 \times 4 \times (6 \times 10^3)^4}{384 \times 210 \times 10^3 \times 12100 \times 10^4} + \frac{25 \times 10^3 \times (6 \times 10^3)^3}{48 \times 210 \times 10^3 \times 12100 \times 10^4}$$

$$= 2.6 + 4.4 \text{ mm}$$

$$= 7 \text{ mm} < \text{allowable} \left(\frac{\text{span}}{350} = \frac{6 \times 10^3}{350} = 17 \text{ mm} \right) \quad \text{OK}$$

Deflection due to permanent and variable loading, w and W, where

$$w = g_k + q_k = 4 + 4 = 8 \text{ kN m}^{-1} = 8 \text{ N mm}^{-1}$$

$$W = Q_k = 25 \text{ kN} = 25\,000 \text{ N}$$

is

$$\delta_{max} = \frac{5 \times 8 (6 \times 10^3)^4}{384 \times 210 \times 10^3 \times 12\,100 \times 10^4} + \frac{25 \times 10^3 (6 \times 10^3)^3}{48 \times 210 \times 10^3 \times 12\,100 \times 10^4}$$

$$= 5.3 + 4.4 \text{ mm}$$

$$= 9.7 \text{ mm} < \text{allowable} \left(\frac{\text{span}}{250} = \frac{6 \times 10^3}{250} = 24 \text{ mm} \right)$$

Example 9.3 Design of a cantilever beam

A cantilever beam is needed to resist the loading shown below. Select a suitable UB section in grade Fe 430 steel to satisfy bending and shear criteria only.

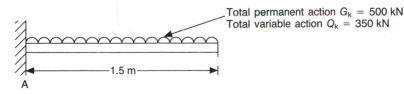

Total permanent action $G_k = 500$ kN
Total variable action $Q_k = 350$ kN

—1.5 m—

A

DESIGN BENDING MOMENT AND SHEAR FORCE

Design action $(F_d) = (\gamma_G G_k + \gamma_Q Q_k)$
$$= (1.35 \times 500 + 1.5 \times 350)$$
$$= 1200 \text{ kN}$$

Design bending moment at A (M_{Sd}) is

$$M_{Sd} = \frac{F_d l}{2} = 1200 \times \frac{1.5}{2} = 900 \text{ kN m}$$

Design shear force at A $(V_{Sd}) = F_d = 1200$ kN

SECTION SELECTION

Since the cantilever beam is relatively short and subject to fairly high shear forces, the bending capacity or the shear strength of the section may be critical and both factors will need to be considered in order to select an appropriate section. Hence, plastic moment of section (assuming beam belongs to class 1), W_{pl}, must exceed the following to satisfy bending:

$$W_{pl} = \frac{M_{pl.Rd}\gamma_{M0}}{f_y} = \frac{900 \times 10^6 \times 1.05}{275}$$

$$= 3436 \times 10^3 \text{ mm}^2 = 3436 \text{ cm}^3$$

The shear area of the section, A_v, must exceed the following to satisfy shear:

$$A_v = \frac{V_{pl.Rd}\gamma_{M0}}{(f_y/\sqrt{3})} = \frac{1200 \times 10^3 \times 1.05}{(275/\sqrt{3})} = 7936 \text{ mm}^2$$

where $A_v = 1.04ht_w$. Hence try $686 \times 254 \times 152$ kg m^{-1} UB section.

CHECK STRENGTH CLASSIFICATION

Flange thickness $(t_f) = 21$ mm, steel grade Fe 430. Hence from *Table 9.3*, $f_y = 275$ N mm^{-2} as assumed.

CHECK SECTION CLASSIFICATION

$$\varepsilon = (235/f_y)^{0.5} = (235/275)^{0.5} = 0.924$$

$$\frac{c}{t_f} = 6.06 < 10\varepsilon = 9.24$$

$$\frac{d}{t_w} = 46.6 < 72\varepsilon = 72 \times 0.924 = 66.5$$

Hence from *Table 9.4*, section belongs to class 1 as assumed.

RESISTANCE OF CROSS-SECTION

Bending moment
Plastic moment of resistance of $686 \times 254 \times 152$ UB is given by

$$M_{pl.Rd} = \frac{W_{pl}f_y}{\gamma_{M0}} = \frac{5000 \times 10^3 \times 275}{1.05}$$

$$= 1309 \times 10^6 \text{ N mm} > M_{Sd} \ (= 900 \text{ kN m}) \quad \text{OK}$$

Shear
For class 1 section, design plastic shear resistance is given by

$$V_{pl.Rd} = A_v(f_y/\sqrt{3})/\gamma_{M0}$$

where

$$A_v = 1.04ht_w = 1.04 \times 687.6 \times 13.2 = 9439 \text{ mm}^2$$

Hence

$$V_{pl.Rd} = 9439 \ (275/\sqrt{3})/1.05 = 1427.3 \times 10^3 \text{ N}$$

$$= 1427.3 \text{ kN} > V_{Sd} \ (= 1200 \text{ kN}) \quad \text{OK}$$

Bending and shear

Since

$$V_{Sd} = 1200 \text{ kN} > 0.5V_{pl.Rd} = 0.5 \times 1427.3 = 713.7 \text{ kN}$$

the section is subject to a 'high shear load' and the design moment of resistance of the section should be reduced to $M_{V.Rd}$ which is given by

$$M_{V.Rd} = f_y \, (W_{pl} - \rho \, A_v^2/4t_w)/\gamma_{M0}$$

where

$$\rho = (2V_{Sd}/V_{pl.Rd} - 1)^2 = (2 \times 1200/1427.3 - 1)^2 = 0.464$$

Hence

$$M_{V.Rd} = 275 \, (5000 \times 10^3 - 0.464 \times 9439^2/4 \times 13.2)/1.05$$

$$= 1104 \times 10^6 = 1104 \text{ kN m} > M_{Sd} \, (= 900 \text{ kN m}) \quad \text{OK}$$

Selected UB section, $686 \times 254 \times 152$ kg m^{-1}, is satisfactory in bending and shear.

Example 9.4 Design of a fully laterally restrained beam with overhang

The figure shows a simply supported beam and cantilever with uniformly distributed loads applied to it. Using grade Fe 430 steel and assuming full lateral restraint, select and check a suitable beam section.

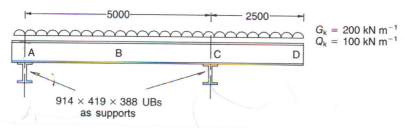

DESIGN BENDING MOMENT AND SHEAR FORCE

Load cases

(a) $1.35G_k + 1.5Q_k$ on span AC, $1.35G_k + 1.5Q_k$ on span CD;
(b) $1.35G_k + 1.5Q_k$ on span AC, $1.0G_k + 0.0Q_k$ on span CD;
(c) $1.0G_k + 0.0Q_k$ on span AC, $1.35G_k + 1.5Q_k$ on span CD.

Shear and bending

Shear force and bending moment diagrams for the three load cases are shown below. Hence $V_{Sd} = 1313$ kN and $M_{Sd} = 1313$ kN m.

STATIC EQUILIBRIUM

Overturning moment = $(1.35 \times 200 + 1.5 \times 100) \, 2.5 \times 1.25 = 1312.5$ kN m
Restorative moment $= (1.1 \times 200) \, 5 \times 2.5 = 2750$ kN m > 1312.5 kN m $\quad$ OK

SECTION SELECTION

$$W_{pl} = \frac{M_{pl.Rd}\gamma_{M0}}{f_y} = \frac{1313 \times 10^6 \times 1.05}{275} = 5.01 \times 10^6 \text{ mm}^3$$

Try $762 \times 267 \times 173$ UB section.

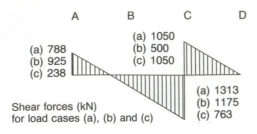

A   B   C   D

(a) 1050
(a) 788   (b) 500
(b) 925   (c) 1050
(c) 238

(a) 1313
(b) 1175
(c) 763

Shear forces (kN)
for load cases (a), (b) and (c)

(a) 1313   (a) 1313
(b) 1313   (b) 625
(c) 625   (c) 1313

Bending moments (kN m)
for load cases (a), (b) and (c)

STRENGTH CLASSIFICATION

Flange thickness = 21.6 mm, steel grade Fe 430. Hence from *Table 9.3*, $f_y = 275$ N mm^{-2}.

SECTION CLASSIFICATION

$$\varepsilon = (235/f_y)^{1/2} = 0.924$$

$$c/t_f = 6.17 < 10\varepsilon = 9.24 \quad d/t_w = 48 < 72\varepsilon = 66.5$$

Hence from *Table 9.4*, section belongs to class 1.

RESISTANCE OF CROSS-SECTIONS

Shear resistance

$$V_{pl.Rd} = A_v\,(f_y/\sqrt{3})/\gamma_{M0}$$

where

$$A_v = 1.04 h t_w = 1.04 \times 762 \times 14.3 = 11\,332 \text{ mm}^2$$

$$V_{pl.Rd} = \frac{11\,332 \times 275}{\sqrt{3} \times 1.05}$$

$$= 1713 \times 10^3 \text{ N} = 1713 \text{ kN} > V_{Sd}\,(= 1313 \text{ kN})$$

Bending and shear

$$V_{Sd} = 1313 \text{ kN} = 0.76 V_{pl.Rd} > 0.5 V_{pl.Rd}$$

Hence beam is subject to high shear load.

$$M_{V.Rd} = f_y(W_{pl} - \rho\,A_v^2/4t_w)/\gamma_{M0}$$

where

$$\rho = (2V_{Sd}/V_{pl.Rd} - 1)^2 = (2 \times 1313/1713 - 1)^2 = 0.284$$

$$M_{V.Rd} = 275\,(6200 \times 10^3 - 0.284 \times 11\,332^2/4 \times 14.3)/1.05$$

$$= 1456 \times 10^6 \text{ N mm} = 1456 \text{ kN m} > M_{Sd}\,(= 1313 \text{ kN m}) \quad \text{OK.}$$

SHEAR BUCKLING RESISTANCE

As will be seen under 'Resistance of web to transverse forces', the web needs stiffeners. Hence, provided d/t_w for the web is less than $30\varepsilon \sqrt{k_\tau}$ no check is required for resistance to shear buckling.

$$k_\tau = 5.34 \quad \text{(assuming stiffeners needed only at supports)}$$

$$\varepsilon = 0.924$$

Hence

$$30\varepsilon \sqrt{k_\tau} = 30 \times 0.924 \times \sqrt{5.34} = 64 > d/t_w \ (= 48.0) \quad \text{OK}$$

FLANGE-INDUCED BUCKLING

Area of web is

$$A_w = (h - 2t_f) \, t_w = (762 - 2 \times 21.6) \, 14.3 = 10\,279 \text{ mm}^2$$

Area of compression flange is

$$A_{fc} = bt_f = 266.7 \times 21.6 = 5761 \text{ mm}^2$$

$$k(E/f_{yf}) \, [\, A_w/A_{fc}]^{1/2} = 0.3 \times (210 \times 10^3/275)[10\,279/5761]^{1/2}$$

$$= 306 > d/t_w = 48.0$$

Hence no check is required.

RESISTANCE OF WEB TO TRANSVERSE FORCES

No check on web buckling is required when concentrated loads are resisted by shear in the web. We must, however, check the crippling and crushing resistance.

Web crippling

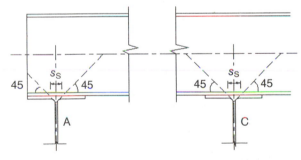

Stiff bearing length, $s_s = 2t_f + t_w + r$

$= 2 \times 36.6 + 21.5 + 24.1 = 118.8$ mm (based on properties of $914 \times 419 \times 388$ UBs as supports)

Thus

$$R_{a.Rd} = 0.5t_w^2(Ef_{yw})^{1/2}[(t_f/t_w)^{1/2} + 3 \, (t_w/t_f) \, (s_s/d)]/\gamma_{M1}$$

$$= 0.5 \times 14.3^2(210 \times 10^3 \times 275)^{1/2}[(21.6/14.3)^{1/2}$$

$$+ 3(14.3/21.6) \, (118.8/685.8)]/1.05$$

$$= 776\,994.7 \, [1.573]/1.05$$

$$= 1164 \times 10^3 \text{ N} = 1164 \text{ kN} < F_{Sd} \quad (= 2363 \text{ kN at C, load case (a)}) \quad \text{Not OK}$$

Where the member is also subject to bending moments, e.g. support C, clause 5.7.4(2) of EC3 recommends the following additional check:

$$\frac{F_{Sd}}{R_{a.Rd}} + \frac{M_{Sd}}{M_{c.Rd}} \leq 1.5 \qquad \frac{2363}{1164} + \frac{1313}{1456} = 3 > 1.5$$

Hence this is also not satisfied.

Web crushing

Support A

$$\sigma_{f.Ed} = 0$$

$$s_y = \frac{2.5\,(h-d)\,[1-(\sigma_{f.Ed}/f_{yf})^2]^{1/2}}{(1+0.8 s_s/(h-d))}$$

$$= \frac{2.5\,(762-685.8)\,[1-0]^{1/2}}{(1+0.8\times118.8/76.2)} = 84.8 \text{ mm}$$

Hence

$$R_{y.Rd} = (s_s + s_y)\,t_w f_{yw}/\gamma_{M1}$$

$$= (118.8 + 84.8/2)\,14.3 \times 275/1.05$$

$$= 603.7 \times 10^3 \text{ N} = 603 \text{ kN} \ (< 925 \text{ kN}) \quad \text{Not OK}$$

Support C

$$\sigma_{f.Ed} = \frac{M_{Sd}\,(h/2)}{I} = \frac{1313 \times 10^6 \times (762/2)}{2.05 \times 10^9} = 244 \text{ N mm}^{-2}$$

$$s_y = \frac{2.5\,(762-685.8)\,[1-(244/275)^2]^{1/2}}{(1+0.8\times118.8/76.2)} = 39.1 \text{ mm}$$

Hence

$$R_{y.Rd} = (118.8 + 39.1)14.3 \times 275/1.05$$

$$= 591.3 \times 10^3 \text{ N} = 591 \text{ kN} \ (< 2363 \text{ kN}) \quad \text{Not OK}$$

Stiffeners are therefore required at A and C with the stiffener at C being the more critical.

Stiffener design at C

(Refer to clause 5.7.6 of EC3 and *section 9.12.1*)
Where loadbearing stiffeners are provided, e.g. at A and C, the effective cross-section of stiffener/web needs to be checked for buckling and crushing resistance.

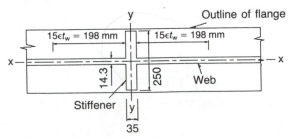

Buckling resistance. When checking the buckling resistance, the effective cross-section of a stiffener should be taken as including a width of web plate equal to $30\varepsilon t_w$, arranged with

$$15\varepsilon\,t_w = 15 \times (235/275)^{0.5}14.3 = 198 \text{ mm each side of the stiffener.}$$

$$\text{Buckling length, } l \geqslant 0.75d = 0.75 \times 685.8 = 514 \text{ mm}$$

Radius of gyration of stiffened section, i_x, is

$$i_x = (I/A)^{1/2} = [(1/12)\ 35 \times 250^3/(2 \times 198 \times 14.3 + 250 \times 35)]^{1/2}$$

$$= (45\ 572\ 916.6/14\ 412.8)^{1/2} = 56.2 \text{ mm}$$

$$\lambda = l/i = 514/56.2 = 9.15$$

$$\lambda_1 = 93.9\varepsilon = 93.9 \times 0.924 = 86.8$$

$$\overline{\lambda} = \lambda/\lambda_1 [\beta_A]^{1/2} = 9.15/86.8 = 0.105$$

where $\beta_A = 1$ (for class 1 section). By substituting these values into equations 9.38 and 9.39 (*section 9.12.1*) and putting $\alpha = 0.49$ gives $\chi_c = 1.0$. Hence, design buckling resistance, $N_{b.Rd}$, is given by (equation 9.37, *section 9.12.1*)

$$N_{b.Rd} = \chi_c \beta_A A_s f_y/\gamma_{M1} = 1 \times 1 \times 14412.8 \times 275/1.05$$

$$= 3.77 \times 10^6 \text{ N} = 3774 \text{ kN}\quad (> 2363 \text{ kN})\quad \text{OK}$$

Crushing resistance

$$\sigma_{f.Ed} = \frac{M_{Sd}\,(h/2)}{I} = \frac{1313 \times 10^6 \times (762/2)}{2.05 \times 10^9} = 244 \text{ N mm}^{-2}$$

$$s_y = \frac{2.5\ (762 - 685.8)\ [1 - (244/275)^2]^{1/2}}{(1 + 0.8 \times 118.8/76.2)} = 39.1 \text{ mm}$$

Area of stiffener cross-section, A, is given by

$$A = (118.8 + 39.1)14.3 + 35(250 - 14.3) = 10\ 507 \text{ mm}^2$$

Stiffener resistance, $N_{c.Rd}$, is given by

$$N_{c.Rd} = A f_y/\gamma_{M0} = 10\ 507 \times 275/1.05 = 2751 \times 10^3 > 2363 \text{ kN}\quad \text{OK}$$

Hence double-sided stiffener made of two plates 118×35 is suitable at support C. A similar stiffener should be provided at support A.

DEFLECTION

For a simply supported span

$$\delta_2 = \frac{5wL^4}{384EI}$$

where w is the unfactored variable load = 100 kN m^{-1} = 100 N mm^{-1},

$$\delta_2 = \frac{5 \times 100 \times 5000^4}{384 \times 210 \times 10^3 \times 2.05 \times 10^9} = 2 \text{ mm} = L/2580 \quad (< L/350)\quad \text{OK}$$

$$\delta_{max} = \frac{5w'L^4}{384EI}$$

where w', the total unfactored permanent and variable load, is $100 + 200 = 300$ kN m^{-1} = 300 N mm^{-1}. Hence

$$\delta_{max} = 6 \text{ mm} = L/833 \quad (< L/250)\quad \text{OK}$$

For a cantilever

$$\delta_2 = \frac{wL^4}{8EI} = \frac{100 \times 2500^4}{8 \times 210 \times 10^3 \times 2.05 \times 10^9}$$

$$= 1.1 \text{ mm} = L/2204 \quad (< L/175) \quad \text{OK}$$

$$\delta_{max} = 3.4 \text{ mm} = L/737 \quad (< L/125) \quad \text{OK}$$

Table 9.6 Values of factor C_1, C_2, and C_3 corresponding to various k factors and end moment or transverse loading combinations (based on Tables F.1.1 & F.1.2, EC3)

Loading and support conditions	Bending moment diagram	Value of k	Values of factors		
			C_1	C_2	C_3
	(a) $\psi = +1$	1.0	1.000		1.000
		0.7	1.000	–	1.113
		0.5	1.000		1.144
M ψM	(b) $\psi = 0$	1.0	1.879		0.939
		0.7	2.092	–	1.473
		0.5	2.150		2.150
	(c) $\psi = -1$	1.0	2.752		0.000
		0.7	3.063	–	0.000
		0.5	3.149		0.000
w	(d)	1.0	1.132	0.459	0.525
		0.5	0.972	0.304	0.980
w	(e)	1.0	1.285	1.562	0.753
		0.5	0.712	0.652	1.070
F	(f)	1.0	1.365	0.553	1.730
		0.5	1.070	0.432	3.050
F	(g)	1.0	1.565	1.267	2.640
		0.5	0.938	0.715	4.800
F F	(h)	1.0	1.046	0.430	1.120
		0.5	1.010	0.410	1.890

Design of beams

9.11.2. LATERAL TORSIONAL BUCKLING OF BEAMS (CLAUSE 5.5.2, EC3)

In order to prevent the possibility of a beam failure due to lateral torsional buckling, the designer needs to ensure that the buckling resistance, $M_{b.Rd}$ exceeds the design moment, M_{Sd}, i.e.

$$M_{b.Rd} \geq M_{Sd} \qquad (9.26)$$

This requires the designer to calculate the values of the following additional parameters:

1. geometrical slenderness ratio λ_{LT}
2. slenderness ratio $\bar{\lambda}_{LT}$
3. buckling factor χ_{LT}
4. buckling resistance $M_{b.Rd}$

9.11.2.1. Geometrical slenderness ratio λ_{LT} (Annex F, clause F.2.2,EC3)

The geometrical slenderness ratio is given by

$$\lambda_{LT} = \frac{L(W_{pl.y}^2/I_z I_y)^{1/4}}{C_1^{1/2}[1+(L^2 GI_t/\pi^2 EI_w)]^{1/4}} \qquad (9.27)$$

where L is the length of beam between points which have lateral restraint, $W_{pl.y}$ the plastic modulus about the major axis, I_z the second moment of area about the minor axis, I_y the second moment of area about the major axis, I_t the torsional constant, I_w the warping constant, E the modulus of elasticity (210 000 N mm^{-2}), G the shear modulus equal to

$$\frac{E}{2(1+\upsilon)} = \frac{210\ 000}{2(1+0.3)} = 80\ 769\ \text{N mm}^{-2}$$

and C_1 the factor depending on the loading and end restraint conditions as indicated by Ψ and k (Table 9.6). Note that Ψ is the ratio of end moments over the length L between lateral restraints, and the effective length factor k varies from 0.5 for full fixity to

1.0 for no fixity, with 0.7 for one end fixed and one end free.

9.11.2.2. Slenderness ratio $\bar{\lambda}_{LT}$

The value of $\bar{\lambda}_{LT}$ may be determined from

$$\bar{\lambda}_{LT} = (\lambda_{LT}/\lambda_1)(\beta_W)^{1/2} \qquad (9.28)$$

where

$$\lambda_1 = \pi(E/f_y)^{1/2} = 93.9\varepsilon \qquad (9.29)$$

where $\varepsilon = (235/f_y)^{1/2}$, $\beta_W = 1$ for class 1 and 2 sections, $\beta_W = W_{el.y}/W_{pl.y}$ for class 3 cross-sections and $\beta_W = W_{eff.y}/W_{pl.y}$ for class 4 cross-sections.

9.11.2.3 Buckling factor χ_{LT} (clause 5.5.2, EC3)

Here χ_{LT} is the reduction factor for lateral torsional buckling which is given by

$$\chi_{LT} = \frac{1}{\phi_{LT}+(\phi_{LT}^2-\bar{\lambda}_{LT}^2)^{1/2}} \leq 1 \qquad (9.30)$$

in which

$$\phi_{LT} = 0.5(1+\alpha_{LT}(\bar{\lambda}_{LT}-0.2)+\bar{\lambda}_{LT}^2) \, (9.31)$$

The imperfection factor, α_{LT}, for lateral torsional buckling should be taken as 0.21 for rolled sections, more for welded sections.

9.11.2.4 Buckling resistance $M_{b.Rd}$ (clause 5.5.2, EC3)

The design buckling resistance moment of a laterally unrestrained beam is given by

$$M_{b.Rd} = \chi_{LT}\beta_W W_{pl.y}f_y/\gamma_{M1} \qquad (9.32)$$

where f_y is the yield strength (*Table 9.4*) and γ_{M1} the partial safety factor for buckling = 1.05 (*section 9.9.3*).

Example 9.5 Analysis of a beam restrained at the supports

Assuming that the beam in *Example 9.1* is only laterally and torsionally restrained at the supports, determine whether a $356 \times 171 \times 51$ kg m^{-1} UB section in grade Fe 430 steel is still suitable .

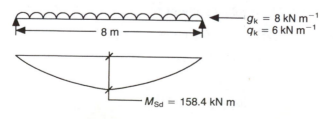

297

SECTION PROPERTIES
From steel tables (*Appendix B*)

Depth of section, h	$= 355.6$ mm
Thickness of flange, t_f	$= 11.5$ mm
Second moment of area about the minor axis, I_z	$= 968 \times 10^4$ mm^4
Radius of gyration about z–z axis, i_z	$= 38.7$ mm
Elastic modulus about the major axis, $W_{el.y}$	$= 796 \times 10^3$ mm^3
Plastic modulus about the major axis $W_{pl.y}$	$= 895 \times 10^3$ mm^3
Warping constant, I_w	$= 286 \times 10^9$ mm^6
Torsional constant, I_t	$= 236 \times 10^3$ mm^4
Yield strength of Fe 430 steel, f_y	$= 275$ N mm^{-2}

$$\text{Shear modulus, } G = \frac{E}{2(1+\upsilon)} = \frac{210 \times 10^3}{2(1+0.3)} \qquad = 80\ 769 \text{ N mm}^{-2}$$

LATERAL BUCKLING RESISTANCE

Geometrical slenderness ratio, λ_{LT}, for lateral torsional buckling.
The effective length factor $k = 1.0$ since compression flange is laterally unrestrained. From *Table 9.6*, factor $C_1 = 1.132$. Length of beam between points which are laterally restrained, $L = 8000$ mm. Buckling slenderness, λ_{LT}, is

$$\lambda_{LT} = \frac{L\left[\dfrac{W_{pl.y}^2}{I_z I_w}\right]^{1/4}}{C_1^{1/2}\left[1 + \dfrac{L^2 G I_t}{\pi^2 E I_w}\right]^{1/4}}$$

$$= \frac{8000\left[\dfrac{(895 \times 10^3)^2}{968 \times 10^4 \times 286 \times 10^9}\right]^{1/4}}{1.132^{1/2}\left[1 + \dfrac{8000^2 \times 80\ 769 \times 236 \times 10^3}{3.142^2 \times 210 \times 10^3 \times 286 \times 10^9}\right]^{1/4}}$$

$$= 131.87$$

Slenderness ratio $\overline{\lambda}_{LT}$ for lateral torsional buckling

$$\beta_w = 1 \quad \text{(for class 1 cross-section)}$$

$$\varepsilon = (235/f_y)^{1/2} = (235/275)^{1/2} = 0.924$$

$$\lambda_1 = 93.9\varepsilon = 93.9 \times 0.924 = 86.8$$

$$\overline{\lambda}_{LT} = (\lambda_{LT}/\lambda_1)\beta_w^{1/2} = (131.87/86.8)1^{1/2} = 1.52$$

Buckling factor χ_{LT}
Imperfection factor $\alpha_{LT} = 0.21$ for rolled sections (clause 5.5.2(3) of EC3). Buckling factor, ϕ_{LT}, is

$$\phi_{LT} = 0.5[1 + \alpha_{LT}(\overline{\lambda}_{LT} - 0.2) + \overline{\lambda}_{LT}^2]$$

$$= 0.5[1 + 0.21(1.52 - 0.2) + 1.52^2] = 1.794$$

$$\chi_{LT} = \frac{1}{\phi_{LT} + [\phi_{LT}^2 - \overline{\lambda}_{LT}^2]^{1/2}} \leq 1$$

$$= \frac{1}{1.794 + [1.794^2 - 1.52^2]^{1/2}} = 0.364 < 1 \quad \text{OK}$$

Buckling resistance

Buckling resistance of beam, $M_{b.Rd}$ is

$$M_{b.Rd} = \chi_{LT}\beta_W W_{pl.y} f_y / \gamma_{M1}$$
$$= 0.364 \times 1 \times 895 \times 10^3 \times 275/1.05$$
$$= 85.3 \times 10^6 \text{ N mm} = 85.3 \text{ kN m} < M_{Sd} = 158.4 \text{ kN m} \quad (Example\ 9.1)$$

Since buckling resistance of beam (85.3 kN m) < design moment (158.4 kN m), the beam section is unsuitable.

Example 9.6 Analysis of a beam restrained at mid-span and supports

Repeat *Example 9.5*, but this time assume that the beam is laterally and torsionally restrained at mid-span and at the supports.

SECTION PROPERTIES
See *Example 9.5*.

LATERAL BUCKLING RESISTANCE

$k = 1.0$ (since compression flange unrestrained between supports and mid-span)

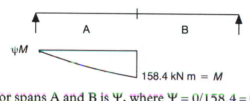

Ratio of end moments for spans A and B is Ψ, where $\Psi = 0/158.4 = 0$. Hence from *Table 9.6(b)*, i.e. $\Psi = 0$ and $k = 1$, $C_1 = 1.879$. Length of beam between points which are laterally restrained, $L = 4000$ mm.

$$\lambda_{LT} = \frac{L\left[\dfrac{W_{pl.y}^2}{I_z I_w}\right]^{1/4}}{C_1^{1/2}\left[1 + \dfrac{L^2 G I_t}{\pi^2 E I_w}\right]^{1/4}}$$

$$= \frac{4000\left[\dfrac{(895 \times 10^3)^2}{968 \times 10^4 \times 286 \times 10^9}\right]^{1/4}}{1.879^{1/2}\left[1 + \dfrac{(4 \times 10^3)^2 \times 80\ 769 \times 236 \times 10^3}{\pi^2 \times 210 \times 10^3 \times 286 \times 10^9}\right]^{1/4}}$$

$$= 61.0$$

Note that clause F.2.2.(4) of EC3 permits the following conservative approximation to be used to determine λ_{LT} for rolled I- or H-sections.

$$\lambda_{LT} = \frac{0.9L/i_z}{C_1^{1/2}\left[1 + \frac{1}{20}\left[\frac{L/i_z}{h/t_f}\right]^2\right]^{1/4}} \tag{9.33}$$

$$= \frac{0.9 \times 4 \times 10^3/38.7}{1.879^{1/2}\left[1 + \frac{1}{20}\left[\frac{4 \times 10^3/38.7}{355.6/11.5}\right]^2\right]^{1/4}}$$

$$= 60.7$$

$$\overline{\lambda}_{LT} = (\lambda_{LT}/\lambda_1)\beta_W^{1/2} = (61/93.9 \times 0.924)\,1^{1/2} = 0.70$$

$$\alpha_{LT} = 0.21$$

$$\phi_{LT} = 0.5[1 + \alpha_{LT}(\overline{\lambda}_{LT} - 0.2) + \overline{\lambda}_{LT}^2]$$

$$= 0.5[1 + 0.21(0.70 - 0.2) + 0.70^2] = 0.80$$

Buckling factor

$$\chi_{LT} = \frac{1}{\phi_{LT} + [\phi_{LT}^2 - \overline{\lambda}_{LT}^2]^{1/2}} \leqslant 1$$

$$= \frac{1}{0.80 + [0.80^2 - 0.70^2]^{1/2}} = 0.84$$

Buckling resistance of beam, $M_{b.Rd}$, is
$$M_{b.Rd} = \chi_{LT}\beta_W\,W_{pl.y}f_y/\gamma_{M1}$$
$$= 0.84 \times 1 \times 895 \times 10^3 \times 275/1.05$$
$$= 196.9 \times 10^6 \text{ N mm} = 196.9 \text{ kN m} > M_{Sd} = 158.4 \text{ kN m}$$

Hence the beam section is now suitable.

9.12 Design of columns

The design of columns is largely covered in Chapter 5 of EC3. However, unlike BS 5950, EC3 does not include the design of cased columns, which is left to Eurocode 4: *Steel Concrete Composite Structures*. EC4 is due to be published as an ENV in 1993 and the early indications are that the design procedures for cased columns in EC4 and BS 5950 will be similar.

The following subsections will consider EC3 requirements in respect of the design of:

1. compression members;
2. members resisting combined axial force and moment;
3. simple column baseplates.

9.12.1 COMPRESSION MEMBERS

According to clause 5.1.4 of EC3, compression members (i.e. struts) should be checked for (1) resistance to compression failure and (2) resistance to buckling.

9.12.1.1 Resistance of cross-sections – compression (clause 5.4.4., EC3)

For members in axial compression, the design value of compressive force N_{Sd} at each cross-section should satisfy

$$N_{Sd} \leqslant N_{c.Rd} \tag{9.34}$$

where $N_{c.Rd}$ is the design compressive resistance of cross-section, taken as the smaller of:

1. the design plastic compressive resistance of the gross section

$$N_{pl.Rd} = Af_y/\gamma_{M0} \tag{9.35}$$

(for classes 1–3 cross-sections);

2. the design local buckling resistance of the gross section

Table 9.7 Imperfection factors (Table 5.5.1, EC3)

Buckling curve	a	b	c	d
Imperfection factor α	0.21	0.34	0.49	0.76

$$N_{o.Rd} = A_{eff}f_y/\gamma_{M1} \qquad (9.36)$$

(for class 4 cross-sections), where A_{eff} is the effective area of section.

9.12.1.2. Buckling resistance of members
(clause 5.5.1, EC3)

The design buckling resistance of a compression member should be taken as

$$N_{b.Rd} = \chi\beta_A A f_y/\gamma_{M1} \qquad (9.37)$$

where $\beta_A = 1$ for classes 1–3 cross-sections, $\beta_A = A_{eff}/A$ for class 4 cross-sections and χ is the reduction factor for the relevant buckling mode, determined as follows or from EC3, Table 5.5.2 (not reproduced here).

For uniform members subjected to axial compression

$$\chi = \frac{1}{\phi + \left[\phi^2 - \overline{\lambda}^2\right]^{1/2}} \leq 1 \qquad (9.38)$$

where

$$\phi = 0.5(1 + \alpha(\overline{\lambda} - 0.2) + \overline{\lambda}^2) \qquad (9.39)$$

and where α is an imperfection factor from *Table 9.7*.

$$\overline{\lambda} = (\beta_A A f_y/N_{cr})^{1/2} = (\lambda/\lambda_1)\,(\beta_A)^{1/2} \qquad (9.40)$$

where λ is the slenderness for the relevant buckling mode (see below).

$$\lambda_1 = \pi(E/f_y)^{1/2} = 93.9\varepsilon \qquad (9.41)$$

where $\varepsilon = (235/f_y)^{1/2}$

Table 9.8 indicates which of the buckling curves is to be used.

The slenderness of a member, λ, is given by

$$\lambda = l/i \qquad (9.42)$$

where l is the buckling length of the member which can be obtained using the assumption in clause 5.5.1.5 of EC3 or from Annex E of EC3, and i the radius of gyration about the relevant axis. Note that buckling length in EC3 is the same as effective length in BS 5950. Clause 5.5.1.5 of EC3 states that the buckling length of a member may conservatively be taken as equal to its system (actual) length provided that both ends of the member are effectively held in position laterally.

Annex E – 'Buckling length of a compression member' goes into considerably more detail than BS 5950 on this subject. Essentially the buckling length is obtained from charts after calculation of distribution factors η_1 and η_2 based on member stiffness coefficients for appropriate subframe models. The buckling lengths are, however, identical to BS 5950 effective lengths, and could therefore be obtained from BS 5950, Table 24 (*Table 4.20*).

9.12.2. MEMBERS SUBJECT TO COMBINED AXIAL FORCE AND MOMENT

Clause 5.1.6 of EC3 states that members subject to combined axial (compression) force and moment should be checked for the following ultimate limit

Table 9.8 Selection of buckling curve for a cross section (Table 5.5.3, EC3)

Cross-section	Limits	Buckling about axis	Buckling curve
Rolled I-sections	h/b > 1.2		
	t_f ≤ 40 mm	y-y	a
		z-z	b
	40mm < t_f ≤ 100 mm	y-y	b
		z-z	c
	h/b ≤ 1.2		
	t_f ≤ 100mm	y-y	b
		z-z	c
	t_f > 100mm	y-y	d
		z-z	d

states: (1) resistance of cross-sections to the combined effects and (2) buckling resistance of member to the combined effects.

9.12.2.1 Resistance of cross-sections – bending and axial force (clause 5.4.8, EC3)

Classes 1 and 2 cross-sections. For classes 1 and 2 cross-sections the criterion to be satisfied in the absence of shear force is

$$M_{Sd} \leq M_{N.Rd} \tag{9.43}$$

where $M_{N.Rd}$ is the reduced design plastic resistance moment allowing for the axial force. No reduction is necessary provided that the axial force does not exceed half the plastic tension resistance of the web, or a quarter of the plastic tension resistance of the whole cross-section, whichever is the smaller.

For larger axial loads the following approximations can be used for standard rolled I- and H-sections:

$$M_{Ny} = M_{pl.y} (1 - n)/(1 - 0.5a) \leq M_{pl.y} \tag{9.44}$$

$$M_{Nz} = M_{pl.z} \left[1 - \left[\frac{n-a}{1-a} r \right]^2 \right] \leq M_{pl.z} \tag{9.45}$$

where

$$n = N_{Sd}/N_{pl.Rd} \tag{9.46}$$

and

$$a = (A - 2bt_f)/A \leq 0.5 \tag{9.47}$$

The above approximations may be further simplified for rolled I- and H- sections only to

$$M_{Ny} = 1.11 M_{pl.y}(1 - n) \leq M_{pl.y} \tag{9.48}$$

$$M_{Nz} = 1.56 M_{pl.z}(1 - n)(n + 0.6) \leq M_{pl.z} \tag{9.49}$$

For biaxial bending, the following approximate criterion can be used:

$$\left[\frac{M_{y.Sd}}{M_{Ny.Rd}} \right]^\alpha + \left[\frac{M_{z.Sd}}{M_{Nz.Rd}} \right]^\beta \leq 1 \tag{9.50}$$

in which $\alpha = 2$ and $\beta = 5n$ but ≥ 1 for I- and H-sections. As a further conservative approximation the following may be used:

$$\frac{N_{Sd}}{Af_y/\gamma_{M0}} + \frac{M_{y.Sd}}{W_{pl.y}f_y/\gamma_{M0}} + \frac{M_{z.Sd}}{W_{pl.z}f_y/\gamma_{M0}} \leq 1 \tag{9.51}$$

Class 3 cross-sections. In the absence of shear force, class 3 cross-sections will be satisfactory if the maximum longitudinal stress does not exceed the design yield strength, i.e.

$$\sigma_{x.Ed} \leq f_{yd} \tag{9.52}$$

where $f_{yd} = f_y/\gamma_{M0}$. For cross-sections without fastener holes, this becomes

$$\frac{N_{Sd}}{Af_{yd}} + \frac{M_{y.Sd}}{W_y f_{yd}} + \frac{M_{z.Sd}}{W_z f_{yd}} \leq 1 \tag{9.53}$$

Class 4 cross-sections. For class 4 cross-sections the above approach should also be used, but calculated using effective, rather than actual, widths of compression elements (clause 5.4.8.3 of EC3).

9.12.2.2 Buckling resistance of members – combined bending and axial compression (clause 5.5.4, EC3)

Again this is presented in a rather cumbersome manner in EC3. We will confine ourselves to the interaction formulae for classes 1 and 2 members:

$$\frac{N_{Sd}}{\chi_{min} Af_y/\gamma_{M1}} + \frac{k_y M_{y.Sd}}{W_{pl.y}f_y/\gamma_{M1}} + \frac{k_z M_{z.Sd}}{W_{pl.z}f_y/\gamma_{M1}} \leq 1 \tag{9.54}$$

in which

$$k_y = 1 - \frac{\mu_y N_{Sd}}{\chi_y Af_y} \leq 1.5 \tag{9.55}$$

$$\mu_y = \bar{\lambda}_y (2\beta_{My} - 4) + \left[\frac{W_{pl.y} - W_{el.y}}{W_{el.y}} \right] \leq 0.9 \tag{9.56}$$

$$k_z = 1 - \frac{\mu_z N_{Sd}}{\chi_z Af_y} \leq 1.5 \tag{9.57}$$

$$\mu_z = \bar{\lambda}_z (2\beta_{Mz} - 4) + \left[\frac{W_{pl.z} - W_{el.z}}{W_{el.z}} \right] \leq 0.9 \tag{9.58}$$

Here χ_{min} is the lesser of χ_y and χ_z, which are the reduction factors for the y–y and z–z axes respectively; β_{My} and β_{Mz} are equivalent uniform moment factors for flexural buckling.

Members for which lateral torsional buckling is a potential failure mode should also satisfy

$$\frac{N_{Sd}}{\chi_z Af_y/\gamma_{M1}} + \frac{k_{LT} M_{y.Sd}}{\chi_{LT} W_{pl.y}f_y/\gamma_{M1}} + \frac{k_z M_{z.Sd}}{W_{pl.z}f_y/\gamma_{M1}} \leq 1 \tag{9.59}$$

in which

Moment diagram	Equivalent uniform moment factor β_M
End moments  M_1 ▨ ψM_1 $-1 \leqslant \psi \leqslant 1$	$\beta_{M,\psi} = 1.8 - 0.7\psi$
Moments due to in-plane lateral loads 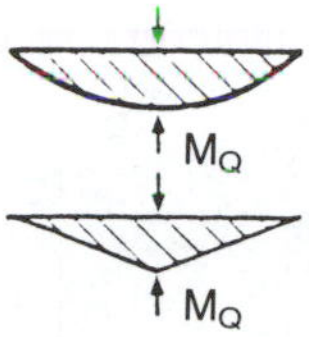 M_Q M_Q	$\beta_{M,Q} = 1.3$ $\beta_{M,Q} = 1.4$

Fig. 9.6 *Equivalent uniform moment factors (based on Fig. 5.5.3 of EC3).*

$$k_{LT} = 1 - \frac{\mu_{LT} N_{Sd}}{\chi_z A f_y} \leqslant 1 \qquad (9.60)$$

$$\mu_{LT} = 0.15 \overline{\lambda}_z \beta_{M.LT} - 0.15 \leqslant 0.90 \qquad (9.61)$$

where $\beta_{M.LT}$ is an equivalent uniform moment factor for lateral torsional buckling and is obtained from *Figure 9.6* according to the shape of the bending moment diagram between braced points.

9.12.3 SIMPLE COLUMN BASEPLATES (ANNEX L, CLAUSE L.1, EC3)

Generally, column baseplates should be checked to ensure (1) the bearing pressure does not exceed the design bearing strength of the foundations and (2) the bending moment in the compression or tension region of the baseplate does not exceed the resistance moment.

9.12.3.1 Bearing pressure and strength

The aim of design is to ensure that the bearing pressure does not exceed the bearing strength of the concrete, f_j, i.e.

$$\text{bearing pressure} = \frac{N_{Sd}}{A_{eff}} \leqslant f_j \qquad (9.62)$$

where N_{Sd} is the design axial force on column and A_{eff} the area in compression under the baseplate. The bearing strength of concrete foundations can be determined using

$$f_j = \beta_j k_j f_{cd} \qquad (9.63)$$

where f_{cd} is the design value of the concrete cylinder compressive strength of the foundation and is given by

$$f_{cd} = f_{ck}/\gamma_m \qquad (9.64)$$

and where β_j is the joint (concrete) coefficient which may generally be taken as two-thirds and k_j, the concentration factor, which may generally be taken as 1.0.

Figure 9.7 shows the effective bearing areas under axially loaded column baseplates. The additional bearing width x is given by

$$x = t[f_y/3f_j\gamma_{M0}]^{1/2} \qquad (9.65)$$

where t is the thickness of the baseplate, f_y the yield strength of baseplate material and f_j the bearing strength of the foundations.

9.12.3.2 Resistance moment

To prevent bending failure, the design bending moment in the baseplate, m_{Sd}, must not exceed the resistance moment, m_{Rd}:

$$m_{Sd} < m_{Rd} \qquad (9.66)$$

The bending moment in the baseplate is given by

$$m_{Sd} = (x^2/2)N_{Sd}/A_{eff} \qquad (9.67)$$

The resistance moment is given by

$$m_{Rd} = \frac{t^2 f_y}{6\gamma_{M0}} \qquad (9.68)$$

303

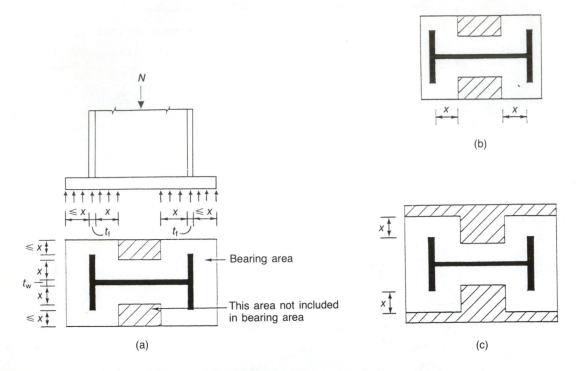

Fig. 9.7 *Area in compression under base plate: (a) general case; (b) short projection; (c) large projetion (Fig. L.1, EC3).*

Example 9.7 Analysis of a column resisting an axial load

Check the suitability of the $203 \times 203 \times 60$ kg m^{-1} UC section in grade Fe 430 steel to resist a design axial compression force of 1400 kN. Assume the column is pinned at both ends and that its height is 6 m.

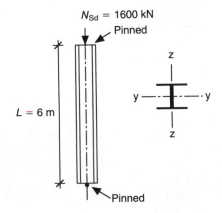

SECTION PROPERTIES
From steel tables (*Appendix B*)

Area of section $(A) = 7580$ mm^2
Thickness of flange $(t_f) = 14.2$ mm

Radius of gyration about the y–y axis (i_y) = 89.6 mm
Radius of gyration about the z–z axis (i_z) = 51.9 mm

STRENGTH CLASSIFICATION
Flange thickness = 14.2 mm, steel grade Fe 430. Hence from *Table 9.3*, f_y = 275 N mm^{-2}.

SECTION CLASSIFICATION

$$\varepsilon = (235/f_y)^{0.5} = (235/275)^{0.5} = 0.924$$

$$c/t_f = 7.23 < 10\varepsilon = 9.24$$

Also

$$d/t_w = 17.3 < 33\varepsilon = 33 \times 0.924 = 30.5$$

Hence from *Table 9.4*, section belongs to class 1.

RESISTANCE OF CROSS-SECTION – COMPRESSION
Plastic compressive resistance of section, $N_{pl.Rd}$, for class 1 section is given by

$$N_{pl.Rd} = \frac{Af_y}{\gamma_{M0}} = 7580 \times 275/1.05$$

$$= 1985 \times 10^3 \text{ N} = 1985 \text{ kN} > N_{Sd} = 1600 \text{ kN}$$

BUCKLING RESISTANCE OF MEMBER
Effective length of column about both axes is given by $L_{eff} = L_{eff\,y} = L_{eff\,z} = 1.0L = 1.0 \times 6000 = 6000$ mm

The column will buckle about the weak (z–z) axis. Slenderness ratio about z–z axis (λ_z) is

$$\lambda_z = \frac{L_{eff\,z}}{i_z} = \frac{6000}{51.9} = 115.6$$

Here $\beta_A = 1$ for class 1 section, $\lambda_1 = 93.9\varepsilon = 93.9 \times 0.924 = 86.76$.

$$\bar{\lambda}_z = (\lambda_z/\lambda_1)(\beta_A)^{1/2} = (115.6/86.76)(1)^{1/2} = 1.332$$

$$\frac{h}{b} = \frac{209.6}{205.2} = 1.02 < 1.2 \quad \text{and} \quad t_f = 14.2 \text{ mm}$$

Hence from *Table 9.8*, for buckling about z–z axis buckling curve c is appropriate and from *Table 9.7*, α=0.49.

$$\phi = 0.5[1 + \alpha(\bar{\lambda}_z - 0.2) + \bar{\lambda}_z^2]$$

$$= 0.5[1 + 0.49(1.332 - 0.2) + 1.332^2] = 1.665$$

$$\chi_z = \frac{1}{\phi + [\phi^2 - \bar{\lambda}_z^2]^{1/2}}$$

$$= \frac{1}{1.665 + [1.665^2 - 1.332^2]^{1/2}} = 0.375$$

Hence design buckling resistance, $N_{b.Rd}$, is given by

$$N_{b.Rd} = \chi_z \beta_A A f_y / \gamma_{M1} = 0.375 \times 1 \times 7580 \times 275/1.05$$

$$= 744 \times 10^3 \, N = 744 \, kN < 1400 \, kN$$

The section is therefore unsuitable to resist the design force.

Example 9.8 Analysis of a column with a tie-beam at mid-height

Recalculate the axial compression resistance of the column in *Example 9.7* if a tie-beam is introduced at mid-height such that in-plane buckling about the z–z axis is prevented (see below).

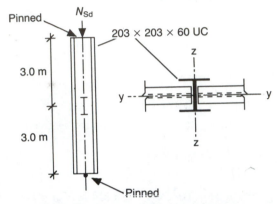

RESISTANCE OF CROSS-SECTION – COMPRESSION
Design plastic compressive resistance of section, $N_{pl.Rd} = 1985 \, kN$, as above.

BUCKLING RESISTANCE OF MEMBER

Buckling about y–y axis
Effective length of column about y–y axis is given by

$$L_{eff\,y} = 1.0L = 1.0 \times 6000 = 6000 \, mm$$

Slenderness ratio about y–y axis (λ_y) is

$$\lambda_y = \frac{L_{eff\,y}}{i_y} = \frac{6000}{89.6} = 67$$

$$\beta_A = 1, \quad \lambda_1 = 86.76 \quad \text{(see above)}$$

$$\overline{\lambda}_y = (\lambda_y/\lambda_1)(\beta_A)^{1/2} = (67/86.76)(1)^{1/2} = 0.77$$

$$h/b = 1.02 \quad \text{and} \quad t_f = 14.2 \, mm$$

Hence from *Table 9.8*, buckling curve b appropriate, and from *Table 9.7* imperfection factor $\alpha = 0.34$.

$$\phi = 0.5[1 + \alpha(\overline{\lambda}_y - 0.2) + \overline{\lambda}_y^2]$$

$$= 0.5[1 + 0.34(0.77 - 0.2) + 0.77^2] = 0.893$$

$$\chi_y = \frac{1}{\phi + [\phi^2 - \overline{\lambda}_y^2]^{1/2}} = \frac{1}{0.893 + [0.893^2 - 0.77^2]^{1/2}} = 0.743$$

Hence design buckling resistance about the y–y axis is given by

$$N_{b.Rd} = \chi_y \beta_A A f_y / \gamma_{M1} = 0.743 \times 1 \times 7580 \times 275/1.05$$

$$= 1475 \times 10^3 \, N = 1475 \, kN$$

Buckling about z–z axis

Effective length of column about z–z axis, $L_{eff\ z}$, is equal to $L_{eff\ z} = 3000$ mm. Slenderness ratio about z–z axis (λ_z) is

$$\lambda_z = \frac{L_{eff\ z}}{i_z} = \frac{3000}{51.9} = 57.8$$

$$\beta_A = 1 \qquad \lambda_1 = 86.76 \quad \text{(see above)}$$

$$\overline{\lambda}_z = (\lambda_z/\lambda_1)(\beta_A)^{1/2} = (57.8/86.76)(1)^{1/2} = 0.666$$

$$h/b = 1.02 \quad \text{and} \quad t_f = 14.2 \, mm$$

Hence from *Table 9.8*, buckling curve c appropriate and, from *Table 9.7*, imperfection factor $\alpha = 0.49$.

$$\phi = 0.5[1 + \alpha(\overline{\lambda}_z - 0.2) + \overline{\lambda}_z^2]$$

$$= 0.5[1 + 0.49(0.666 - 0.2) + 0.666^2] = 0.836$$

$$\chi_z = \frac{1}{\phi + [\phi^2 - \overline{\lambda}_z^2]^{1/2}}$$

$$= \frac{1}{0.836 + [0.836^2 - 0.666^2]^{1/2}} = 0.746$$

$$N_{b.Rd} = \chi_z \beta_A A f_y / \gamma_{M1} = 0.746 \times 1 \times 7580 \times 275/1.05$$

$$= 1481 \times 10^3 \, N = 1481 \, kN$$

Hence compressive resistance of column is 1475 kN > 1400 kN   OK.

Example 9.9 Analysis of a column resisting an axial load and moment

A 305 × 305 × 137 kg m^{-1} UC section extends through a height of 3.5 m and is pinned at both ends. Check whether this member is suitable to support a design axial permanent load of 600 kN together with a major axis variable bending moment of 300 kN m applied at the top of the element. Assume grade Fe 430 steel is to be used and that all effective length factors are unity.

ACTIONS

Factored axial loading is

$$N_{Sd} = 600 \times 1.35 = 810 \, kN$$

Factored bending moment at top, middle and bottom of column is

$$M_{Sd.t} = 300 \times 1.5 = 450 \, kN \, m$$

$$M_{Sd.m} = 225 \, kN \, m \qquad M_{Sd.b} = 0 \, kN \, m$$

STRENGTH CLASSIFICATION

Flange thickness $t_f = 21.7$ mm, steel grade = Fe 430. Hence from *Table 9.3*, $f_y = 275$ N mm^{-2}.

SECTION CLASSIFICATION

$$c/t_f = 7.11 < 10\varepsilon = 9.24 \qquad d/t_w = 17.9 < 33\varepsilon = 30.5$$

Hence from *Table 9.4*, section belongs to class 1.

RESISTANCE OF CROSS-SECTIONS: BENDING AND AXIAL FORCE

Squash load, $N_{pl.Rd}$, is

$$N_{pl.Rd} = Af_y/\gamma_{M0} = \frac{17\,500 \times 275}{1.05}$$

$$= 4583\,333 \text{ N} = 4583 \text{ kN}$$

Full plastic moment of resistance of section, $M_{pl.Rd}$, is

$$M_{pl.Rd} = W_{pl}f_y/\gamma_{M0} = \frac{2300 \times 10^3 \times 275}{1.05}$$

$$= 602 \times 10^6 \text{ N mm} = 602 \text{ kN m}$$

$$n = N_{Sd}/N_{pl.Rd} = 810/4583 = 0.177$$

$$a = (A - 2bt_f)/A = \frac{(17500 - 2(308.7 \times 21.7)}{17\,500} = 0.234$$

$$M_{Ny} = M_{pl.y}(1 - n)/(1 - 0.5a)$$

$$= 602(1 - 0.177)/(1 - 0.5 \times 0.234)$$

$$= 561 \text{ kN m} > M_{Sd} (= 450 \text{ kN m}) \quad \text{OK}$$

RESISTANCE OF MEMBER: COMBINED BENDING AND AXIAL COMPRESSION

Effective length of column about both axes is given by

$$L_{eff} = L_{eff\,y} = L_{eff\,z} = 1.0L = 1.0 \times 3500 = 3500 \text{ mm}$$

The column will buckle about the weak (z–z) axis. Slenderness ratio about z–z axis, λ_z is

$$\lambda_z = \frac{L_{eff\,z}}{i_z} = \frac{3500}{78.2} = 44.7$$

Here $\beta_A = 1$ for class 1 section and

$$\lambda_1 = 93.9\varepsilon = 93.9 \times 0.924 = 86.76$$

$$\bar{\lambda}_z = (\lambda_z/\lambda_1)(\beta_A)^{1/2} = (44.7/86.76)(1)^{1/2} = 0.515$$

$$\frac{h}{b} = 1.04 < 1.2 \quad \text{and} \quad t_f = 21.7 \text{ mm}$$

From *Table 9.8*, for buckling about z–z axis use buckling curve c. Hence from *Table 9.7*, $\alpha = 0.49$.

$$\phi = 0.5[1 + \alpha(\bar{\lambda}_z - 0.2) + \bar{\lambda}_z{}^2]$$

$$= 0.5[1 + 0.49(0.515 - 0.2) + 0.515^2] = 0.71$$

$$\chi_z = \frac{1}{\phi + [\phi^2 - \bar{\lambda}_z{}^2]^{1/2}}$$

$$= \frac{1}{0.71 + [0.71^2 - 0.515^2]^{1/2}} = 0.834$$

Slenderness ratio about y–y axis, λ_y, is

$$\lambda_y = \frac{L_{\text{eff y}}}{i_y} = \frac{3500}{137} = 25.6$$

$$\bar{\lambda}_y = (\lambda_y/\lambda_1)(\beta_A)^{1/2} = (25.6/86.76)\,(1)^{1/2} = 0.294$$

$$\mu_y = \bar{\lambda}_y(2\beta_{My} - 4) + \left(\frac{W_{\text{pl.y}} - W_{\text{el.y}}}{W_{\text{el.y}}}\right)$$

$$= 0.294(2 \times 1.8 - 4) + \frac{(2300 - 2050)}{2050} = 0 \text{ approx.}$$

where $\beta_{My} = 1.8$ from *Fig. 9.6*. Thus

$$k_y = 1 - \frac{\mu_y N_{Sd}}{\chi_y A f_y} = 1$$

From *Table 9.8*, since $h/b < 1.2$ and $t_f < 100$ mm, for buckling about y–y axis use buckling curve b. Hence from *Table 9.7*, $\alpha = 0.34$.

$$\phi = 0.5[1 + \alpha(\bar{\lambda}_y - 0.2) + \bar{\lambda}_y{}^2]$$

$$= 0.5[1 + 0.34(0.294 - 0.2) + 0.294^2] = 0.56$$

$$\chi_y = \frac{1}{\phi + [\phi^2 - \bar{\lambda}_y{}^2]^{1/2}}$$

$$= \frac{1}{0.56 + [0.56^2 - 0.294^2]^{1/2}} = 0.964$$

χ_{min} is the smaller of $\chi_x (= 0.834)$ and $\chi_y (= 0.964)$.

Furthermore,

$$N_{c.Rd} = N_{pl.Rd} = 4583 \text{ kN} \qquad M_{pl.y.Rd} = M_{pl.Rd} = 602 \text{ kN m}$$

Check using equation 9.54, i.e.

$$\frac{N_{Sd}}{\chi_{min} N_{c.Rd}} + \frac{k_y.M_{y.Sd}}{M_{pl.y.Rd}} \leqslant 1$$

$$\frac{810}{0.834 \times 4583} + \frac{1 \times 450}{602} = 0.96$$

Hence the selected section is suitable.

Example 9.10 Analysis of a column baseplate

Check that the column baseplate shown below is suitable to resist an axial design load, N_{Sd}, of 2200 kN. Assume that the foundations are of concrete of compressive cylinder strength, f_{ck}, of 30 N mm^{-2} and that the baseplate is made of steel grade Fe 430.

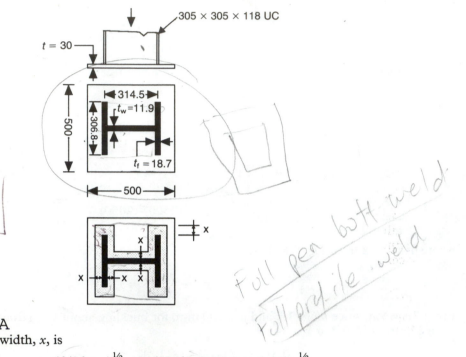

EFFECTIVE AREA
Additional bearing width, x, is

$$x = t(f_y/3f_j\gamma_{M0})^{1/2} = 30(275/3 \times 13.3 \times 1.05)^{1/2}$$
$$= 77 \text{ mm} < 0.5(500 - 314.5) = 92 \text{ mm} \quad \text{OK}$$

where

$$f_j = \beta_j k_j f_{cd} = 2/3 \times 1 \times 30/1.5 = 13.3 \text{ N mm}^{-2}$$
$$A_{eff} = (2x + h)(2x + b) - 2(c - t_w/2)(h - 2t_f - 2x)$$
$$= (2 \times 77 + 314.5)(2 \times 77 + 306.8)$$
$$- 2(153.4 - 11.9/2)(314.5 - 2 \times 18.7 - 2 \times 77)$$
$$= 215\ 885 - 36\ 302 = 179\ 583 \text{ mm}^2$$

AXIAL LOAD CAPACITY
Axial load capacity of baseplate is equal to

$$A_{eff}f_j = 179\ 583 \times 13.3 = 2388 \times 10^3 \text{ N} > N_{Sd}(= 2200 \text{ kN}) \quad \text{OK}$$

BENDING IN BASEPLATE
Bending moment per unit length in baseplate, m_{Sd}, is

$$m_{Sd} = (x^2/2)N_{Sd}/A_{eff} = (77^2/2)2200/179\ 583 = 36.3 \text{ kN mm mm}^{-1}$$

Moment of resistance, m_{Rd}, is

$$m_{Rd} = t^2 f_y/6\gamma_{M0} = 30^2 \times 275/6 \times 1.05 = 39.3 \times 10^3 \text{ N mm mm}^{-1}$$
$$= 39.3 \text{ kN mm mm}^{-1} > m_{Sd} \quad \text{OK}$$

9.13 Connections

In EC2, Chapter 6 on connections is more comprehensive than BS 5950, but the results seem broadly similar and the principles are essentially the same. One exception may be the design of friction grip fasteners, where the slip factor for untreated surfaces may have to be taken as 0.2 rather than 0.45 in BS 5950. In general the results for bolting and welding seem slightly more conservative than BS 5950. This is largely because of the larger material safety factors for connections $\gamma_M = 1.25$.

To help comparison of the design methods in BS 5950 and EC3 with regards to connections, the material in this section is presented under the following headings:

1. material properties
2. clearances in holes for fasteners
3. positioning of holes for bolts
4. bolted connections
5. high-strength bolts in slip-resistant connections
6. welded connections
7. design of connections.

9.13.1 MATERIAL PROPERTIES

9.13.1.1 Nominal bolt strengths (clause 3.3, EC3)

The nominal values of the yield strength, f_{yb}, and the ultimate tensile strength, f_{ub}, of bolts are shown in *Table 9.9*. According to clause 3.3.2.2 of EC3, high-strength bolts may be used as preloaded bolts with controlled tightening provided that they conform with the requirements in reference 3 of Annex B. With regard to welded connections, EC3 requires that the specified yield strength, ultimate tensile strength, etc. should be equal to or greater than the values specified for the steel grade being welded.

9.13.1.2 Partial safety factors

Partial safety factors, γ_M, are taken as follows:

Resistance of bolts, $\gamma_{Mb} = 1.25$

Resistance of welds, $\gamma_{Mw} = 1.25$

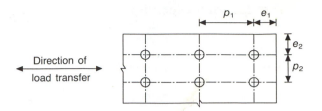

Fig. 9.8 *Spacing of fasteners.*

9.13.2 CLEARANCES IN HOLES FOR FASTENERS (CLAUSE 7.5.2, EC3)

The nominal clearance in standard holes for bolted connections should be as follows:

1. 1 mm for M12 and M14 bolts
2. 2 mm for M16 and M24 bolts
3. 3 mm for M27 and larger bolts.

The nominal clearance in oversize holes for slip-resistant connections should be:

1. 3 mm for M12 bolts
2. 4 mm for M14 to M22 bolts
3. 6 mm for M24 bolts
4. 8 mm for M27 and larger bolts.

9.13.3 POSITIONING OF HOLES FOR BOLTS (CLAUSE 6.5.1, EC3)

9.13.3.1 Minimum end distance

The end distance e_1 from the centre of a fastener hole to the adjacent end of any part, measured in the direction of load transfer (*Fig. 9.8*), should be not less than $1.2d_0$, where d_0 is the hole diameter.

9.13.3.2 Minimum edge distance

The edge distance e_2 from the centre of a fastener hole to the adjacent edge of any part, measured at right angles to the direction of load transfer (*Fig. 9.8*), should not generally be not less than $1.5d_0$.

9.13.3.3 Maximum end and edge distances

Under normal conditions, the end and edge distance should not exceed $12t$ or 150 mm, whichever is the larger, where t is the thickness of the thinner outer connected part.

Table 9.9 Nominal values of f_{yb} and f_{ub} for bolts (Table 3.3, EC3)

Bolt grade	4.6	4.8	5.6	5.8	6.8	8.8	10.9
f_{yb} (N mm^{-2})	240	320	300	400	480	640	900
f_{ub} (N mm^{-2})	400	400	500	500	600	800	1000

9.13.3.4 Minimum spacing

The spacing p_1 between centres of fasteners in the direction of load transfer (*Fig. 9.8*), should be not less than $2.2d_0$. The spacing p_2 between rows of fasteners, measured perpendicular to the direction of load transfer (*Fig. 9.8*), should normally be not less than $3d_0$.

9.13.4 BOLTED CONNECTIONS (CLAUSE 6.5.5, EC3)

9.13.4.1 Design shear resistance per shear plane

If the shear plane passes through the threaded portion of the bolt, the design shear resistance per shear plane, $F_{v.Rd}$, for strength grades 4.6, 5.6 and 8.8 bolts, is given by

$$F_{v.Rd} = \frac{0.6f_{ub}A_s}{\gamma_{Mb}} \qquad (9.69)$$

and for strength grades 4.8, 5.8 and 6.8 and 10.9 bolts, is given by

$$F_{v.Rd} = \frac{0.5f_{ub}A_s}{\gamma_{Mb}} \qquad (9.70)$$

If the shear plane passes through the unthreaded portion of the bolt, the design shear resistance is given by

$$F_{v.Rd} = \frac{0.6f_{ub}A}{\gamma_{Mb}} \qquad (9.71)$$

where A is the gross cross-section area of the bolt, A_s the tensile stress area of the bolt, d the bolt diameter and d_0 the hole diameter. Note that these values for design shear resistance apply only where the bolts are used in holes with nominal clearances specified in *section 9.13.2*.

9.13.4.2 Bearing resistance

The design bearing resistance, $F_{b.Rd}$, is given by

$$F_{b.Rd} = \frac{2.5\alpha f_u dt}{\gamma_{Mb}}$$

$$(9.72)$$

where α is the smallest of:

$$\frac{e_1}{3d_0}; \qquad \frac{p_1}{3d_0} - \frac{1}{4}; \qquad \frac{f_{ub}}{f_u} \quad \text{or} \quad 1.0$$

Note that the values of the design bearing resistance only apply where the edge distance e_2 is not less than $1.5d_0$ and the spacing p_2 is not less than $3.0d_0$.

9.13.5 HIGH-STRENGTH BOLTS IN SLIP-RESISTANT CONNECTIONS (CLAUSE 6.5.8, EC3)

9.13.5.1 Slip resistance

The design slip resistance of a preloaded high-strength bolt, $F_{s.Rd}$, is given by

$$F_{s.Rd} = \frac{k_s n \mu F_{p.Cd}}{\gamma_{Ms}} \qquad (9.73)$$

where $k_s = 1.0$ where the holes in all the plies have standard nominal clearances as outlined in *section 9.13.2* above, n is the number of friction interfaces, μ the slip factor (see below) and γ_{Ms} the partial safety factor. For bolt holes in standard nominal clearance holes, $\gamma_{Ms} = 1.25$ and 1.10 for the ultimate and serviceability limit states respectively. Finally, $F_{p.Cd}$ is the design preloading force. It is given by

$$F_{p.Cd} = 0.7f_{ub}A_s \qquad (9.74)$$

9.13.5.2 Slip factor

The value of the slip factor μ is dependent on the class of surface treatment. The value of μ should be taken as follows:

$$\mu = 0.5 \text{ for class A surfaces}$$

$$\mu = 0.4 \text{ for class B surfaces}$$

$$\mu = 0.3 \text{ for class C surfaces}$$

$$\mu = 0.2 \text{ for class D surfaces}$$

The surface descriptions are as follows:

Class A: surfaces blasted with shot or grit, with any loose rust removed, no pitting; surfaces blasted with shot or grit, and spray-metallized with aluminium; surface blasted with shot or grit, and spray-metallized with a zinc-based coating certified to provide a slip factor of not less than 0.5.

Class B: surfaces blasted with shot or grit, and painted with an alkali–zinc silicate paint to produce a coating thickness of 50–80 μm.

Class C: surfaces cleaned by wire brushing or flame cleaning, with any loose rust removed;

Class D: surfaces not treated.

9.13.6 WELDED CONNECTIONS (CLAUSE 6.6, EC3)

9.13.6.1 Design resistance of a fillet weld

The design resistance per unit length of a fillet weld, $F_{w.Rd}$, is given by

$$F_{w.Rd} = f_{vw}a \qquad (9.75)$$

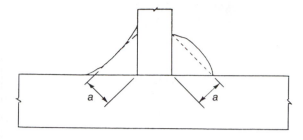

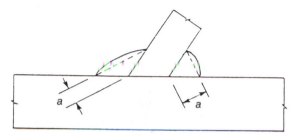

Fig. 9.9 *Throat thickness of a fillet weld.*

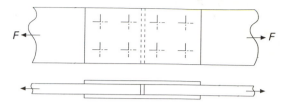

Fig. 9.10 *Splice connection.*

where a is the throat thickness of the weld and is taken as the height of the largest triangle which can be inscribed within the fusion faces and weld surface, measured perpendicular to the outer side of this triangle (*Fig. 9.9*). Note that a should not be less than 3 mm. Here f_{vw} is the design shear strength of the weld and is given by

$$f_{vw} = \frac{f_u/\sqrt{3}}{\beta_w \gamma_{Mw}} \qquad (9.76)$$

where f_u is the nominal ultimate tensile strength of the weaker part joined and β_w is a correlation factor whose value should be taken as follows. Linear interpolation for intermediate values of f_u is allowed.

EN10025 steel grade	Ultimate tensile strength f_u (N mm^{-2})	Correlation factor β_w
Fe 360	360	0.8
Fe 430	430	0.85
Fe 510	510	0.9

9.13.7 DESIGN OF CONNECTIONS

The use of the above equations is illustrated by means of the following design examples:

1. tension splice connection;
2. welded end plate to beam connection;
3. bolted beam-to-column connection using end plates;
4. bolted beam-to-column connection using web cleats.

9.13.7.1 Splice connections

The design of splice connections (*Fig. 9.10*) in EC3 is essentially the same as that used in BS 5950 and involves determining the design values of the following parameters:

1. design shear resistance of fasteners (*section 9.13.4* or *9.13.5*);
2. critical bearing resistance (*section 9.13.4*);
3. critical tensile resistance (section below).

Tension resistance of cross-sections (clause 5.4.3, EC3) For members in axial tension, the design value of the tensile force, N_{Sd}, at each cross-section should satisfy the following:

$$N_{Sd} \leq N_{t.Rd} \qquad (9.77)$$

where $N_{t.Rd}$ is the design tension resistance of the cross-section taken as the design ultimate resistance of the net section at holes for fasteners, $N_{u.Rd}$, which is given by

$$N_{u.Rd} = 0.9 A_{net} f_u / \gamma_{M2} \qquad (9.78)$$

Example 9.11 Analysis of tension splice connections

Calculate the design resistance of the splice connection shown below. The cover plates are made of grade Fe 430 steel and connected with either (a) non-preloaded bolts of diameter 20 mm and grade 4.6 or (b) prestressed bolts also of diameter 20 mm and grade 4.6. Assume that in both cases, the shear plane passes through the unthreaded portions of the bolts.

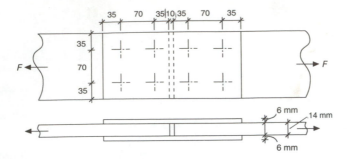

NON-PRELOADED BOLTS

Design shear resistance

Design shear resistance per shear plane, $F_{v.Rd}$, is given by

$$F_{v.Rd} = \frac{0.6 f_{ub} A}{\gamma_{Mb}} = \frac{0.6 \times 400 \times 314.16}{1.25}$$

$$= 60\ 318\ N = 60\ kN$$

where $f_{ub} = 400\ N\ mm^{-2}$ (*Table 9.9*), $\gamma_{Mb} = 1.25$ and $A = \pi 20^2/4 = 314.16\ mm^2$. All four bolts are in double shear. Hence, shear resistance, F_{sd}, of connection is

$$F_{sd} = 4 \times (2 \times 60) = 480\ kN$$

Bearing resistance

Bearing failure will tend to take place in the cover plates since they are thinner. According to clause 6.1.1(2) of EC3, α is the smallest of

$$e_1/3d_0 = 35/3 \times 22 = 0.530$$

$$p_1/3d_0 - 1/4 = 70/3 \times 22 - 1/4 = 0.810$$

$$f_{ub}/f_u = 400/430 = 0.930 \quad (\textit{Tables 9.3 and 9.9})$$

$$\alpha = 1.000$$

Hence, $\alpha = 0.530$ and the bearing resistance of one shear plane, $F_{b.Rd}$ is given by

$$F_{b.Rd} = \frac{2.5 \alpha f_u d t}{\gamma_{Mb}} = \frac{2.5 \times 0.530 \times 430 \times 20 \times 6}{1.25}$$

$$= 54\ 696\ N = 54.7\ kN$$

Bearing resistance of double shear plane is

$$2 \times 54.7 = 109.4\ kN$$

Bearing resistance of bolt group is

$$4 \times 109.4 = 437.6\ kN$$

Tensile resistance of cover plates

Net area of cover plate, A_{net}, is

$$A_{net} = 6 \times 140 - 2 \times 6 \times 22 = 576 \text{ mm}^2$$

Design ultimate resistance, $N_{u.Rd}$, is given by

$$N_{u.Rd} = \frac{0.9 A_{net} f_u}{\gamma_{M2}} = \frac{0.9 \times 576 \times 430}{1.25}$$

$$= 178\,330 \text{ N} = 178 \text{ kN}$$

Total ultimate resistance of connection, i.e. two cover plates = $(2 \times 178) = 356$ kN. Hence design resistance of the connection with grade 4.6, 20 mm diameter non-preloaded bolts is 356 kN.

PRESTRESSED BOLTS

Slip resistance

Preloading force, $F_{p.Cd}$, is given by

$$F_{p.Cd} = 0.7 f_{ub} A_s = 0.7 \times 400 \times 245$$

$$= 68\,600 \text{ N} = 68.6 \text{ kN}$$

Assuming the surfaces have been shot blasted, i.e. class A, take $\mu=0.5$. For two surfaces $n=2$, standard clearances $k_s=1.0$ and $\gamma_{Ms}=1.25$. Hence slip resistance for each bolt, $F_{s.Rd}$, is given by

$$F_{s.Rd} = \frac{k_s n \mu}{\gamma_{M2}} F_{p.Cd} = \frac{1 \times 2 \times 0.5}{1.25} \times 68.6 = 54.9 \text{ kN}$$

Hence design resistance = $4 \times 54.9 = 219.6$ kN.

Tensile resistance of cover plate

Tensile resistance of cover plates=356 kN (from above). Hence design resistance of connection with grade 4.6, 20 mm diameter prestressed bolts = 219 kN.

9.13.7.2 Welded end plate to beam connection

The relevant equations for the design of such con- nections were discussed in *section 9.13.6*.

Example 9.12 Welded end plate to beam connection

Calculate the shear resistance of the welded end plate-to-beam connection shown below. Assume the throat thickness of the fillet weld is 4 mm and the steel grade is Fe 430.

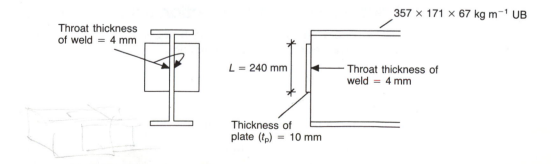

Ultimate tensile strength of Fe 430 steel (f_u) = 430 N mm^{-2} (*Table 9.3*); correlation factor (β_w) = 0.85; throat thickness of weld (a) = 4 mm; partial safety factor for welds (γ_{Mw}) = 1.25. Design shear strength of weld, f_{vw}, is given by

$$f_{vw} = \frac{f_u/\sqrt{3}}{\beta_w \gamma_{Mw}} = \frac{430/\sqrt{3}}{0.85 \times 1.25} = 233.6 \text{ N mm}^{-2}$$

Design resistance of weld per unit length, $F_{w.Rd}$, is given by

$$F_{w.Rd} = f_{vw}a = 233.6 \times 4 = 934.4 \text{ N mm}^{-1}$$

Hence for weld length of 2×240 mm the shear resistance of the weld, $V_{w.Rd}$, is

$$V_{w.Rd} = 2 \times (F_{w.Rd}L) = 2 \times (934.4 \times 240) = 448\ 512 \text{ N}$$
$$= 448 \text{ kN}$$

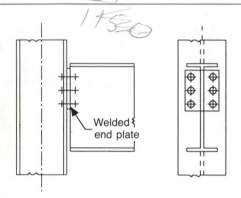

Fig. 9.11 *Typical bolted beam-to-column connection.*

9.13.7.3 Bolted beam-to-column connection using end plate

The design of such connections (*Fig. 9.11*) involves determining the design values of the following parameters:

1. design shear resistance of fasteners (*section 9.13.4*);
2. bearing resistance of fasteners (*section 9.13.4*);
3. resistance of welded connection (*section 9.13.6*),
4. shear resistance of end plate (section below);

5. local shear resistance of beam web (*section 9.11.1.2*).

Shear resistance of end plate (clause 5.4.6, EC3). The design value of the shear force, V_{Sd}, at each cross-section should satisfy the following:

$$V_{Sd} \leqslant V_{pl.Rd} \tag{9.79}$$

where $V_{pl.Rd}$ is the design plastic shear resistance given by

$$V_{pl.Rd} = A_v (f_y/\sqrt{3}/\gamma_{M0}) \tag{9.80}$$

where A_v is the shear area and may be taken as being equal to the cross-sectional area, A, for plate sections.

Clause 5.4.6(8) of EC3 also states that fastener holes in webs need not be allowed for provided that

$$A_{v.net}/A_v \geqslant f_y/f_u \tag{9.81}$$

However, when $A_{v.net}$ is less than this limit, the following value of the effective shear area may used:

$$\text{Effective shear area} = A_{v.net}f_u/f_y \tag{9.82}$$

Example 9.13 Bolted beam-to-column connection using end plate

If the beam in *Example 9.12* is to be connected to a column using 8 no. grade 4.6, M16 bolts as shown below, calculate the maximum shear resistance of the connection.

CHECK POSITIONING OF HOLES FOR BOLTS

Diameter of bolt, $d = 16$ mm

Diameter of bolt hole, $d_0 = 18$ mm

End distance, $e_1 = 30$ mm

Edge distance, $e_2 = 35$ mm

Spacing between centres of bolts in the direction of load transfer, $p_1 = 60$ mm

Spacing between rows of bolts, $p_2 = 115$ mm

Thickness of end plate, $t_p = 10$ mm

The following conditions need to be met:

End distance, $e_1 \not< 1.2d_0 = 1.2 \times 18 = 21.6$ mm < 30 mm   OK

Edge distance, $e_2 \not< 1.5\ d_0 = 1.5 \times 18 = 27$ mm < 35 mm   OK

e_1 and $e_2 \not> $ larger of $12t$ ($= 12 \times 10 = 120$ mm) or 150 mm $> 30, 35$ OK

Spacing, $p_1 \geqslant 2.2d_0 = 2.2 \times 18 = 39.6 < 60$ OK

Spacing, $p_2 \geqslant 3d_0 = 3 \times 18 = 54 < 115$ OK

p_1 and $p_2 \geq$ lesser of $14t$ ($= 14 \times 10 = 140$ mm) or 200 mm $> 60, 115$ OK

SHEAR RESISTANCE OF BOLT GROUP

Cross-sectional area of bolt $(A) = \dfrac{\pi 16^2}{4} = 201$ mm^2

Shear resistance per shear plane $(F_{v.Rd})$ is given by

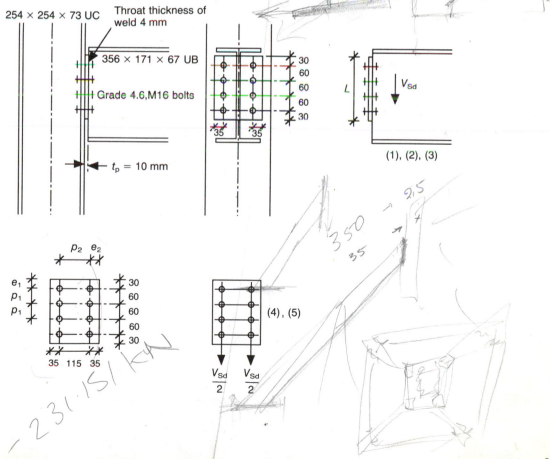

$$F_{v.Rd} = \frac{0.6 f_{ub} A}{\gamma_{Mb}} = \frac{0.6 \times 400 \times 201}{1.25}$$

$$= 38\ 592\ N$$

Hence, shear resistance of bolt group is

$$V_{v.Rd} = 8 \times F_{v.Rd} = 8 \times 38.6 = 308.7\ kN$$

BEARING RESISTANCE OF BOLT GROUP

Bolt diameter $(d) = 16$ mm, hole diameter $(d_0) = 18$ mm and α is the smallest of the following:

$$e_1/3d_0 = 30/3 \times 18 = 0.555$$

$$p_1/3d_0 - 1/4 = 60/3 \times 18\ - 1/4 = 0.861$$

$$f_{ub}/f_u = 400/430 = 0.930 \quad (Tables\ 9.3\ \text{and}\ 9.9) \quad \text{or} \quad 1.000$$

Hence $\alpha = 0.555$.

Bearing resistance of one bolt

$$F_{b.Rd} = \frac{2.5\ \alpha f_u d t_p}{\gamma_{Mb}} = \frac{2.5 \times 0.555 \times 430 \times 16 \times 10}{1.25}$$

$$= 76\ 368\ N = 76.3\ kN$$

Bearing resistance of bolt group is

$$= 8 F_{b.Rd} = 8 \times 76.3 = 610.4\ kN$$

RESISTANCE OF WELDED CONNECTION BETWEEN BEAM AND END PLATE

$$V_{w.Rd} = 448\ kN \quad (Example\ 9.12)$$

SHEAR RESISTANCE OF END PLATE

Bolt holes in plate do not need to be taken into account provided that

$$A_{v.net}/A_v \geqslant f_y/f_u \quad \text{where} \quad A_v = 10 \times 240 = 2400\ mm^2$$

and

$$A_{v.net} = 2400 - 4 \times 18 \times 10 = 1680\ mm^2$$

Substituting above gives

$$1680/2400 = 0.7 > f_y/f_u = \frac{275}{430} = 0.64$$

Hence shear resistance of end plate per section is given by

$$V_{pl.Rd} = A_v(f_y/\sqrt{3})/\gamma_{M0}$$

$$= 2400(275/\sqrt{3})/1.05 = 362\ 906\ N = 362.9\ kN$$

For failure, two planes have to shear. Hence shear resistance of end plate is

$$2 \times V_{pl.Rd} = 2 \times 362.9 = 725.8\ kN$$

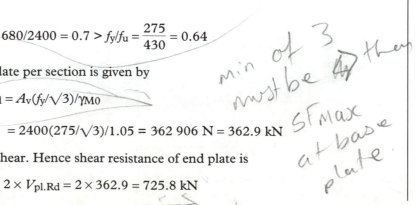

LOCAL SHEAR RESISTANCE OF BEAM WEB

Shear area of web is

$$(A_v)_{web} = Lt_{web} = 240 \times 9.1 = 2184 \text{ mm}^2$$

where t_{web} = beam web thickness = 9.1 mm. Hence local shear resistance of web, $V_{pl.Rd}$, is given by

$$V_{pl.Rd} = (A_v)_{web}(f_y/\sqrt{3})/1.05$$

$$= 2184 \times (275/\sqrt{3})/1.05 = 330\ 244 \text{ N} = 330 \text{ kN}$$

Hence the maximum shear resistance of the connection is controlled by the shear resistance of the bolt group and is equal to 308.7kN $> V_{sd}$ ∴ O.K

9.13.7.4 Bolted beam-to-column connection using web cleats

The design of such connections (*Fig. 9.12*) involves determining the design values of the following parameters:

1. design shear resistance of fasteners (*section 9.13.4*);
2. bearing resistance of fasteners (*section 9.13.4*);
3. shear resistance of cleats (see previous example);
4. distribution of shear forces between fasteners (see below);
5. bearing resistance of cleats (based on (2) and (4));
6. bearing resistance of beam web (*section 9.13.4*);
7. shear rupture strength (see below).

Distribution of forces between fasteners (clause 6.5.4, EC3). The distribution of internal forces between fasteners at the ultimate limit state can be assumed to be proportional to the distance from the centre of rotation (*Fig. 9.13*) where the design shear resistance of a fastener, $F_{v.Rd}$, is less than the design bearing resistance $F_{b.Rd}$, i.e.

$$F_{v.Rd} < F_{b.Rd} \qquad (9.83)$$

Thus, for the connection detail shown in *Fig. 9.13*, the maximum horizontal shear force on the bolts, $F_{h.Sd}$, is given by

$$F_{h.Sd} = M_{Sd}/5p \qquad (9.84)$$

and the vertical shear force per bolt is = $V_{Sd}/5$. The design shear force, $F_{v.Sd}$, is given by

$$F_{v.Sd} = \left[\left(\frac{M_{Sd}}{5p}\right)^2 + \left(\frac{V_{Sd}}{5}\right)^2\right]^{1/2} \qquad (9.85)$$

where M_{Sd} is the design bending moment, V_{Sd} the design shear force and p the spacing between fasteners.

Design shear rupture resistance (clause 6.5.2.2, EC3) 'Block shear' failure at a group of fastener

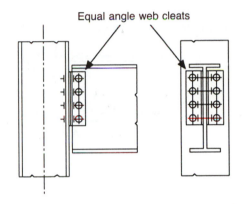

Equal angle web cleats

Fig. 9.12 *Typical bolted beam-to-column connection using wet cleat.*

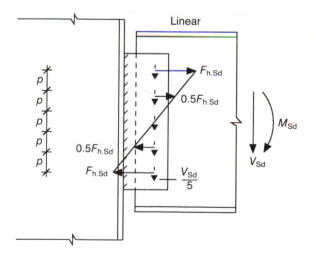

Fig. 9.13 *Distribution of loads between fasteners (Fig. 6.5.7, EC3).*

holes near the end of a beam web may occur as shown in *Fig. 9.14*. The design value of the effective resistance to block shear, $V_{eff.Rd}$, is given by

$$V_{eff.Rd} = (f_y/\sqrt{3})A_{v.eff}/\gamma_{M0}$$

where $A_{v.eff}$ is the effective shear area and is given by

$$A_{v.eff} = tL_{v.eff} \quad where \quad L_{v.eff} = L_v + L_1 + L_2 \leqslant L_3$$

in which

$$L_1 = a_1 \leqslant 5d \qquad L_2 = (a_2 - kd_{0.t})(f_u/f_y)$$

and

$$L_3 = L_v + a_1 + a_3 \leqslant (L_v + a_1 + a_3 - nd_{0.v})(f_u/f_y)$$

where a_1, a_2, a_3 and L_v are as indicated in *Fig. 9.14*, d is the nominal diameter of fasteners, $d_{0.t}$, $d_{0.v}$ is generally the hole diameter, n the number of fastener holes on the shear face, t is the thickness of the web or bracket and k is a coefficient with values as follows:

For a single row of bolts $k = 0.5$

For two rows of bolts $k = 2.5$

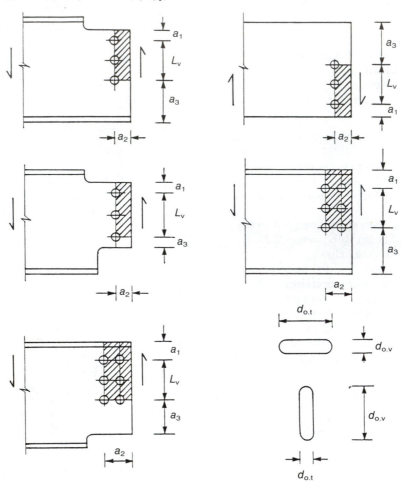

Figure 9.14 *Block shear–effective shear areas (Fig. 6.5.5, EC3).*

Example 9.14 Bolted beam-to-column connection using web cleats

Show that the double angle web cleat beam-to-column connection detail below is suitable to resist the design shear force, V_{Sd}, of 225kN. Assume the steel grade is Fe 430 and the bolts are of grade 5.8 and diameter 16 mm.

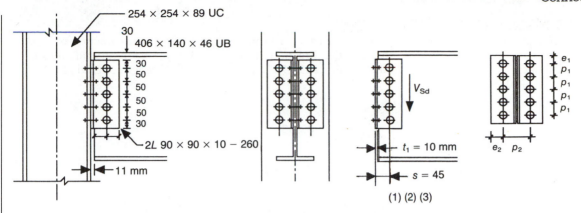

(1) (2) (3)

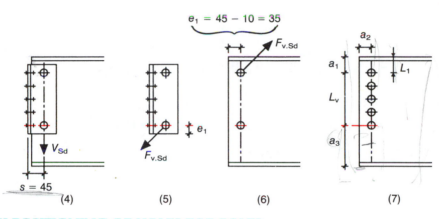

(4)　　　　(5)　　　　(6)　　　　(7)

CHECK POSITIONING OF HOLES FOR BOLTS

Diameter of bolt, $d = 16$ mm　　　　Diameter of bolt hole, $d_0 = 18$ mm

End distance, $e_1 = 30$ mm　　　　Edge distance, $e_2 = 45$ mm

Spacing between centres of bolts in the direction of load transfer, $p_1 = 50$ mm

Thickness of angle cleat, $t_p = 10$ mm

The following conditions need to be met:

End distance, $e_1 \not< 1.2d_0 = 1.2 \times 18 = 21.6$ mm < 30 mm　　　OK

Edge distance, $e_2 \not< 1.5d_0 = 1.5 \times 18 = 27$ mm < 45 mm　　　OK

e_1 and $e_2 \not> $ larger of $12t$ ($= 12 \times 10 = 120$ mm) or 150 mm $> 30, 45$　　　OK

Spacing, $p_1 \geqslant 2.2d_0 = 2.2 \times 18 = 39.6 < 50$　　　OK

Spacing, $p_1 \geq$ lesser of $14t$ ($= 14 \times 10 = 140$ mm) or 200 mm > 50　　　OK

SHEAR RESISTANCE OF BOLT GROUP

Assume that the shear plane passes through threaded portion of the bolt. Hence tensile stress area of bolt $A_s = 157$ mm^2. Shear resistance per bolt, $F_{v.Rd}$, is

$$F_{v.Rd} = \frac{0.5f_{ub}A_s}{\gamma_{Mb}} = \frac{0.5 \times 500 \times 157}{1.25} = 31\,400 \text{ N} = 31.4 \text{ kN}$$

where $f_{ub} = 500$ N mm^{-2} (Table 9.9) and $\gamma_{Mb} = 1.25$. Shear resistance for bolt group is

$$10 \times F_{v.Rd} = 10 \times 31.4 = 314 \text{ kN} > V_{Sd} = 225 \text{ kN}　OK$$

321

BEARING RESISTANCE OF BOLT GROUP

Diameter of bolts $(d) = 16$ mm; hole diameter $(d_0) = 18$ mm; end distance in the direction of load transfer $(e_1) = 30$; spacing between bolts in the direction of load transfer $(p_1) = 50$; ultimate tensile strength of grade 5.8 bolts $(f_{ub}) = 500$ N mm^{-2} (*Table 9.9*); ultimate tensile strength of grade Fe 430 steel $(f_u) = 430$ N mm^{-2} (*Table 9.3*).

Here α is the smallest of

$$\frac{e_1}{3d_0} = \frac{30}{3 \times 18} = 0.555$$

$$\frac{p_1}{3d_0} - \frac{1}{4} = \frac{50}{3 \times 18} - \frac{1}{4} = 0.676$$

$$\frac{f_{ub}}{f_u} = \frac{500}{430} = 1.163 \quad \text{or} \quad 1.000$$

Take $\alpha = 0.555$:

$$F_{b.Rd} = 2.5\alpha \, f_u dt_c / \gamma_{Mb}$$
$$= \frac{2.5 \times 0.555 \times 430 \times 16 \times 10}{1.25} = 76\,368 \text{ N} = 76.3 \text{ kN}$$

Bearing resistance of bolt group is

$$10 \times F_{b.Rd} = 10 \times 76.3 = 763 \text{ kN} > V_{Sd} = 225 \text{ kN} \quad \text{OK.}$$

SHEAR RESISTANCE OF LEGS OF CLEATS

Fastener holes need not be allowed for provided that

$$\frac{A_{v.net}}{A_v} \geqslant \frac{f_y}{f_u} = \frac{1700}{2600} = 0.65 > \frac{275}{430} = 0.64$$

where

$$A_{v.net} = 10(260 - 5 \times 18) = 1700 \text{ mm}^2$$
$$A_v = 10 \times 260 = 2600 \text{ mm}^2$$

and for steel Fe 430 $f_u = 430$ N mm^{-2} and $f_y = 275$ N mm^{-2} (*Table 9.3*). Hence design shear resistance of each leg of cleat, $V_{pl.Rd}$, is

$$V_{pl.Rd} = A_v(f_y/\sqrt{3})/\gamma_{MO}$$
$$= 2600(275/\sqrt{3})/1.05 = 393\,148 \text{ N} = 393.1 \text{ kN}$$

where $\gamma_{MO} = 1.05$. Total shear resistance of both legs of cleats is

$$2 \times V_{pl.Rd} = 2 \times 393.1 = 786.2 \text{ kN} > V_{Sd} = 225 \text{ kN} \quad \text{OK.}$$

SHEAR ACTION ON BOLTS DUE TO SHEAR FORCE AND RESULTANT BENDING MOMENT

Shear force per bolt in vertical direction, $F_{v.Sd}$, is

$$F_{v.Sd} = \frac{V_{Sd}}{n} = \frac{225}{5} = 45 \text{ kN}$$

Maximum shear force on bolt assembly in horizontal direction, $F_{h.Sd}$, is

$$F_{h.Sd} = \frac{M_{Sd}}{5p_1} = \frac{V_{Sd}s}{5p_1} = \frac{225 \times 45}{5 \times 50} = 40.5 \text{ kN}$$

Resultant force, $F_{V.Sd}$, is

$$F_{V.Sd} = (F_{v.Sd}^2 + F_{h.Sd}^2)^{1/2} = (45^2 + 40.5^2)^{1/2} = 60.5 \text{ kN}$$

Shear resistance per bolt, $F_{v.Rd}$, is 31.4 kN (see 'Shear resistance of bolt group' above). Since the resultant shear force is resisted by bolts in double shear, the total shear resistance is

$$F_{v.Rd} = 2 \times 31.4 = 62.8 \text{ kN} > F_{v.Sd} = 60.5 \text{ kN} \quad \text{OK}$$

BEARING RESISTANCE OF LEGS OF CLEATS CONNECTED TO WEB OF BEAM
Here α is the smallest of

$$\frac{e_1}{3d_0} = \frac{30}{3 \times 18} = 0.555$$

$$\frac{p_1}{3d_0} - \frac{1}{4} = \frac{50}{3 \times 18} - \frac{1}{4} = 0.676$$

$$\frac{f_{ub}}{f_u} = \frac{500}{430} = 1.163 \quad \text{or} \quad 1.000$$

Take $\alpha = 0.555$:

$$F_{b.Rd} = 2.5\alpha f_u d t_c / \gamma_{Mb}$$

$$= \frac{2.5 \times 0.555 \times 430 \times 16 \times 10}{1.25} = 76\,368 \text{ N} = 76.4 \text{ kN}$$

$$F_{v.Sd} = 60.5 \text{ kN} < 2F_{b.Rd} = 2 \times 76.4 = 152.8 \text{ kN} \quad \text{OK}$$

BEARING RESISTANCE OF WEB OF BEAMS
Here α is the smallest of

$$e_1/3d_0 = 35/3 \times 18 = 0.648$$
$$p_1/3d_0 - 1/4 = 50/3 \times 18 - 1/4 = 0.676$$
$$f_{ub}/f_u = 500/430 = 1.163 \quad \text{or} \quad 1.000$$

Hence $\alpha = 0.648$. Bearing resistance per bolt, $F_{b.Rd}$, is

$$F_{b.Rd} = \frac{2.5\alpha f_u d t_w}{\gamma_{Mb}} = \frac{2.5 \times 0.648 \times 430 \times 16 \times 6.9}{1.25}$$

$$= 61\,523 \text{ N} = 61.5 \text{ kN} > F_{v.Sd} = 60.5 \text{ kN}$$

BLOCK SHEAR RESISTANCE

$$t = \text{thickness of web of beam} = 6.9 \text{ mm}$$
$$L_1 = a_1 = 60 \text{ mm} < 5d = 5 \times 16 = 80 \text{ mm}$$
$$L_v = 200 \text{ mm and } a_3 = 95.6 \text{ mm}$$
$$L_2 = (a_2 - kd_{0.t})(f_u/f_y)$$
$$= (35 - 0.5 \times 18)(430/275) = 40.6$$
$$L_3 = L_v + a_1 + a_3 \leqslant (L_v + a_1 + a_3 - nd_{0.v})(f_u/f_y)$$
$$= 200 + 60 + 95.6 = 355.6$$
$$< (200 + 60 + 95.6 - 5 \times 18)(430/275) = 415.3$$
$$L_{v.eff} = L_v + L_1 + L_2 \leqslant L_3$$
$$= 200 + 60 + 40.6 = 300.6 \text{ mm} < L_3$$
$$A_{v.eff} = tL_{v.eff}$$
$$= 6.9 \times 300.6 = 2074 \text{ mm}^2$$
$$V_{eff.Rd} = (f_y/\sqrt{3})A_{v.eff}/\gamma_{M0}$$
$$= (275/\sqrt{3})2074/1.05$$
$$= 313.6 \times 10^3 \text{ N} = 313.6 \text{ kN} > V_{Sd} = 225 \text{ kN} \quad \text{OK}$$

Chapter 10

Eurocode 6: Design of masonry structures

This chapter briefly describes the purpose, scope and progress to date on Eurocode 6: Design of Masonry Structures. The chapter highlights developments in drafting the document and summarizes the layout and content of Part 1.1 of Eurocode 6, which should be ready for publication as an ENV by 1995. The chapter also discusses the advantages of using Eurocode 6 and its future direction.

10.1 Introduction

Eurocode 6 will be the new structural European standard for the design of masonry structures. Eurocode 6 is similar in scope to BS 5628 (*Chapter 5*), which it will eventually replace. Of the four structural Eurocodes discussed in this book, work on Eurocode 6 is the least advanced. Latest estimates suggest that Part 1.1 of Eurocode 6 will not be ready for publication as an ENV until 1995, at the earliest. This delay has been partly attributed to problems with the CEN mandate and agreement of production time-scales during the transition of Eurocode 6 from Commission responsibility to CEN responsibility, and partly to financial difficulties. These problems have now mainly been resolved. In common with the other structural Eurocodes, there have also been a number of technical difficulties in drafting the document, some of which are still ongoing and will be discussed more fully in *section 10.2*.

Like BS 5628, Eurocode 6 is based on limit state principles and will be published in a number of parts as follows:

Part 1.1: General rules – rules for reinforced and unreinforced masonry; crack and deflection control
Part 1.2: Detailed rules on lateral loading
Part 1.3: Complex shapes in masonry structures
Part 2: Other aspects of masonry
Part 3: Simplified and simple rules for masonry structures

Part 4: Constructions with lesser requirements for reliability and durability
Part 10: Fire design of masonry structures.

Work on Parts 1.1 and 10 is most advanced. Part 1.1, hereafter referred to as EC6, will be similar in scope to Parts 1 and 2 of BS 5628 except that only some aspects of flexural design will be covered, perhaps just the principle or possibly expanded into a design philosophy with methodology. The remainder will be included in Part 1.2 of Eurocode 6. Some probable reasons why it has been decided to devote a separate part to flexural design are discussed in *section 10.2.2*. Part 10 was issued for public comment during 1991/92 and then referred back to the CEN Committee for development towards ENV status. The time-scale for Parts 1.1 and 10 of Eurocode 6 are broadly similar.

Work on Part 2 of Eurocode 6 began in January 1993. This will attempt to produce a European equivalent of BS 5628: Part 3. As yet, there is very little or no activity on the remaining parts of Eurocode 6.

10.2 Development work

As mentioned above, the drafting panel of EC6, CEN Technical Committee 250, Sub-Committee SC6 has faced various difficulties in drafting the document. These principally relate to (a) the diverse range of masonry units produced throughout Europe and (b) national differences in design methodologies. The way in which these difficulties were tackled is discussed next.

10.2.1 CEN SUPPORTING STANDARDS FOR PRODUCTS

There is a wide variety of masonry units used throughout Europe. The units differ in many respects including appearance, size, strength and disposition and percentage of perforations. One of the stated aims of the Single Act is that there should be no barriers to trade, which has generally been interpreted as meaning that design rules must not disadvantage available products from EC countries. Thus, the EC6 drafting panel has had to develop design methods which enable the majority of masonry products currently available throughout Europe to be used in design.

The first step towards solving this problem was to agree a European specification for masonry units which is the responsibility of CEN Technical Committee 125. In the case of clay bricks (and some other units) this has now largely been achieved. The European standard for clay masonry units, reference no. prEN 777-1, uses a declaration system for specifying product characteristics. This is somewhat similar to that already adopted in BS 3921 (*Chapter 5*) and covers six essential requirements, all of which can be assessed by means of standard tests. These requirements are: compressive strength, water absorption, dimensional tolerance, frost resistance, soluble salt content and density. The latter is an additional parameter to those mentioned in BS 3921 and has been included because of its correlation with noise attenuation and thermal insulation. Note that appearance is not mentioned as it requires subjective judgement and cannot be supported with an appropriate test method. It is worth remembering that the masonry product standards have not been developed by the EC6 drafting panel and, therefore, they contain general details on unit specifications in terms of individual product types and will be relevant to all uses, not just structural applications.

Masonry units which have been characterized according to the appropriate European standard could be used to size masonry elements. However, it is considered by the EC6 drafting panel that this would result in a large volume of design data covering the range of masonry units produced throughout Europe which would be impractical to include in a working design standard. Therefore, to simplify the process, it has been decided that masonry units with similar characteristics should be grouped together into unit groups, for which typical design parameters, e.g. characteristic compressive strengths, would be available. In all, it is expected there will be three unit groups, with the superior structural use masonry units being placed in group 1. (Note that BS 5268 for the design of timber structures uses a similar approach, as discussed in *Chapter 6*.) At this stage it is certain that all the EC6 design rules will cover group 1 units.

It would appear that all UK brick units and probably most concrete block units will fit into the group 1 unit specification. Furthermore, UK bricks compare very favourably from an appearance point of view with those available in many of the other EC countries. Both these factors may benefit the UK brick industry by increasing the potential for export.

10.2.2 DESIGN PROCEDURE

Although there are differences over certain aspects of compressive design, e.g. slender members, shear resistance and concentrated load design, EC6 committee members are optimistic that they will be able to reach a consensus view. However, flexural design (i.e. design of laterally loaded panel walls) remains a particular difficulty for the reasons discussed below.

In the UK code of practice, the design of panel walls for lateral loading is based on 'yield line' principles. This assumes that the wall behaves plastically when subject to lateral loading, an assumption which correlates well with the available test data based on the BS 5628 panel design range. These tests were used to establish the characteristic flexural strength values given in Table 3 of BS 5628 (*Table 5.12*). Many continental standards, however, either do not contain a method for flexural design as their walls tend to be much thicker than those normally used in the UK, or they are based on alternative design principles. In both cases, the increased thickness of wall will lead to increased costs of masonry construction. It will therefore not be easy to achieve agreement on a pan-European basis.

While it is difficult to be certain, some commentators have suggested that a compromise solution may be to base the new European method on yield line principles but use lower characteristic flexural strength than that specified in BS 5628. If this is accepted, the next step will be to produce the actual design data. This is, of course, available for UK masonry units, and practice and a substantial data bank exist for wallette and full size wall panel flexure testing which means that UK units to group 1 CEN specification can be used in flexural design. However, much of the design data for other than UK masonry units is not available as yet and, bearing in mind the range of masonry units which the new European specifications will allow, may take some time to produce.

The above are probable reasons for deciding to deal with flexure design methods in a separate part of Eurocode 6. However, very little information is available. The only thing that can be said with any degree of certainty is that Eurocode 6 will contain some information on flexure design although the extent is not currently known as far as Part 1.1 is concerned. There is no work in Part 1.2 yet and the scope of this part is also uncertain at present.

10.3   Layout and contents

The latest and only draft of EC6 was published and issued for public comment in 1988. (A draft for reinforced and prestressed design was produced and issued for public comment later but this is not discussed here.) The 1988 draft of EC6 was based on a number of documents drawn up by experts from Europe. Obviously many changes have been made to the text since then and it would be rather pointless to describe in detail the contents of the draft. However, it is considered instructive to briefly summarize in general terms the layout and contents of EC6, as this follows the style adopted in the other structural Eurocodes and is therefore unlikely to undergo any major revisions.

EC6 is divided into six sections as described under the headings below.

SECTION 1:   INTRODUCTION
Section 1 will describe the scope of EC6. As with the other Eurocodes the material in EC6 will be divided into principles and application rules.

SECTION 2:   BASIS OF DESIGN
Section 2 will describe the overall aims of design and how these aims can be achieved in practice using limit state principles. It will also set out the basis of the method and include the basic equations for calculating the design actions and design values of material properties. (It should be noted that Eurocode 1 which is due to be published around 1993 will contain a 'basis of design' common to all Eurocodes. At present EC6 is still using its own pending agreement on an EC1 document.)

SECTION 3:   MATERIALS
Section 3 will describe the types and general properties of masonry units, mortars, reinforcement and infill concrete for reinforced and post-tensioned design. It will also include the characteristic compressive strengths based on a 'Normalized' value to accommodate different unit shapes. Mortars will be specified (classified) by strength, i.e. M5, M10. A flexure classification method may be given.

In addition it will describe the type and properties of the ancillary materials used in masonry construction, e.g. damp-proof courses, wall ties and other metal ancillary components. Reference will be made to the harmonized product standards and testing specifications.

SECTION 4:   DESIGN OF MASONRY
Section 4 will contain the main design fundamentals including compressive design, concentrated load design, shear, and probably flexure and cover design concepts such as effective height, effective thickness, slenderness ratio and reduction factor for slenderness ratio and eccentricity of loading. Some of the design procedures covered in EC6 are expected to be similar in approach to BS 5628; others are likely to be different.

SECTION 5:   STRUCTURAL DETAILING
Section 5 provides typical construction details of supports which satisfy the assumptions made in design, e.g. enhanced or simple supports and free, partially restrained and completely restrained edge support conditions. It will also include typical details of bonding, connections, recesses and chases, damp-proof courses and movement joints in masonry walls.

SECTION 6:   CONSTRUCTION
Section 6 provides guidance on good site practice and workmanship and will cover such aspects as supply and storage of masonry units, properties of masonry and maintenance of newly constructed masonry.

APPENDICES
There will be a number of appendices although the subject coverage remains incomplete at present.

10.4   Advantages and future direction

Preliminary work which was undertaken by the Building Research Establishment suggested that generally design solutions achieved by following the recommendations contained in BS 5628 and EC6 should be similar. In some aspects the design procedures in EC6 are perhaps a little more conservative than BS 5628 and in others less conservative. A new calibration exercise of EC6 versus BS 5628 is now needed in the light of EC6 drafting progress since the issue of the original draft for public comment.

Assuming that the design procedures in the two codes are not too dissimilar, it should not take very long for engineers in this country to become proficient with the design techniques in EC6 which will obviously help in tendering for work in other EC countries. Furthermore, as suggested above, since UK masonry products are generally of high quality, this may encourage more widespread use of masonry construction throughout Europe and open up new markets for UK producers.

At this stage it is too early to talk about supporting documents but it is highly likely that organizations such as the Brick Development Association will produce a number of design aids which will encourage and assist with the use of EC6. However, it is worth remembering that the ENV is only supposed to be valid for three years, with work starting on the ENV's development to EN status after two years from ENV issue. Therefore, organizations will probably be wary of rushing in to produce a whole range of design aids for the ENV period; they will be more interested in the longer-term EN.

Finally, it is interesting to note that the many difficulties faced by the EC6 drafting panel, which have inevitably delayed its development, will mean that in all probability BS 5628 will continue in existence at least until the early part of the new century.

Chapter 11

Eurocode 5: Design of timber structures

This chapter briefly describes the content of Part 1.1 of Eurocode 5, the new European standard for the design of buildings in timber, which is scheduled for publication as an ENV in the late summer of 1993. The chapter highlights the principal differences between the standard and its British equivalent, BS 5268:Part 2. It also includes a number of worked examples to illustrate the new procedures for designing flexural and compression members.

11.1   Introduction

Eurocode 5 applies to the design of building and civil engineering structures in timber. It is based on limit state principles and comes in several parts as shown in *Table 11.1*.

Part 1-1 of Eurocode 5, which is the subject of this discussion, gives a general basis for the design of buildings and civil engineering works in timber. It is largely similar in scope to Part 2 of BS 5268, which was discussed in *Chapter 6*. Part 1-1 of Eurocode 5, hereafter referred to as EC5, is scheduled to be published as a preliminary standard, reference no: DD ENV 1995-1-1, probably in the late summer of 1993. Work on Part 1-2 of Eurocode 5 on fire is also quite advanced and it too should be ready for publication as a preliminary standard, reference no: DD ENV 1995-1-2, in the latter half of 1993.

In contrast, work on Part 2 of Eurocode 5 on bridges only began in January 1993 and is expected to take several years to complete. This delay relates partly to problems in issuing the CEN mandate and

Table 11.1   Overall scope of Eurocode 5

Part	Subject
1–1	General rules and rules for buildings
1–2	Structural fire design
2	Bridges

the fact that there is not the same commercial pressure for timber bridges as those made from concrete and steel.

At the time of writing, EC5 is only available as a draft pre-standard, prEN 1995-1-1, and the comments that follow are based on the July 1992 document and incorporates the November amendments.

11.2   Layout

In common with the other structural Eurocodes, EC5 was drafted by a panel of experts drawn from the various EC member state countries. It is based on studies carried out by Working Commission W18: *Timber Structures of the CIB International Council for Building Research Studies and Documentation*, in particular on *CIB Structural Timber Design Code: Report No. 66*, published in 1983. The following subjects are covered in EC5:

Chapter 1:   Introduction
Chapter 2:   Basis of design
Chapter 3:   Material properties
Chapter 4:   Serviceability limit states
Chapter 5:   Ultimate limit states
Chapter 6:   Joints
Chapter 7:   Structural detailing and control
Annex A: Determination of 5-percentile characteristic values from test results and acceptance criteria for a sample
Annex B: Mechanically jointed beams
Annex C: Built-up columns
Annex D: The design of trusses with punched metal plate fasteners

As can be appreciated from the above contents list, the organization of material is different from that used in BS 5268 but follows the layout adopted in the other structural Eurocodes. Generally, the design rules in EC5 are sequenced on the basis of action effects rather than on the type of member, as in BS 5268.

All the annexes in EC5 are labelled 'Normative'. As explained in *section 7.4*, this signifies that this material has the same status as the rest of the code but appears here rather than in the body of the code in order to make the document easier to use.

This chapter briefly describes the contents of EC5, in so far as it is relevant to the design of flexural and compression members in solid timber. The rules governing the design of joints are not discussed.

11.3 Principles/application rules, shaded values, symbols

As with the other structural Eurocodes and for the reasons discussed in *Chapter 7*, the clauses in EC5 have been divided into principles and application rules. Principles comprise general statements, definitions, requirements and models for which no alternative is permitted. Principles are preceded by the letter P. The application rules are generally recognized rules which follow the statements and satisfy the requirements given in the principles.

Some of the numerical values in EC5, particularly partial safety coefficients, are shaded. These values are meant to be for guidance only and the appropriate UK values will be specified in the UK National Application Document (NAD) which will be published in conjunction with the ENV.

Chapter 1 of EC5 lists the symbols used in EC5. Those relevant to this discussion are reproduced below.

GEOMETRICAL PROPERTIES

b	breadth of beam
h	depth of beam
A	area
i	radius of gyration
I	second moment of area
Z	section modulus

BENDING

l	span
M	bending moment
G	permanent action
Q	variable action
$\sigma_{m,d}$	design normal bending stress
$f_{m,k}$	characteristic bending strength
$f_{m,d}$	design bending strength
k_{mod}	modification factor for strength values

k_{ls}	load-sharing factor
k_{inst}	instability factor for lateral buckling
$E_{0,05}$	characteristic modulus of elasticity (parallel) to grain
$E_{0,mean}$	mean modulus of elasticity (parallel) to grain
G_{mean}	mean shear modulus $= E_{0,mean}/16$
γ_G	partial coefficient for permanent actions
γ_Q	partial coefficient for variable actions
γ_m	partial coefficient for material properties

DEFLECTION

$u_{2,inst}$	instantaneous deflection due to variable loads
$u_{2,fin}$	final deflection due to variable loads
$u_{net,fin}$	net, final deflection due to total load plus precamber (if applied)
k_{def}	deformation factor

VIBRATION

f_1	fundamental frequency of vibration
b	floor width
l	floor length
v	unit impulse velocity
ζ	damping coefficient
n_{40}	number of first-order modes with natural frequencies below 40 Hz

SHEAR

V_d	design shear force
τ_d	design shear stress
$f_{v,k}$	characteristic shear strength
$f_{v,d}$	design shear strength

BEARING

$F_{90,d}$	design bearing force
l	length of bearing
$\sigma_{c,90,d}$	design compression stress perpendicular to grain
$f_{c,90,k}$	characteristic compression strength perpendicular to grain
$f_{c,90,d}$	design compression strength perpendicular to grain

COMPRESSION

l_{ef}	effective length of column
λ_y, λ_z	slenderness ratios about y–y and z–z axes
$\lambda_{rel,y}, \lambda_{rel,z}$	relative slenderness ratios about y–y and z–z axes
N	design axial force
$\sigma_{c,0,d}$	design compression stress parallel to grain

$f_{c,0,k}$ characteristic compression strength parallel to grain

$f_{c,0,d}$ design compression strength parallel to grain

$\sigma_{m,y,d}$, $\sigma_{m,z,d}$ design bending stresses parallel to grain

$f_{m,y,d}$, $f_{m,z,d}$ design bending strengths parallel to grain

k_c compression factor

11.4 Basis of design

As pointed out above, EC5, unlike BS 5268, is based on limit state principles. However, in common with other limit state codes, EC5 recommends that the two principal categories of limit states to be considered in design are the ultimate and serviceability limit states. The terms ultimate state and serviceability state apply in the same way as is understood in other limit state codes. Thus ultimate limit states are those associated with collapse or with other forms of structural failure which may endanger the safety of people, while serviceability limit states correspond to states beyond which specific service criteria are no longer met.

The serviceability limit states which must be checked in EC5 are deflection and vibration. The ultimate limit states, which must be checked singly or in combination, are bending, shear, compression and tension. The various design rules for checking these limit states are discussed later.

11.4.1 ACTIONS

Actions is the Eurocode terminology for loads and imposed deformations. Permanent actions, G, are all the dead loads acting on the structure, including the finishes, fixtures and self-weight of the structure. Variable actions, Q, include the imposed, wind and snow loads.

The characteristic permanent actions, G_k, and variable actions, Q_k, are specified in Eurocode 1: *Basis of Design and Actions on Structures*, which is due to be published as an ENV late in 1993. In the mean time, therefore, designers should continue using BS 648, BS 6399:Part 1 and CP 3: Chapter 5:Part 2 for characteristic values of actions.

The design values of actions, F_d, are obtained by multiplying the characteristic actions, F_k, by the appropriate partial safety factor for actions, γ_F:

$$F_d = \gamma_F F_k \qquad (11.1)$$

According to EC5, the partial safety factors for permanent actions, γ_G, and variable actions, γ_Q,

Table 11.2 Partial safety factors for actions in building structures for persistent and transient design situations (based on Table 2.3.3.1, EC5)

	Permanent actions (γ_G)	Variable actions (γ_Q)	
		One with its characteristic value	Others with their combination values
Favourable effect	1.0	0	0
Unfavourable effect	1.35	1.5	1.5

should normally be taken as 1.35 and 1.5 respectively (*Table 11.2*).

11.4.2 MATERIAL PROPERTIES

EC5, unlike BS 5268, does not contain the material properties, e.g. bending and shear strengths, necessary for sizing members. This information is to be found in a CEN supporting standard for timber products, namely provisional standard prEN 338. Like BS 5268, prEN 338 provides for a number of strength classes and gives typical characteristic strength and stiffness values and densities for each (*Table 11.3*). It should be noted that the European standard specifies 15 strength classes, rather than the nine identified in BS 5268 (*Table 6.3*). Presumably, this reflects the wider range of timber species found on the Continent. Further, the bending strengths are indicated by the strength class designations.

It is rather difficult to directly compare European and UK strength classes since the former refer to characteristic values which are generally fifth percentile values derived directly from laboratory tests of five minutes' duration, and the latter refer to grade stress which have been reduced for long-term duration and already include a safety factor. The UK NAD will, however, include a table by which the various combinations of UK grades and species may be assigned to their appropriate European strength classes. Meanwhile, as a rough guide, it can be assumed that timber strength classes C22 (*Table 11.3*) and SC4 (*Table 6.3*) have similar structural properties.

One benefit of using characteristic values of material properties rather than grade stresses is that it will make it easier to sanction the use of new materials and components for structural purposes, since such values can be utilized immediately, without first

Table 11.3 Structural timber strength classes (Table 1, prEN338)

		Poplar and conifer species									Deciduous species					
		C14	C16	C18	C22	C24	C27	C30	C35	C40	D30	D35	D40	D50	D60	D70
Strength properties (N mm^{-2})																
Bending	$f_{m,k}$	14	16	18	22	24	27	30	35	40	30	35	40	50	60	70
Tension parallel	$f_{t,0,k}$	8	10	11	13	14	16	18	21	24	18	21	24	30	36	42
Tension perpendicular	$f_{t,90,k}$	0.3	0.3	0.3	0.3	0.4	0.4	0.4	0.4	0.4	0.6	0.6	0.6	0.6	0.7	0.9
Compression parallel	$f_{c,0,k}$	16	17	18	20	21	22	23	25	26	23	25	26	29	32	34
Compression perpendicular	$f_{c,90,k}$	4.3	4.6	4.8	5.1	5.3	5.6	5.7	6.0	6.3	8.0	8.4	8.8	9.7	10.5	13.5
Shear	$f_{v,k}$	1.7	1.8	2.0	2.4	2.5	2.8	3.0	3.4	3.8	3.0	3.4	3.8	4.6	5.3	6.0
Stiffness properties (kN mm^{-2})																
Mean modulus of elasticity parallel	$E_{0,mean}$	7	8	9	10	11	12	12	13	14	10	10	11	14	17	20
5% modulus of elasticity parallel	$E_{0,05}$	4.7	5.4	6.0	6.7	7.4	8.0	8.0	8.7	9.4	8.0	8.7	9.4	11.8	14.3	6.8
Mean modulus of elasticity perpendicular	$E_{90,mean}$	0.23	0.27	0.30	0.33	0.37	0.40	0.40	0.43	0.47	0.64	0.69	0.75	0.93	1.13	1.33
Mean shear modulus	G_{mean}	0.44	0.50	0.56	0.63	0.69	0.75	0.75	0.81	0.88	0.60	0.65	0.70	0.88	1.06	1.25
Density (kgm^{-2})																
Density	ρ_k	290	310	320	340	350	370	380	400	420	530	560	590	650	700	900
Average density	ρ_{mean}	350	370	380	410	420	450	460	480	500	640	670	700	780	840	1080

Table 11.4 Partial coefficients for material properties, γ_m (based on Table 2.3.3.2, EC5)

Limit states	γ_m
Ultimate limit states	
Fundamental combinations	
Timber- and wood-based materials	1.3
Steel used in joints	1.1
Accidental combinations	1.0
Serviceability limit states	1.0

having to determine what reduction factors are needed to convert them to permissible or working values.

The characteristic strength values given in *Table 11.3* are related to a depth in bending and width in tension of solid timber of 150 mm. For depths in bending or widths in tension of solid members, h, less than 150 mm the characteristic bending or tension strengths may be increased by the factor k_h which is given by

$$k_h = (150/h)^{0.2} \quad (11.2)$$

The characteristic strengths, X_k, are converted to design values, X_d, by dividing them by a partial coefficient for material properties, γ_m, generally taken from *Table 11.4*, and multiplying by a factor k_{mod}, taken from Table 11.5.

$$X_d = k_{mod}X_k/\gamma_m \quad (11.3)$$

EC5, like BS 5268, allows the design strengths calculated using equation 11.3 to be multiplied by a load-sharing factor, k_{ls}, where several equally spaced similar members are able to resist a common load. Typical members which fall into this category may include joists in flat roofs or floors with a maximum span of 6 m and wall studs with a maximum height of 4 m. According to clause 5.4.6 of EC5, a value of $k_{ls} = 1.1$ may generally be assumed.

Here k_{mod} is a modification factor which takes

Table 11.5 Values of k_{mod} for solid and glued laminated timber plywood (based on Table 3.1.7, EC5)

| Load duration class | Service class | | |
	1	2	3
Permanent	0.60	0.60	0.50
Long term	0.70	0.70	0.55
Medium term	0.80	0.80	0.65
Short term	0.90	0.90	0.70
Instantaneous	1.10	1.10	0.90

Table 11.6 Service classes

Service class	Moisture content	Typical service conditions
1	≤12%	20°C, 65% RH
2	≤20%	20°C, 85% RH
3	> 20%	Climatic conditions leading to a higher moisture content than in service class 2

into account the effect on the strength parameters of the duration of loading and the climatic conditions that the structure will experience in service. EC5 defines three service classes and five load duration classes as summarized in *Tables 11.6* and *11.7* respectively.

Where a load combination consists of actions belonging to different load duration classes the value of k_{mod} should correspond to the action with the shortest duration. For example, for members subject to permanent and variable (imposed) loading, a value of k_{mod} corresponding to the long-term load duration class should be used.

Having discussed these more general aspects it is now possible to describe in detail EC5 rules governing the design of flexural and compression members.

11.5 Design of flexural members

The design of flexural members to EC5 principally involves checking the effect of the following actions as discussed next:

1. bending

Table 11.7 Load duration classes (Table 3.1.6, EC5)

Load duration class	Order of duration	Examples of loading
Permanent	> 10 years	Self-weight
Long term	6 months–10 years	Imposed storage load
Medium term	1 week–6 months	Imposed occupational loads
Short term	< 1 week	Snow[a] and wind
Instantaneous		Accidental impact

[a] Depending on local conditions – in parts of Scotland snow may be medium term.

2. deflection
3. vibration
4. lateral buckling
5. shear
6. bearing.

11.5.1 BENDING (CLAUSE 5.1.6, EC5)

If members are not to fail in bending, the following conditions should be satisfied:

$$k_m \frac{\sigma_{m,y,d}}{f_{m,y,d}} + \frac{\sigma_{m,z,d}}{f_{m,z,d}} \leq 1 \qquad (11.4)$$

$$\frac{\sigma_{m,y,d}}{f_{m,y,d}} + k_m \frac{\sigma_{m,z,d}}{f_{m,z,d}} \leq 1 \qquad (11.5)$$

where $\sigma_{m,y,d}$ and $\sigma_{m,z,d}$ are the design bending stresses about axes y–y and z–z as shown in Figure 11.1, $f_{m,y,d}$ and $f_{m,z,d}$ the design bending strengths from equation 11.3 and k_m the bending factor. It should be noted that in EC5 the x–x axis is the axis along the member and that axes y–y and z–z are the major and minor axes respectively. These definitions are consistent with the other structural Eurocodes.

For a beam with rectangular cross-sections

$$\sigma_{m,y,d} = \frac{M_y}{Z_y} = \frac{M_y}{bh^2/6} \qquad (11.6)$$

$$\sigma_{m,z,d} = \frac{M_z}{Z_z} = \frac{M_z}{hb^2/6} \qquad (11.7)$$

where M_y and M_z are the design bending moments about axes y–y and z–z, Z_y and Z_z the moduli of elasticity about axes y–y and z–z, b the breadth of beam and h the depth of beam. The value of the factor k_m assumes the following values: for rectangular sections $k_m = 0.7$; for other cross-sections $k_m = 1.0$.

11.5.2 DEFLECTION (CLAUSE 4.3, EC5)

To prevent the possibility of damage to surfacing materials, ceilings, partitions and finishes, and to the functional needs as well as aesthetic requirements, EC5 recommends various limiting values of deflection for beams. The components of deflection are shown

Fig. 11.1 Beam axes.

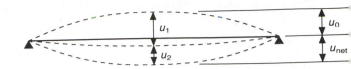

Fig. 11.2 Components of deflection.

in Figure 11.2, where the symbols are defined as

u_0 = precamber (if applied)
u_1 = deflection due to permanent loads
u_2 = deflection due to variable loads

EC5 recommends that the following limiting values will normally need to be observed:

1. Instantaneous deflection due to variable load, $u_{2,inst}$, should not exceed
   $u_{2,inst} \leq 1/300 \times span$
   $u_{2,inst} \leq 1/150 \times span$ (for cantilever)
2. Final deflection due to variable load only, $u_{2,fin}$, should not exceed
   $u_{2,fin} \leq 1/200 \times span$
   $u_{2,fin} \leq 1/100 \times span$ (for cantilever)
3. Final deflection due to all the loads and any precamber, $u_{net,fin}$, should not exceed
   $u_{net,fin} \leq 1/200 \times span$
   $u_{net,fin} \leq 1/100 \times span$ (for cantilever)

The instantaneous deflection due to the variable loads, $u_{2,inst}$, and the final deflection due to the total load, $u_{net,fin}$, can be calculated using the formulae given in *Table 6.9* and should be based on $E_{0,mean}$ or $E_{90,mean}$. The final deflection due to variable loading, $u_{2,fin}$, is derived from the instantaneous deflection using the following expression:

$$u_{fin} = u_{inst}(1 + k_{def}) \qquad (11.8)$$

where k_{def} is the deformation factor which takes into account the increase in deformation with time due to the combined effect of creep and moisture. Values of k_{def} are given in *Table 11.8*.

11.5.3 VIBRATION (CLAUSE 4.4, EC5)

EC5, unlike BS 5268, gives a procedure for calculating the vibrational characteristics of residential floors. These must satisfy certain requirements otherwise the vibrations may impair the proper functioning of the structure or cause unacceptable discomfort to the users. The method given in EC5 assumes that the floor is supported on four edges which is rarely achieved in UK design; there is some doubt as to the relevance of the EC5 design method to UK floors.

Table 11.8 Values of k_{def} for solid timber and glue-laminated timber (based on Table 4.1, EC5)

Load duration class	Service class		
	1	*2*	*3*
Permanent	0.80	0.80	2.00
Long term	0.50	0.50	1.50
Medium term	0.25	0.25	0.75
Short term	0.00	0.00	0.30

The fundamental frequency of vibration of a rectangular residential floor supported on four edges, f_1, can be estimated using

$$f_1 = \frac{\pi}{2l^2} = \sqrt{\frac{(EI)_l}{m}} \qquad (11.9)$$

where m is the mass equal to the self-weight of the floor and other permanent actions per unit area (kg m^{-2}), l the floor span (m) and $(EI)_l$ the equivalent bending stiffness in the beam direction (N m^2 m^{-1}).

For residential floors with a fundamental frequency greater than 8 Hz the following conditions should be satisfied:

$$u/F \leq 1.5 \text{ mm kN}^{-1} \qquad (11.10)$$

and

$$v \leq 100^{(f_1 \zeta - 1)} \qquad (11.11)$$

where ζ is the damping coefficient, normally taken as 0.01, u the maximum vertical deflection caused by a concentrated static force $F = 1$ kN and v the unit impulse velocity.

In carrying out the deflection check in equation 11.10, the transverse distribution of load can be taken as 50%, i.e. 0.5 kN on the loaded joist and 25% on the adjacent ones. The value of v may be estimated from

$$v = 4(0.4 + 0.6n_{40})/(mbl + 200) \text{ m N}^{-1} \text{ s}^{-2} \qquad (11.12)$$

where b is the floor width (m) and n_{40} the number of first-order modes with natural frequencies below 40 Hz given by

$$n_{40} = \left\{ \left[\left(\frac{40}{f_1} \right)^2 - 1 \right] \left(\frac{b}{l} \right)^4 \frac{(EI)_l}{(EI)_b} \right\}^{0.25} \qquad (11.13)$$

where $(EI)_b$ is the equivalent plate bending stiffness parallel to the beams.

11.5.4 LATERAL BUCKLING (CLAUSE 5.2.2, EC5)

Clause 5.1.6(2) of EC5 also requires that beams should be checked for lateral instability. Generally, it will be necessary to show that the following condition is satisfied:

$$\sigma_{m,d} \leq k_{inst} f_{m,d} \qquad (11.14)$$

where $\sigma_{m,d}$ is the design bending stress, $f_{m,d}$ the design bending strength and k_{inst} the instability factor. Here k_{inst} is given by

$$k_{inst} = 1 \quad \text{for } \lambda_{rel,m} \leq 0.75 \qquad (11.15)$$

$$k_{inst} = 1.56 - 0.75\lambda_{rel,m} \quad \text{for } 0.75 < \lambda_{rel,m} \leq 1.4 \qquad (11.16)$$

$$k_{inst} = 1/\lambda^2_{rel,m} \text{ for } 1.4 < \lambda_{rel,m} \qquad (11.17)$$

where $\lambda_{rel,m}$ is the relative slenderness ratio for bending. For beams with rectangular cross-section, $\lambda_{rel,m}$ can be calculated from the following expression:

$$\lambda_{rel,m} = \sqrt{\left[\frac{l_{ef}hf_{m,k}}{\pi b^2 E_{0,k05}} \sqrt{\left(\frac{E_{0,mean}}{G_{mean}} \right)} \right]} \qquad (11.18)$$

where l_{ef} is the effective length of the beam and is obtained from *Fig. 11.3*, b the width of beam, h the depth of beam, $f_{m,k}$ the characteristic bending strength (*Table 11.3*), $E_{0,k05}$ the characteristic modulus of elasticity parallel to grain (*Table 11.3*), $E_{0,mean}$ the mean modulus of elasticity parallel to grain (*Table 11.3*) and G_{mean} the mean shear modulus = $E_{0,mean}/16$.

11.5.5 SHEAR (CLAUSE 5.1.7, EC5)

If flexural members are not to fail in shear, the following condition should be satisfied:

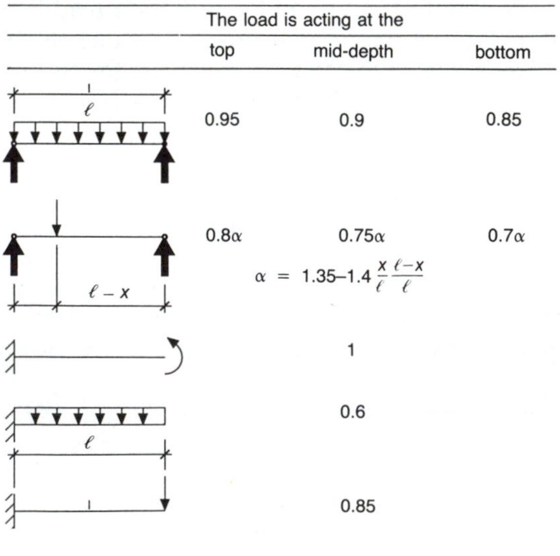

	The load is acting at the		
	top	mid-depth	bottom
	0.95	0.9	0.85
	0.8α	0.75α	0.7α
		$\alpha = 1.35 - 1.4 \frac{x}{l} \frac{l-x}{l}$	
		1	
		0.6	
		0.85	

Fig. 11.3 Ratios of l_{ef}/l

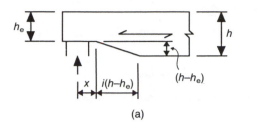

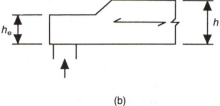

Fig. 11.4 *End notched beams: (a) notch on loaded side; (b) notch on unloaded side (Fig. 5.1.7.2, EC5).*

$$\tau_d \leqslant f_{v,d} \qquad (11.19)$$

where τ_d is the design shear stress and $f_{v,d}$ the design shear strength.

For beams with a rectangular cross-section, the design shear stress occurs at the neutral axis and is given by

$$\tau_d = \frac{3V_d}{2A} \qquad (11.20)$$

where V_d is the design shear force and A the cross-sectional area. The design shear strength, $f_{v,d}$, is given by

$$f_{v,d} = \frac{k_{mod}f_{v,k}}{\gamma_m} \qquad (11.21)$$

where $f_{v,k}$ is the characteristic shear strength (*Table 11.3*).

For beams notched at the ends as shown in *Fig. 11.4*, the following condition should be checked:

$$\tau_d \leqslant k_v f_{v,d} \qquad (11.22)$$

where k_v is the shear factor which may attain the following values: (a) for beams notched on the unloaded side $k_v = 1$ and (b) for beams of solid timber notched on the loaded side k_v is taken as the lesser of $k_v = 1$ and

$$k_v = \frac{5(1 + \frac{1.1i^{1.5}}{\sqrt{h}})}{\sqrt{h}[\sqrt{\alpha}(1 - \alpha) + 0.8\frac{x}{h}\sqrt{(\frac{1}{\alpha} - \alpha^2)}]} \qquad (11.23)$$

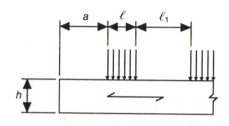

Fig. 11.5 *Compression perpendicular to grain (Fig. 5.1.5(a), EC5).*

where h is the beam depth (mm), x the distance from line of action to the corner, $\alpha = h_e/h$ (*Fig. 11.4*) and i is defined in *Fig. 11.4*.

11.5.6 COMPRESSION PERPENDICULAR TO GRAIN (CLAUSE 5.1.4, EC5)

For compression perpendicular to the grain the following condition should be satisfied:

$$\sigma_{c,90,d} \leqslant k_{c,90}f_{c,90,d} \qquad (11.24)$$

where $\sigma_{c,90,d}$ is the design compressive stress perpendicular to grain, $f_{c,90,d}$ the design compressive strength perpendicular to grain from equation 11.3 and $k_{c,90}$ the compressive strength factor. Here, $k_{c,90}$ takes into account that the load can be increased if the loaded length, l in *Fig. 11.5*, is short. *Table 11.9* shows values for $k_{c,90}$ for various combinations of a, l and l_1.

Table 11.9 Values of $k_{c,90}$ (Table 5.1.5, EC5)

	$l_1 \leq 150\ mm$	$l_1 > 150\ mm$ $a \geqslant 100\ mm$	$a < 100\ mm$
$l \geqslant 150$ mm	1	1	1
150 mm $> l \geqslant 15$ mm	1	$1 + \dfrac{150 - l}{170}$	$1 + \dfrac{a(150 - l)}{17000}$
15 mm $> l$	1	1.8	$1 + a/125$

Example 11.1 Design of timber floor joists

Design the timber floor joists for a domestic dwelling using timber of strength class C22 given that the:

1. Floor width, b, is 3.6 m and floor span, l, is 3.4 m.
2. Joists are spaced at 600 mm centres.
3. Flooring is tongue and groove boarding of thickness 21 mm and has a self-weight of 0.1 kN m^{-2}.
4. Ceiling is of plasterboard with a self-weight of 0.2 kN m^{-2}.
5. The bearing length is 100 mm.

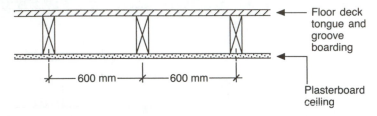

DESIGN LOADING

Permanent loading, G_k

Tongue and groove boarding	= 0.10 kN m^{-2}
Ceiling	= 0.20 kN m^{-2}
Joists (say)	= 0.10 kN m^{-2}
Total characteristic permanent load	= 0.40 kN m^{-2}

Variable load, Q_k

Imposed floor load for domestic dwelling (BS 6399:Part 1) is 1.50 kN m^{-2}.

Design load

Total design load is

$$\gamma_G G_k + \gamma_Q Q_k = 1.35 \times 0.40 + 1.5 \times 1.5 = 2.79 \text{ kN m}^{-2}$$

Design load/joist, F_d, is

$$F_d = \text{joist spacing} \times \text{effective span} \times \text{load} = 0.6 \times 3.4 \times 2.79 = 5.7 \text{ kN}$$

CHARACTERISTIC STRENGTHS AND MODULUS OF ELASTICITY FOR TIMBER OF STRENGTH CLASS C22

Values in N mm^{-2} are given as follows:

Bending strength ($f_{m,k}$)	Compression perpendicular to grain ($f_{c,90,k}$)	Shear parallel to grain ($f_{v,k}$)	Modulus of elasticity ($E_{0,\text{mean}}$)
22.0	5.1	2.4	10 000

BENDING

Bending moment

$$(M) = \frac{Wl}{8} = \frac{5.7 \times 3.4}{8} = 2.42 \text{ kN m}$$

Assuming that the average moisture content of the timber joists does not exceed 20% during the life of the structure, the design should be based on service class 2 (*Table 11.6*). Further, since the joists are required to carry permanent and variable (imposed) loads, the critical load duration class is 'medium term' (*Table 11.7*). Hence from *Table 11.5*, $k_{mod} = 0.8$. From *Table 11.4*, γ_m (for ultimate limit state) = 1.3. Since the joists satisfy the conditions for a load-sharing system outlined in clause 5.4.6 of EC5, the design strengths can be multiplied by a load-sharing factor, $k_{ls} = 1.1$. Assuming $k_h = 1$, design bending strength about the y–y axis is

$$f_{m,y,d} = k_h k_{ls} k_{mod} f_{m,k}/\gamma_m = 1.0 \times 1.1 \times 0.8 \times 22/1.3 = 14.9 \text{ N mm}^{-2}$$

The design bending stress is obtained by substituting into equation 11.5. Hence

$$\frac{\sigma_{m,y,d}}{f_{m,y,d}} + k_m \frac{\sigma_{m,z,d}}{f_{m,z,d}} \leqslant 1$$

$$\frac{\sigma_{m,y,d}}{14.9} + 0.7 \frac{0}{14.9} \leqslant 1$$

$$\sigma_{m,y,d} = 14.9 \text{ N mm}^{-2}$$

$$Z_y \text{ req} \geqslant \frac{M_y}{\sigma_{m,y,d}} = \frac{2.42 \times 10^6}{14.9} = 162 \times 10^3 \text{ mm}^3$$

From *Table 6.8* a 50×200 mm joist would be suitable ($Z_y = 333 \times 10^3$ mm^3, $I_y = 33.3 \times 10^6$ mm^4, $A = 10 \times 10^3$ mm^2).

Since $h > 150$ mm, $k_h = 1$ (as assumed).

DEFLECTION

Check $u_{2,inst}$

From *Table 11.4*, γ_m (for serviceability limit state) is 1.0 and factored variable load, Q, is

$$Q = \gamma_m Q_k = 1.0 \times 1.5 = 1.5 \text{ kN m}^{-2}$$

Factored variable load per joist is

Total load × joist spacing × span length = $1.5 \times 0.6 \times 3.4 = 3.06$ kN

From *Table 6.9*, instantaneous deflection due to variable load, $u_{2,inst}$, is given by

$$u_{2,inst} = \text{bending deflection + shear deflection}$$

$$= \frac{5}{384} \times \frac{Wl^3}{EI} + \frac{12}{5} \times \frac{Wl}{EA}$$

$$= \frac{5}{384} \times \frac{3.06 \times 10^3 \times (3.4 \times 10^3)^3}{10 \times 10^3 \times 33.3 \times 10^6} + \frac{12}{5} \times \frac{3.06 \times 10^3 \times 3.4 \times 10^3}{10 \times 10^3 \times 10 \times 10^3}$$

$$= 4.7 + 0.2 = 4.9 \text{ mm}$$

Permissible instantaneous deflection is $1/300 \times$ span = $3.4 \times 10^3/300 = 11.3$ mm > 4.9 mm   OK

Check $u_{2,fin}$

From *Table 11.8*, for solid timber members subject to service class 2 and medium-term loading, $k_{def} = 0.25$. Final deflection due to variable load only is given by

$$u_{2,fin} = u_{inst} (1 + k_{def}) = 4.9(1 + 0.25) = 6.1 \text{ mm}$$

Permissible final deflection is $1/200 \times$ span = $1/200 \times 3.4 \times 10^3 = 17$ mm > 6.1 mm   OK

Check $u_{net, fin}$

Factored permanent load, $G = \gamma_m G_k = 1.0 \times 0.40 = 0.40$ kN m^{-2} and factored permanent load per joist is

$$\text{Total load} \times \text{joist spacing} \times \text{span length} = 0.4 \times 0.6 \times 3.4 = 0.82 \text{ kN}$$

Instantaneous deflection due to permanent load is given by

$$u_{1,inst} = \text{bending deflection} + \text{shear deflection}$$

$$= \frac{5}{384} \times \frac{Wl^3}{EI} + \frac{12}{5} \times \frac{Wl}{EA}$$

$$= \frac{5}{384} \times \frac{0.82 \times 10^3 \times (3.4 \times 10^3)^3}{10 \times 10^3 \times 33.3 \times 10^6} + \frac{12}{5} \times \frac{0.82 \times 10^3 \times 3.4 \times 10^3}{10 \times 10^3 \times 10 \times 10^3}$$

$$= 1.26 + 0.07 = 1.3 \text{ mm}$$

From *Table 11.8*, for solid timber members subject to service class 2 and permanent loading, $k_{def} = 0.8$. Hence final deflection due to permanent loading, $u_{1,fin}$, is given by

$$u_{1,fin} = u_{1,inst} (1 + k_{def}) = 1.3(1 + 0.8) = 2.3 \text{ mm}$$

and final deflection due to permanent and variable loading is given by

$$u_{net,fin} = u_{1,fin} + u_{2,fin} = 2.3 + 6.1 = 8.4 \text{ mm}$$

Permissible deflection is

$$1/200 \times \text{span} = 1/200 \times 3.4 \times 10^3 = 17 \text{ mm} > 8.4 \text{ mm} \quad \text{OK}$$

Therefore at 50×200 mm joist is adequate in deflection.

VIBRATION

Assuming that $f_1 > 8$ Hz, check that $u/F \leqslant 1.5$ mm kN^{-1} and $v \leqslant 100^{(f_1 \zeta - 1)}$

Check u/F **ratio**

Assuming that due to transverse distribution, 50% of the concentrated load occurs on the joist, i.e. $W = F/2 = 0.5$ kN, from *Table 6.9* the maximum deflection, u, is given by

$$u = \frac{1}{48} \times \frac{Wl^3}{EI} + \frac{24}{5} \times \frac{Wl}{EA}$$

$$= \frac{1}{48} \times \frac{0.50 \times 10^3 \times (3.4 \times 10^3)^3}{10 \times 10^3 \times 33.3 \times 10^6} + \frac{24}{5} \times \frac{0.50 \times 10^3 \times 3.4 \times 10^3}{10 \times 10^3 \times 10 \times 10^3}$$

$$= 1.23 + 0.08 = 1.3 \text{ mm}$$

Hence $u/F = 1.3/1 = 1.3$ mm kN^{-1} < permissible = 1.5 mm kN^{-1} OK

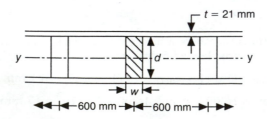

Check impulse velocity

Floor width, $b = 3.4$ m and floor span, $l = 3.6$ m. I_y = second moment of area of joist (ignore tongue and groove boarding unless a specific shear calculation at the interface of joist and board is made):

$$I_y = 33.3 \times 10^6 \text{ mm}^4 = 33.3 \times 10^{-6} \text{ m}^4$$

$$E_{0,\text{ mean}} = 10\ 000 \text{ N mm}^{-2} = 10 \times 10^9 \text{ N m}^{-2} \quad (\textit{Table 11.3})$$

$$(EI)_l = E_{0,\text{mean}} I_y / \text{joist spacing}$$

$$= 10 \times 10^9 \times 33.3 \times 10^{-6} / 0.6 = 555 \times 10^3 \text{ Nm}^2 \text{ m}^{-1}$$

Mass due to permanent actions per unit area, m, is

$$m = \text{permanent action/gravitational constant} = 0.40 \times 10^3 / 9.81 = 40.8 \text{ kg m}^{-2}$$

$$f_1 = \frac{\pi}{2\lambda^2} \, 2l^2 \sqrt{\left[\frac{(EI)_l}{m} \right]}$$

$$= \frac{\pi}{2 \times 3.4^2} \sqrt{\left(\frac{555 \times 10^3}{40.8} \right)}$$

$$= 15.8 \text{ Hz} > 8 \text{ Hz as assumed}$$

I (parallel to beam) is tongue and groove boarding and is

$$I = bt^3/12 = 1000 \times 21^3/12$$

$$= 0.772 \times 10^6 \text{ mm}^4 = 0.772 \times 10^{-6} \text{ m}^4$$

$$(EI)_b = (E_{0,\text{mean}} I) = 10 \times 10^9 \times 0.772 \times 10^{-6} = 7.72 \times 10^3 \text{ Nm}^2 \text{ m}^{-1}$$

$$n_{40} = \left\{ \left[\left(\frac{40}{f_1} \right)^2 - 1 \right] \left(\frac{b}{l} \right)^4 \frac{(EI)_l}{(EI)_b} \right\}^{0.25}$$

$$= \left\{ \left[\left(\frac{40}{15.8} \right)^2 - 1 \right] \left(\frac{3.4}{3.6} \right)^4 \frac{555 \times 10^3}{7.72 \times 10^3} \right\}^{0.25} = 4.19$$

$$v = 4(0.4 + 0.6 n_{40})/(mbl + 200) \text{ m N}^{-1} \text{ s}^{-2}$$

$$= 4(0.4 + 0.6 \times 4.19)/(40.8 \times 3.4 \times 3.6 + 200)$$

$$= 0.017 \text{ m N}^{-1} \text{ s}^{-2}$$

Assume damping coefficient $\zeta = 0.01$. Permissible floor velocity is

$$100^{(f_1 \zeta - 1)} = 100^{(15.8 \times 0.01 - 1)} = 0.02 \text{ m N}^{-1} \text{ s}^{-2} > v = 0.017 \text{ m N}^{-1} \text{ s}^{-2} \quad \text{OK}$$

LATERAL BUCKLING

From *Figure 11.3*, $l_{ef}/l = 0.95$. Hence

$$l_{ef} = 0.95l = 0.95 \times 3.4 \times 10^3 = 3230 \text{ mm}$$

$$\lambda_{rel,m} = \sqrt{\left[\frac{l_{ef}h}{\pi b^2} \times \frac{f_{m,k}}{E_{0,05}} \sqrt{\left(\frac{E_{0,mean}}{G_{mean}} \right)} \right]}$$

$$= \sqrt{\left[\frac{3230 \times 200}{\pi \times 50^2} \times \frac{22}{6700} \sqrt{\left(\frac{E_{0,mean}}{E_{0,mean}|16} \right)} \right]}$$

$$= 1.04$$

For $0.75 < \lambda_{rel,m} < 1.4$

$$k_{inst} = 1.56 - 0.75\lambda_{rel,m} = 1.56 - 0.75 \times 1.04 = 0.78$$

Buckling strength is

$$k_{inst}f_{m,d} = 0.78 \times 14.9 = 11.6 \text{ N mm}^{-2}$$

Buckling stress is

$$\sigma_{m.d} = \frac{M}{Z} = \frac{2.42 \times 10^6}{333 \times 10^3} = 7.3 \text{ N mm}^{-2} < 11.6 \text{ N mm}^{-2} \quad \text{OK}$$

SHEAR

Design shear strength is

$$f_{v,d} = k_{ls}k_{mod}f_{v,k}/\gamma_m = 1.1 \times 0.8 \times 2.4/1.3 = 1.62 \text{ N mm}^{-2}$$

Maximum shear force is

$$F_v = W/2 = \frac{5.7 \times 10^3}{2} = 2.85 \times 10^3 \text{ N}$$

Design shear stress at neutral axis is

$$\tau_d = \frac{3}{2} \times \frac{F_v}{A} = \frac{3}{2} \times \frac{2.85 \times 10^3}{10 \times 10^3}$$

$$= 0.43 \text{ N mm}^{-2} < \text{permissible}$$

Therefore joist is adequate in shear.

BEARING

Design compressive stress

Design bearing force is

$$F_{90,d} = W/2 = 5.7 \times 10^3/2 = 2.85 \times 10^3 \text{ N}$$

100 mm

Assuming that the floor joists span on to 100 mm wide walls as shown above, the bearing stress is given by

$$\sigma_{c,90,d} = \frac{F_{90,d}}{bl} = \frac{2.85 \times 10^3}{50 \times 100} = 0.57 \text{ N mm}^{-2}$$

Design compressive strength

Design compressive strength perpendicular to grain, $f_{c,90,d}$, is given by

$$f_{c,90,d} = k_{ls}k_{mod}f_{c,90,k}/\gamma_m = 1.1\times0.8\times5.1/1.3 = 3.4 \text{ N mm}^{-2}$$

Factor $k_{c,90}$

By comparing the above diagram with *Fig. 11.5*, it can be seen that $a = 0$, $l = 100$ mm and $l_1 > 150$ mm. From *Table 11.9*

$$k_{c,90} = 1 + \frac{a(150 - l)}{17\ 000} = 1$$

From above,

$$\sigma_{c,90,d}\ (= 0.57 \text{ N mm}^{-2}) < k_{c,90}f_{c,90,d} = 1.0\times3.4 = 3.4 \text{ N mm}^{-2}$$

Hence the joist is adequate in bearing.

CHECK ASSUMED SELF-WEIGHT OF JOISTS

From *Table 11.3*, density of timber of strength class C22 is 410 kg m^{-3}. Hence self-weight of the joists is

$$\frac{50 \times 200 \times 10^{-6} \times 410 \text{ kg m}^{-3} \times 9.81 \times 10^{-3}}{0.6} = 0.07 \text{ kN m}^{-2} < \text{assumed}\quad \text{OK}$$

Example 11.2 Design of a notched floor joist

The joists in *Example 11.1* are to be notched at the bearings with a 75 mm deep notch as shown below. Check the notched section is still adequate.

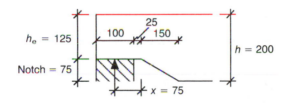

The presence of the notch only affects the shear stresses in the joists.

FACTOR k_v

For beams notched on the loaded side, k_v is taken as the lesser of 1 and the value calculated using equation 11.23. Comparing the above diagram with *Fig. 11.4(a)* gives

$$i = 2 \qquad x = 775 \text{ mm}$$

$$\alpha = h_e/h = 125/200 = 0.625$$

$$k_v = \frac{5(1 + \frac{1.1i^{1.5}}{\sqrt{h}})}{\sqrt{h}\,[\sqrt{\alpha}(1 - \alpha) + 0.8\frac{x}{h}\sqrt{(\frac{1}{\alpha} - \alpha^2)}]}$$

$$k_v = \frac{5(1 + \frac{1.1 \times 2^{1.5}}{\sqrt{200}})}{\sqrt{200}\,[\sqrt{0.625}\,(1 - 0.625) + 0.8\frac{75}{200}\sqrt{(\frac{1}{0.625} - 0.625^2)}]} = 0.53$$

The shear strength, $f_{v,d}$, is

$$f_{v,d} = k_{ls}k_{mod}f_{v,k}/\gamma_m = 1.1\times0.8\times2.4/1.3 = 1.62 \text{ N mm}^{-2}$$

For a notched member, the design shear strength is given by $k_v f_{v,d} = 0.53\times1.62 = 0.86 \text{ N mm}^{-2}$

Design shear stress is $\tau_d = 1.5 V_d/bh_e = 1.5\times2.85\times10^3/50\times125$

$$= 0.68 \text{ N mm}^{-2} < 0.86 \text{ N mm}^{-2}$$

Therefore the section is also adequate when notched with a 75 mm deep bottom edge notch at the bearing.

11.6 Design of compression members

11.6.1 MEMBERS SUBJECT TO AXIAL COMPRESSION ONLY (CLAUSE 5.1.4, EC5)

Members subject to axial compression only should be designed according to the following expression provided there is no tendency for buckling to occur:

$$\sigma_{c,0,d} \leq f_{c,0,d} \qquad (11.25)$$

where $f_{c,0,d}$ is the design compressive strength parallel to the grain obtained from equation 11.3 and $\sigma_{c,0,d}$ the design compressive stress parallel to the grain given by

$$\sigma_{c,0,d} = \frac{N}{A} \qquad (11.26)$$

in which N is the axial load and A is the cross-sectional area.

11.6.2 COLUMNS SUBJECT TO BENDING AND AXIAL COMPRESSION

EC5 gives two sets of conditions for designing columns resisting combined bending and axial compression. Provided that the relative slenderness ratios about both the y–y and z–z axes of the column, $\lambda_{rel,y}$ and $\lambda_{rel,z}$ respectively, are not greater than 0.5, i.e.

$$\lambda_{rel,y} \leq 0.5 \quad \text{and} \quad \lambda_{rel,z} \leq 0.5$$

the suitability of the design can be assessed using the more stringent of the following conditions:

$$\left(\frac{\sigma_{c,0,d}}{f_{c,0,d}}\right)^2 + \frac{\sigma_{m,y,d}}{f_{m,y,d}} + k_m\frac{\sigma_{m,z,d}}{f_{m,z,d}} \leq 1 \qquad (11.27)$$

$$\left(\frac{\sigma_{c,0,d}}{f_{c,0,d}}\right)^2 + k_m\frac{\sigma_{m,y,d}}{f_{m,y,d}} + \frac{\sigma_{m,z,d}}{f_{m,z,d}} \leq 1 \qquad (11.28)$$

where $\sigma_{c,0,d}$ is the design compressive stress from equation 11.26, $f_{c,0,d}$ the design compressive strength from equation 11.3 and $k_m = 0.7$ for rec-

tangular sections and 1.0 for other cross-sections. Note that

$$\lambda_{rel,y} = \sqrt{\left(\frac{f_{c,0,k}}{\sigma_{c,crit,y}}\right)} \qquad (11.29)$$

and

$$\lambda_{rel,z} = \sqrt{\left(\frac{f_{c,0,k}}{\sigma_{c,crit,z}}\right)} \qquad (11.30)$$

where

$$\sigma_{c,crit,y} = \frac{\pi^2 E_{0,05}}{\lambda_y^2} \qquad (11.31)$$

$$\sigma_{c,crit,z} = \frac{\pi^2 E_{0,05}}{\lambda_z^2} \qquad (11.32)$$

in which λ is the slenderness ratio given by

$$\lambda = \frac{l_{ef}}{i} \qquad (11.33)$$

where l_{ef} is the effective length and i the radius of gyration. EC5 does not include a method for determining the effective length of a column. Therefore, designers should follow the recommendation contained in BS 5268: Part 2, as discussed in *section 6.7.1*.

In all other cases the stresses should satisfy the more stringent of the following conditions:

$$\frac{\sigma_{c,0,d}}{k_{c,y}f_{c,0,d}} + \frac{\sigma_{m,y,d}}{f_{m,y,d}} + k_m\frac{\sigma_{m,z,d}}{f_{m,z,d}} \leq 1 \qquad (11.34)$$

$$\frac{\sigma_{c,0,d}}{k_{c,z}f_{c,0,d}} + k_m\frac{\sigma_{m,y,d}}{f_{m,y,d}} + \frac{\sigma_{m,z,d}}{f_{m,z,d}} \leq 1 \qquad (11.35)$$

where σ_m is the bending stress due to any lateral or eccentric loads

$$k_c = \frac{1}{k + \sqrt{(k^2 - \lambda_{rel}^2)}}$$

where $k = 0.5(1 + \beta_c(\lambda_{rel} - 0.5) + \lambda_{rel}^2)$ and $\beta_c = 0.2$ (for solid timber).

Example 11.3 Analysis of a column resisting an axial load

A visually graded timber column of strength class C16 consists of a 100 mm square section which is restrained at both ends in position but not in direction. Assuming that the service conditions comply with service class 2 and the actual height of the column is 3.75 m, calculate the design axial medium-term load that the column can support.

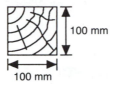

100 mm

100 mm

SLENDERNESS RATIO

$$\lambda_y = \lambda_z = \frac{l_{ef}}{i}$$

$$l_{ef} = 1.0 \times h = 1.0 \times 3750 = 3750 \text{ mm} \quad (Table\ 6.11)$$

$$i = \sqrt{(I/A)} = \sqrt{\left(\frac{db^3/12}{db}\right)} = \sqrt{\frac{b^2}{12}} = \frac{100}{\sqrt{12}} = 28.867$$

$$\lambda_y = \lambda_z = \frac{3750}{28.87} = 129.9$$

CHARACTERISTIC STRENGTH AND STIFFNESSES FOR GRADE C16 TIMBER

Values in N mm^{-2} are as follows:

Compressive strength parallel to grain $f_{c,0,k}$	Modulus of elasticity (5-percentile) $E_{0,05}$
17	5400

(EULER) CRITICAL STRESS

$$\sigma_{c,crit,y} = \sigma_{c,crit,z} = \frac{\pi^2 E_{0,05}}{\lambda_y^2} = \frac{\pi^2 5400}{129.9^2} = 3.16$$

$$\lambda_{rel,y} = \lambda_{rel,z} = \sqrt{\left(\frac{f_{c,0,k}}{\sigma_{c,crit,y}}\right)} = \sqrt{\left(\frac{17}{3.16}\right)} = 2.32$$

Since $\lambda_{rel,y}$ and $\lambda_{rel,z}$ > 0.5, use equations 11.34 or 11.35.

AXIAL LOAD CAPACITY

$$k = 0.5(1 + \beta_c (\lambda_{rel} - 0.5) + \lambda_{rel}^2)$$

$$= 0.5(1 + 0.2(2.32 - 0.5) + 2.32^2) = 3.37$$

$$k_{c,y} = k_{c,z} = \frac{1}{k + \sqrt{(k^2 - \lambda_{rel}^2)}} = \frac{1}{3.37 + \sqrt{(3.37^2 - 2.32^2)}}$$

$$= 0.17$$

Design compressive strength parallel to grain is given by

$$f_{c,0,d} = k_{mod} f_{c,0,k}/\gamma_m = 0.8 \times 17/1.3 = 10.46 \text{ N mm}^{-2}$$

where $\gamma_m = 1.3$ (*Table 11.4*) and $k_{mod} = 0.8$ (service class 2 and medium-term loading). Since column is axially loaded $\sigma_{m,y,d} = \sigma_{m,z,d} = 0$.

Substituting into equation 11.34 or 11.35 gives $\sigma_{c,0,d} = k_{c,y}f_{c,0,d} = 0.17 \times 10.46 = 1.77 \text{ N mm}^{-2}$

Hence axial load capacity of column, N, is given by $N = \sigma_{c,0,d}A = 1.77 \times 10^4 = 17.7 \times 10^3 \text{ N} = 17.7$ kN

Example 11.4 Analysis of an eccentrically loaded column

Check the adequacy of the column in *Example 11.3* to resist a medium-term design (ultimate) axial load of 10 kN applied 35 mm eccentric to its y–y axis.

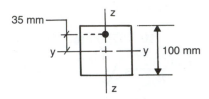

Plan showing column loading

SLENDERNESS RATIO

$$\lambda_y = \lambda_z = 129.9 \quad (\textit{Example 11.3})$$

CHARACTERISTIC STRENGTHS AND MODULUS OF ELASTICITY

Values in N mm^{-2} for machine-graded timber of strength class C16:

Bending parallel to grain	Compression parallel to grain	Modulus of elasticity 5-percentile value
$f_{m,k}$	$f_{c,0,k}$	$E_{0,05}$
16	17	5400

COMPRESSION AND BENDING STRESSES AND STRENGTHS

Design compression stress is

$$\sigma_{c,0,d} = \frac{\text{design axial load}}{A} = \frac{10 \times 10^3}{10^4} = 1 \text{ N mm}^{-2}$$

Design compression strength is

$$f_{c,0,d} = k_{mod}f_{c,0,k}/\gamma_m = 0.8 \times 17/1.3 = 10.46 \text{ N mm}^{-2}$$

Design bending moment about y–y axis, M_y, is

$$M_y = \text{Axial load} \times \text{eccentricity} = 10 \times 35 = 350 \text{ kN mm}$$

Design bending stress about y–y axis, $\sigma_{m,y,d}$,

$$\sigma_{m,y,d} = \frac{M_y}{Z_y} = \frac{350 \times 10^3}{167 \times 10^3} = 2.10 \text{ N mm}^{-2}$$

Design bending strength about y–y axis is $f_{m,d}$ and

$$f_{m,d} = k_h k_{mod} f_{m,k} / \gamma_m = 1.08 \times 0.8 \times 16/1.3 = 10.63 \text{ N mm}^{-2}$$

where $k_h = (150/h)^{0.2} = (150/100)^{0.2} = 1.08$. Design bending stress about z–z axis is $\sigma_{m,y,d} = 0$. Compression factor is $k_{c,y} = 0.17$ (*Example 11.3*). Check the suitability of the column by using equation 11.34:

$$\frac{\sigma_{c,0,d}}{k_{c,y} f_{c,0,d}} + \frac{\sigma_{m,y,d}}{f_{m,y,d}} + k_m \frac{\sigma_{m,z,d}}{f_{m,z,d}} \leqslant 1$$

$$\frac{1}{0.17 \times 10.46} + \frac{2.1}{10.63} + k_m \frac{0}{f_{m,z,d}} = 0.56 + 0.20 = 0.76 < 1$$

Therefore a 100×100 mm column is adequate.

Appendix A

Permissible stress and load factor design

The purpose of this appendix is to illustrate the salient features and highlight essential differences between the following philosophies of structural design:

1. permissible stress approach, i.e. elastic design;

2. load factor approach, i.e. plastic design.

The reader is referred to *Chapter 2* for revision of some basic concepts of structural analysis.

Example A.1

Consider the case of a simply supported, solid rectangular beam (*Figure A.1*), depth (d) 200 mm, span (l) 10 m and subject to a uniformly distributed load (w) of 12 kN m^{-1}. Calculate the minimum width of beam (b) using permissible stress and load factor approaches to design assuming the following:

$$\sigma_{yield} = 265 \text{ N mm}^{-2}$$

Factor of safety (f.o.s.) = 1.5

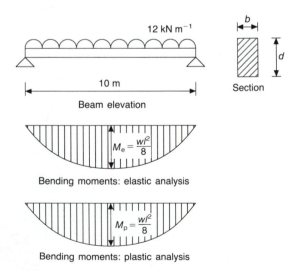

Fig. A.1 *Simply supported beam.*

PERMISSIBLE STRESS APPROACH

Elastic design moment, M_e

$$M_e = \frac{wl^2}{8} = \frac{12 \times 10^2}{8} \quad (Table\ 2.4)$$

$$= 150 \text{ kN m} = 150 \times 10^6 \text{ N mm}$$

LOAD FACTOR APPROACH

Plastic design moment, M_p

Plastic section modulus, $S = \dfrac{bd^2}{4}$

Elastic section modulus, $Z = \dfrac{bd^2}{6}$ *(Table 2.4)*

Shape factor (s.f.) = S/Z
Hence

$$\text{s.f.} = (bd^2/4)/(bd^2/6) = 1.5$$

Load factor   = s.f. × f.o.s.

$$= 1.5 \times 1.5 = 2.25$$

Actual load (w)   = 12 kN m^{-1}
Factored load (w')   = 12 × 2.25 = 27 kN m^{-1}

$$M_p = \frac{w'l^2}{8} = \frac{27 \times 10^2}{8}$$

$$= 337.5 \text{ kN m} = 337.5 \times 10^6 \text{ N mm}$$

Moment of resistance, M_r

Permissible stress, σ_{perm}, is

$$\sigma_{perm} = \frac{\sigma_{yield}}{\text{f.o.s.}} = \frac{265}{1.5}$$

$M_r = \sigma_{perm} Z$
where Z is the elastic section modulus = $bd^2/6$
(Table 2.4). Hence

$$M_r = \frac{(265)}{1.5} \times \frac{200^2 b}{6}$$

$$= 1.178 \times 10^6 b \text{ N mm}^{-2}$$

Moment of resistance, M_r

$$M_r = \sigma_{yield} S = \frac{265 \times bd^2}{4}$$

$$= \frac{265 \times 200^2 b}{4} = 2.65 \times 10^6 b$$

Breadth of beam
At equilibrium, $M_e = M_r$

$$150 \times 10^6 = 1.178 \times 10^6 b$$

Hence breadth of beam, b, is

$$b = 150/1.178 = 127 \text{ mm}$$

Breadth of beam
At equilibrium, $M_p = M_r$

$$337.5 \times 10^6 = 2.65 \times 10^6 b$$

Hence breadth of beam, b, is

$$b = 337.5/2.65 = 127 \text{ mm}$$

It can therefore be seen that for the case of a simply supported beam, provided all the factors are taken into account, both approaches will give the same result and this will remain true irrespective of the shape of the section.

Example A.2

Repeat *Example A.1* but this time assume that the beam is built in at both ends as shown in *Figure A.2*.

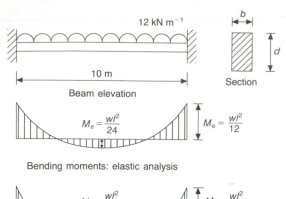

Fig. A.2 *Beam with built-in supports.*

PERMISSIBLE STRESS APPROACH

Elastic design moment, M_e

$$M_e = \frac{wl^2}{12} = \frac{12 \times 10^2}{12}$$
$$= 100 \text{ kN m} = 100 \times 10^6 \text{ N mm}$$

LOAD FACTOR APPROACH

Plastic design moment, M_p

Shape factor (s.f.) = S/Z

Hence
s.f. = $(bd^2/4)/(bd^2/6)$
   = 1.5
Load factor = s.f. × f.o.s.
       = 1.5×1.5 = 2.25
Actual load (w)   = 12 kN m^{-1}
Factored load (w')   = 12×2.25
              = 27 kN m^{-1}

$$M_p = \frac{w'l^2}{16} = \frac{27 \times 10^2}{16}$$
$$= 168.7 \text{ kN m} = 168.7 \times 10^6 \text{ N mm}$$

Moment of resistance, M_r

$$M_r = \sigma_{perm} Z = \frac{(265)}{1.5} \times \frac{200^2 b}{6}$$

$$= 1.178 \times 10^6 b \text{ N mm}^{-2}$$

Moment of resistance, M_r

$$M_r = \sigma_{yield} S = 265 \times \frac{bd^2}{4}$$

$$= 265 \times \frac{200^2 b}{4} = 2.65 \times 10^6 b$$

Breadth of beam

At equilibrium, $M_e = M_r$
$100 \times 10^6 = 1.178 \times 10^6 b$
Hence breadth of beam, b, is
$b = 100/1.178 = 85$ mm

Breadth of beam

At equilibrium, $M_p = M_r$
$168.7 \times 10^6 = 2.65 \times 10^6 b$
Hence breadth of beam, b, is
$b = 168.7/2.65 = 75$ mm

Hence, it can be seen that the load factor approach gives a more conservative estimate for the breadth of the beam, a fact which will be generally found to hold for other sections and indeterminate structures.

The basic difference between these two approaches to design is that while the permissible stress method models behaviour of the structure under working loads, and realistic predictions of behaviour in service can be calculated, the load factor method only models failure, and no information on behaviour in service is obtained.

Appendix B

Dimensions and properties of steel universal beams and columns

Table B1 Dimensions and properties of steel universal beams (structural sections to BS 4: Part 1 and BS 4848: Part 4)

Serial size (mm)	Mass per metre (kg)	Depth of section D (mm)	Width of section B (mm)	Web t_w (mm)	Flange T (mm)	Root radius r (mm)	Depth between fillets d (mm)	Flange b/T	Web d/t_w	2nd moment Axis x-x (cm⁴)	2nd moment Axis y-y (cm⁴)	Radius gyr. Axis x-x (cm)	Radius gyr. Axis y-y (cm)	Elastic mod. Axis x-x (cm³)	Elastic mod. Axis y-y (cm³)	Plastic mod. Axis x-x (cm³)	Plastic mod. Axis y-y (cm³)	Buckling parameter u	Torsional index x	Warping constant H (dm⁶)	Torsional constant J (cm⁴)	Area of section A (cm²)
914×419	388	920.5	420.5	21.5	36.6	24.1	799.1	5.74	37.2	719 000	45 400	38.1	9.58	15 600	2 160	17 700	3 340	0.884	26.7	88.7	1730	494
	343	911.4	418.5	19.4	32.0	24.1	799.1	6.54	41.2	625 000	39 200	37.8	9.46	13 700	1870	15 500	2 890	0.883	30.1	75.7	1190	437
914×305	289	926.6	307.8	19.6	32.0	19.1	824.5	4.81	42.1	505 000	15 600	37.0	6.51	10 900	1010	12 600	1 600	0.867	31.9	31.2	929	369
	253	918.5	305.5	17.3	27.9	19.1	824.5	5.47	47.7	437 000	13 300	36.8	6.42	9 510	872	10 900	1 370	0.866	36.2	26.4	627	323
	224	910.3	304.1	15.9	23.9	19.1	824.5	6.36	51.9	376 000	11 200	36.3	6.27	8 260	738	9 520	1 160	0.861	41.3	22.0	421	285
	201	903.0	303.4	15.2	20.2	19.1	824.5	7.51	54.2	326 000	9 430	35.6	6.06	7 210	621	8 360	983	0.853	46.8	18.4	293	256
838×292	226	850.9	293.8	16.1	26.8	17.8	761.7	5.48	47.3	340 000	11 400	34.3	6.27	7 990	773	9 160	1 210	0.87	35.0	19.3	514	289
	194	840.7	292.4	14.7	21.7	17.8	761.7	6.74	51.8	279 000	9 070	33.6	6.06	6 650	620	7 650	974	0.862	41.6	15.2	307	247
	176	834.9	291.6	14.0	18.8	17.8	761.7	7.76	54.4	246 000	7 790	33.1	5.90	5 890	534	6 810	842	0.856	46.5	13.0	222	224
762×267	197	769.6	268.0	15.6	25.4	16.5	685.8	5.28	44.0	240 000	8 170	30.9	5.71	6 230	610	7 170	959	0.869	33.2	11.3	405	251
	173	762.0	266.7	14.3	21.6	16.5	685.8	6.17	48.0	205 000	6 850	30.5	5.57	5 390	513	6 200	807	0.864	38.1	9.38	267	220
	147	753.9	265.3	12.9	17.5	16.5	685.8	7.58	53.2	169 000	5 470	30.0	5.39	4 480	412	5 170	649	0.857	45.1	7.41	161	188
686×254	170	692.9	255.8	14.5	23.7	15.2	615.1	5.40	42.4	170 000	6 620	28.0	5.53	4 910	518	5 620	810	0.872	31.8	7.41	307	217
	152	687.6	254.5	13.2	21.0	15.2	615.1	6.06	46.6	150 000	5 780	27.8	5.46	4 370	454	5 000	710	0.871	35.5	6.42	219	194
	140	683.5	253.7	12.4	19.0	15.2	615.1	6.68	49.6	136 000	5 180	27.6	5.38	3 990	408	4 560	638	0.868	38.7	5.72	169	179
	125	677.9	253.0	11.7	16.2	15.2	615.1	7.81	52.6	118 000	4 380	27.2	5.24	3 480	346	4 000	542	0.862	43.9	4.79	116	160
610×305	238	633.0	311.5	18.6	31.4	16.5	537.2	4.96	28.9	208 000	15 800	26.1	7.22	6 560	1 020	7 460	1 570	0.886	21.1	14.3	788	304
	179	617.5	307.0	14.1	23.6	16.5	537.2	6.50	38.1	152 000	11 400	25.8	7.08	4 910	743	5 520	1 140	0.886	27.5	10.1	341	228
	149	609.6	304.8	11.9	19.7	16.5	537.2	7.74	45.1	125 000	9 300	25.6	6.99	4 090	610	4 570	937	0.886	32.5	8.09	200	190
610×229	140	617.0	230.1	13.1	22.1	12.7	547.3	5.21	41.8	112 000	4 510	25.0	5.03	3 630	392	4 150	612	0.875	30.5	3.99	217	178
	125	611.9	229.0	11.9	19.6	12.7	547.3	5.84	46.0	98 600	3 930	24.9	4.96	3 220	344	3 680	536	0.873	34.0	3.45	155	160
	113	607.3	228.2	11.2	17.3	12.7	547.3	6.60	48.9	87 400	3 440	24.6	4.88	2 880	301	3 290	470	0.87	37.9	2.99	112	144
	101	602.2	227.6	10.6	14.8	12.7	547.3	7.69	51.6	75 700	2 910	24.2	4.75	2 510	256	2 880	400	0.863	43.0	2.51	77.2	129
533×210	122	544.6	211.9	12.8	21.3	12.7	476.5	4.97	37.2	76 200	3 390	22.1	4.67	2 800	320	3 200	501	0.876	27.6	2.32	180	156
	109	539.5	210.7	11.6	18.8	12.7	476.5	5.60	41.1	66 700	2 940	21.9	4.60	2 470	279	2 820	435	0.875	30.9	1.99	126	139
	101	536.7	210.1	10.9	17.4	12.7	476.5	6.04	43.7	61 700	2 690	21.8	4.56	2 300	257	2 620	400	0.874	33.1	1.82	102	129
	92	533.1	209.3	10.2	15.6	12.7	476.5	6.71	46.7	55 400	2 390	21.7	4.51	2 080	229	2 370	356	0.872	36.4	1.60	76.2	118
	82	528.3	208.7	9.6	13.2	12.7	476.5	7.91	49.6	47 500	2 010	21.3	4.38	1 800	192	2 060	300	0.865	41.6	1.33	51.3	104
457×191	98	467.4	192.8	11.4	19.6	10.2	407.9	4.92	35.8	45 700	2 340	19.1	4.33	1 960	243	2 230	378	0.88	25.8	1.17	121	125
	89	463.6	192.0	10.6	17.7	10.2	407.9	5.42	38.5	41 000	2 090	19.0	4.28	1 770	217	2 010	338	0.879	28.3	1.04	90.5	114
	82	460.2	191.3	9.9	16.0	10.2	407.9	5.98	41.2	37 100	1 870	18.8	4.23	1 610	196	1 830	304	0.877	30.9	0.923	69.2	105
	74	457.2	190.5	9.1	14.5	10.2	407.9	6.57	44.8	33 400	1 670	18.7	4.19	1 460	175	1 660	272	0.876	33.9	0.819	52.0	95.0
	67	453.6	189.9	8.5	12.7	10.2	407.9	7.48	48.0	29 400	1 450	18.5	4.12	1 300	153	1 470	237	0.873	37.9	0.706	37.1	85.4
457×152	82	465.1	153.5	10.7	18.9	10.2	407.0	4.06	38.0	36 200	1 140	18.6	3.31	1 560	149	1 800	235	0.872	27.3	0.569	89.3	104
	74	461.3	152.7	9.9	17.0	10.2	407.0	4.49	41.1	32 400	1 010	18.5	3.26	1 410	133	1 620	209	0.87	30.0	0.499	66.6	95.0
	67	457.2	151.9	9.1	15.0	10.2	407.0	5.06	44.7	28 600	878	18.3	3.21	1 250	116	1 440	182	0.867	33.6	0.429	47.5	85.4
	60	454.7	152.9	8.0	13.3	10.2	407.7	5.75	51.0	25 500	794	18.3	3.23	1 120	104	1 280	163	0.869	37.5	0.387	33.6	75.9
	52	449.8	152.4	7.6	10.9	10.2	407.7	6.99	53.6	21 300	645	17.9	3.11	949	84.6	1 090	133	0.859	43.9	0.311	21.3	66.5
406×178	74	412.8	179.7	9.7	16.0	10.2	360.5	5.62	37.2	27 300	1 540	17.0	4.03	1 320	172	1 500	267	0.881	27.6	0.608	63.0	95.0
	67	409.4	178.8	8.8	14.3	10.2	360.5	6.25	41.0	24 300	1 360	16.9	4.00	1 190	153	1 350	237	0.88	30.5	0.533	46.0	85.5
	60	406.4	177.8	7.8	12.8	10.2	360.5	6.95	46.2	21 500	1 200	16.8	3.97	1 060	135	1 190	208	0.88	33.9	0.464	32.9	76.0
	54	402.6	177.6	7.6	10.9	10.2	360.5	8.15	47.4	18 600	1 020	16.5	3.85	925	114	1 050	177	0.872	38.5	0.39	22.7	68.4
406×140	46	402.3	142.4	6.9	11.2	10.2	359.7	6.36	52.1	15 600	539	16.3	3.02	778	75.7	888	118	0.87	38.8	0.206	19.2	59.0
	39	397.3	141.8	6.3	8.6	10.2	359.7	8.24	57.1	12 500	411	15.9	2.89	627	58.0	721	91.1	0.859	47.4	0.155	10.6	49.4
356×171	67	364.0	173.2	9.1	15.7	10.2	312.3	5.52	34.3	19 500	1 360	15.1	3.99	1 070	157	1 210	243	0.887	24.4	0.413	55.5	85.4
	57	358.6	172.1	8.0	13.0	10.2	312.3	6.62	39.0	16 100	1 110	14.9	3.92	896	129	1 010	199	0.884	28.9	0.331	33.1	72.2
	51	355.6	171.5	7.3	11.5	10.2	312.3	7.46	42.8	14 200	968	14.8	3.87	796	113	895	174	0.882	32.2	0.286	23.6	64.6
	45	352.0	171.0	6.9	9.7	10.2	312.3	8.81	45.3	12 100	812	14.6	3.78	687	95.0	774	147	0.875	36.9	0.238	15.7	57.0
356×127	39	352.8	126.0	6.5	10.7	10.2	311.2	5.89	47.9	10 100	357	14.3	2.69	572	56.6	654	88.7	0.872	35.3	0.104	14.9	49.4
	33	348.5	125.4	5.9	8.5	10.2	311.2	7.38	52.7	8 200	280	14.0	2.59	471	44.7	540	70.2	0.864	42.2	0.081	8.68	41.8
305×165	54	310.9	166.8	7.7	13.7	8.9	265.7	6.09	34.5	11 700	1 060	13.1	3.94	753	127	845	195	0.889	23.7	0.234	34.5	68.4
	46	307.1	165.7	6.7	11.8	8.9	265.7	7.02	39.7	9 950	897	13.0	3.90	648	108	723	166	0.89	27.2	0.196	22.3	58.9
	40	303.8	165.1	6.1	10.2	8.9	265.7	8.09	43.6	8 520	763	12.9	3.85	561	92.4	624	141	0.888	31.1	0.164	14.7	51.5
305×127	48	310.4	125.2	8.9	14.0	8.9	264.6	4.47	29.7	9 500	460	12.5	2.75	612	73.5	706	116	0.874	23.3	0.101	31.4	60.8
	42	306.6	124.3	8.0	12.1	8.9	264.6	5.14	33.1	8 140	388	12.4	2.70	531	62.5	610	98.2	0.872	26.5	0.0842	21.0	53.2

Table B1 Continued

Designation		Dimensions						Ratios for local buckling		Second moment of area		Radius of gyration		Elastic modulus		Plastic modulus		Buckling parameter	Torsional index	Warping constant	Torsional constant	Area of section
Serial size	Mass per metre	Depth of section D	Width of section B	Thickness Web t_w	Thickness Flange T	Root radius r	Depth between fillets d	Flange b/T	Web d/t	Axis x-x	Axis y-y	Axis x-x	Axis y-y	Axis x-x	Axis y-y	Axis x-x	Axis y-y	u	x	H	J	A
(mm)	(kg)	(mm)	(mm)	(mm)	(mm)	(mm)	(mm)			(cm^4)	(cm^4)	(cm)	(cm)	(cm^3)	(cm^3)	(cm^3)	(cm^3)			(dm^6)	(cm^4)	(cm^2)
305×102	37	303.8	123.5	7.2	10.7	8.9	264.6	5.77	36.7	7 160	337	12.3	2.67	472	54.6	540	85.7	0.871	29.6	0.0724	14.9	47.5
	33	312.7	102.4	6.6	10.8	7.6	275.9	4.74	41.8	6 490	193	12.5	2.15	415	37.8	480	59.8	0.866	31.7	0.0441	12.1	41.8
	28	308.9	101.9	6.1	8.9	7.6	275.9	5.72	45.2	5 420	157	12.2	2.08	351	30.8	407	48.9	0.858	37.0	0.0353	7.63	36.3
	25	304.8	101.6	5.8	6.8	7.6	275.9	7.47	47.6	4 390	120	11.8	1.96	288	23.6	338	38.0	0.844	43.8	0.0266	4.65	31.4
254×146	43	259.6	147.3	7.3	12.7	7.6	218.9	5.80	30.0	6 560	677	10.9	3.51	505	92.0	568	141	0.889	21.1	0.103	24.1	55.1
	37	256.0	146.4	6.4	10.9	7.6	218.9	6.72	34.2	5 560	571	10.8	3.47	434	78.1	485	120	0.889	24.3	0.0858	15.5	47.5
	31	251.5	146.1	6.1	8.6	7.6	218.9	8.49	35.9	4 440	449	10.5	3.35	353	61.5	396	94.5	0.879	29.4	0.0662	8.73	40.0
254×102	28	260.4	102.1	6.4	10.0	7.6	225.1	5.10	35.2	4 010	178	10.5	2.22	308	34.9	353	54.8	0.873	27.5	0.0279	9.64	36.2
	25	257.0	101.9	6.1	8.4	7.6	225.1	6.07	36.9	3 410	148	10.3	2.14	265	29.0	306	45.8	0.864	31.4	0.0228	6.45	32.2
	22	254.0	101.6	5.8	6.8	7.6	225.1	7.47	38.8	2 870	120	10.00	2.05	226	23.6	262	37.5	0.854	35.9	0.0183	4.31	28.4
203×133	30	206.8	133.8	6.3	9.6	7.6	172.3	6.97	27.3	2 890	384	8.72	3.18	279	57.4	313	88.1	0.882	21.5	0.0373	10.2	38.0
	25	203.2	133.4	5.8	7.8	7.6	172.3	8.55	29.7	2 360	310	8.54	3.10	232	46.4	260	71.4	0.876	25.4	0.0295	6.12	32.3
203×102	23	203.2	101.6	5.2	9.3	7.6	169.4	5.46	32.6	2 090	163	8.49	2.37	206	32.1	232	49.5	0.89	22.6	0.0153	6.87	29.0
178×102	19	177.8	101.6	4.7	7.9	7.6	146.8	6.43	31.2	1 360	138	7.49	2.39	153	27.2	171	41.9	0.889	22.6	0.00998	4.37	24.2
152×89	16	152.4	88.9	4.6	7.7	7.6	121.8	5.77	26.5	838	90.4	6.40	2.10	110	20.3	124	31.4	0.889	19.5	0.00473	3.61	20.5
127×76	13	127.0	76.2	4.2	7.6	7.6	96.6	5.01	23.0	477	56.2	5.33	1.83	75.1	14.7	85	22.7	0.893	16.2	0.002	2.92	16.8

Properties

Table B2 Dimensions and properties of steel universal columns (structural sections to BS 4: Part 1 and BS 4848: Part 4)

Serial size (mm)	Designation Mass per metre (kg)	Depth of section D (mm)	Width of section B (mm)	Thickness Web t_w (mm)	Thickness Flange T (mm)	Root radius r (mm)	Depth between fillets d (mm)	Ratios for local buckling Flange b/T	Ratios for local buckling Web d/t	Second moment of area Axis x-x (cm⁴)	Second moment of area Axis y-y (cm⁴)	Radius of gyration Axis x-x (cm)	Radius of gyration Axis y-y (cm)	Elastic modulus Axis x-x (cm³)	Elastic modulus Axis y-y (cm³)	Plastic modulus Axis x-x (cm³)	Plastic modulus Axis y-y (cm³)	Buckling parameter u	Torsional index x	Warping constant H (dm⁶)	Torsional constant J (cm⁴)	Area of section (cm²)
356×406	634	474.7	424.1	47.6	77.0	15.2	290.2	2.75	6.10	275 000	98 200	18.5	11.0	11 600	4 630	14 200	7 110	0.843	5.46	38.8	13 700	808
	551	455.7	418.5	42.0	67.5	15.2	290.2	3.10	6.91	227 000	82 700	18.0	10.9	9 960	3 950	12 100	6 060	0.841	6.05	31.1	9 240	702
	467	436.6	412.4	35.9	58.0	15.2	290.2	3.56	8.08	183 000	67 900	17.5	10.7	8 390	3 290	10 000	5 040	0.839	6.86	24.3	5 820	595
	393	419.1	407.0	30.6	49.2	15.2	290.2	4.14	9.48	147 000	55 400	17.1	10.5	7 000	2 720	8 230	4 160	0.837	7.86	19.0	3 550	501
	340	406.4	403.0	26.5	42.9	15.2	290.2	4.70	11.0	122 000	46 800	16.8	10.4	6 030	2 320	6 990	3 540	0.836	8.85	15.5	2 340	433
	287	393.7	399.0	22.6	36.5	15.2	290.2	5.47	12.8	100 000	38 700	16.5	10.3	5 080	1 940	5 820	2 950	0.835	10.2	12.3	1 440	366
	235	381.0	395.0	18.5	30.2	15.2	290.2	6.54	15.7	79 100	31 000	16.2	10.2	4 150	1 570	4 690	2 380	0.834	12.1	9.54	812	300
COLCORE	477	427.0	424.4	48.0	53.2	15.2	290.2	3.99	6.05	172 000	68 100	16.8	10.6	8 080	3 210	9 700	4 980	0.815	6.91	23.8	5 700	607
356×368	202	374.7	374.4	16.8	27.0	15.2	290.2	6.93	17.3	66 300	23 600	16.0	9.57	3 540	1 260	3 980	1 920	0.844	13.3	7.14	560	258
	177	368.3	372.1	14.5	23.8	15.2	290.2	7.82	20.0	57 200	20 500	15.9	9.52	3 100	1 100	3 460	1 670	0.844	15.0	6.07	383	226
	153	362.0	370.2	12.6	20.7	15.2	290.2	8.94	23.0	48 500	17 500	15.8	9.46	2 680	944	2 960	1 430	0.844	17.0	5.09	251	195
	129	355.6	368.3	10.7	17.5	15.2	290.2	10.5	27.1	40 200	14 600	15.6	9.39	2 260	790	2 480	1 200	0.843	19.9	4.16	153	165
305×305	283	365.3	321.8	26.9	44.1	15.2	246.6	3.65	9.17	78 800	24 500	14.8	8.25	4 310	1 530	5 100	2 340	0.855	7.65	6.33	2 030	360
	240	352.6	317.9	23.0	37.7	15.2	246.6	4.22	10.7	64 200	20 200	14.5	8.14	3 640	1 270	4 250	1 950	0.854	8.73	5.01	1 270	306
	198	339.9	314.1	19.2	31.4	15.2	246.6	5.00	12.8	50 800	16 200	14.2	8.02	2 990	1 030	3 440	1 580	0.854	10.2	3.86	734	252
	158	327.2	310.6	15.7	25.0	15.2	246.6	6.21	15.7	38 700	12 500	13.9	7.89	2 370	806	2 680	1 230	0.852	12.5	2.86	379	201
	137	320.5	308.7	13.8	21.7	15.2	246.6	7.11	17.9	32 800	10 700	13.7	7.82	2 050	691	2 300	1 050	0.851	14.1	2.38	250	175
	118	314.5	306.8	11.9	18.7	15.2	246.6	8.20	20.7	27 600	9 010	13.6	7.75	1 760	587	1 950	892	0.851	16.2	1.97	160	150
	97	307.8	304.8	9.9	15.4	15.2	246.6	9.90	24.9	22 200	7 270	13.4	7.68	1 440	477	1 590	723	0.85	19.3	1.55	91.1	123
254×254	167	289.1	264.5	19.2	31.7	12.7	200.3	4.17	10.4	29 900	9 800	11.9	6.79	2 070	741	2 420	1 130	0.852	8.49	1.62	625	212
	132	276.4	261.0	15.6	25.3	12.7	200.3	5.16	12.8	22 600	7 520	11.6	6.67	1 630	576	1 870	879	0.85	10.3	1.18	322	169
	107	266.7	258.3	13.0	20.5	12.7	200.3	6.30	15.4	17 500	5 900	11.3	6.57	1 310	457	1 490	695	0.848	12.4	0.894	173	137
	89	260.4	255.9	10.5	17.3	12.7	200.3	7.40	19.1	14 300	4 850	11.2	6.52	1 100	379	1 230	575	0.849	14.4	0.716	104	114
	73	254.0	254.0	8.6	14.2	12.7	200.3	8.94	23.3	11 400	3 870	11.1	6.46	894	305	989	462	0.849	17.3	0.557	57.3	92.9
203×203	86	222.3	208.8	13.0	20.5	10.2	160.9	5.09	12.4	9 460	3 120	9.27	5.32	851	299	979	456	0.85	10.2	0.317	138	110
	71	215.9	206.2	10.3	17.3	10.2	160.9	5.96	15.6	7 650	2 540	9.16	5.28	708	246	802	374	0.852	11.9	0.25	81.5	91.1
	60	209.6	205.2	9.3	14.2	10.2	160.9	7.23	17.3	6 090	2 040	8.96	5.19	581	199	652	303	0.847	14.1	0.195	46.6	75.8
	52	206.2	203.9	8.0	12.5	10.2	160.9	8.16	20.1	5 260	1 770	8.90	5.16	510	174	568	264	0.848	15.8	0.166	32.0	66.4
	46	203.2	203.2	7.3	11.0	10.2	160.9	9.24	22.0	4 560	1 540	8.81	5.11	449	151	497	230	0.846	17.7	0.142	22.2	58.8
152×152	37	161.8	154.4	8.1	11.5	7.6	123.5	6.71	15.2	2 220	709	6.84	3.87	274	91.8	310	140	0.848	13.3	0.04	19.5	47.4
	30	157.5	152.9	6.6	9.4	7.6	123.5	8.13	18.7	1 740	558	6.75	3.82	221	73.1	247	111	0.848	16.0	0.0306	10.5	38.2
	23	152.4	152.4	6.1	6.8	7.6	123.5	11.2	20.2	1 260	403	6.51	3.68	166	52.9	184	80.9	0.837	20.4	0.0214	4.87	29.8

References and further reading

References

BRITISH STANDARDS

BS 4: *Structural Steel Sections*; Part 1: *Specification for Hot-rolled Sections*

BS 18: *Method of Tensile Testing of Metals Including Aerospace Materials* (now superseded by BS EN10002: Part 1)

BS 449: *The Use of Structural Steel in Buildings*: Part 2: 1969

BS 639: *Specification for Covered Carbon and Manganese Steel Electrodes for Manual Metal Arc Welding*

BS 648: *Schedule of Weights of Building Materials*

BS 709: *Methods of Destructive Testing Fusion Welded Joints and Weld Metal in Steel*

BS 1243: *Specification for Metal Ties for Cavity Wall Construction*

BS 1881: *Methods of Testing Concrete*

BS 3921: *Specification for Clay Bricks*

BS 4360: *Specification for Weldable Structural Steels*

BS 4848: *Specification for Hot-rolled Structural Steel Sections*

BS 4978: *Specification for Softwood Grades for Structural Use*

BS 5268: *Structural Use of Timber*; Part 2: *1991 Code of Practice for Permissible Stress Design, Materials and Workmanship*

BS 5400: *Code of Practice for the Design of Steel, Concrete and Composite Bridges*

BS 5628: *Code of Practice for Use of Masonry*; Part 1: *1992: Unreinforced Masonry*

BS 5950: *Structural Use of Steelwork in Buildings*; Part 1: *1990 Code of Practice for Design in Simple and Continuous Construction: Hot Rolled Sections*

BS 6073: *Precast Concrete Masonry Units*. Part 1: *Specification for Precast Concrete Masonry Units*

BS 6399: *Design Loading for Buildings*; Part 1: *Code of Practice for Dead and Imposed Loads*

BS 8007: *Code of Practice for the Design of Concrete Structures for Retaining Aqueous Liquids*

BS 8110: *Structural Use of Concrete*;

Part 1: 1985 *Code of Practice for Design and Construction*;
Part 2: 1985 *Code of Practice for Special Circumstances*;
Part 3: 1985 *Design Charts for Singly Reinforced Beams, Doubly Reinforced Beams and Rectangular Columns*

CP 3: *Code of Basic Data for the Design of Buildings*; Chapter V: Part 2: *Wind Loads* (will shortly be superseded by BS 6399: Part 2)

CP 114: *Structural Use of Reinforced Concrete in Buildings* (now withdrawn)

EUROCODES AND EUROPEAN STANDARDS

ENV 1991: Eurocode 1: *Basis of Design and Actions on Structures*

ENV 1992-1-1: Eurocode 2: *Design of Concrete Structures*; Part 1-1: 1992 *General Rules and Rules for Buildings*

ENV 1993-1-1: Eurocode 3: *Design of Steel Structures*; Part 1-1: 1992 *General Rules and Rules for Buildings*

prEN 1995-1-1: Eurocode 5: *Design of Timber Structures*; Part 1-1: 1992 *General Rules and Rules for Buildings*

ENV 1996-1-1: Eurocode 6: *Design of Masonry Structures*; Part 1-1: *General Rules and Rules for Buildings* (unpublished)

ENV 206: *Concrete – Performance, Production, Placing and Compliance Criteria*: 1992

prEN 338: *Structural Timber; Strength Classes*: 1990

prEN 77-1: *Specification for Masonry Units*; Part 1: 1992 *Clay Masonry Units*

BS EN 10002: *Tensile Testing of Metallic Materials*; Part 1: *Method of Test at Ambient Temperature*

prEN 10080: *Steel for the Reinforcement of Concrete Weldable Ribbed Reinforcing Steel B 500; Technical Delivery Condition for Bars, Coils and Weldable Fabrics*

GENERAL

British Cement Association *Concise Eurocode for the Design of Concrete Building Structures*

British Cement Association *Shearheads*

British Standards Institution *Extracts from British Standards for Students of Structural Design*

British Steel PLC *Structural Sections to BS 4: Part 1 and BS 4848: Part 4*

CIB (International Council for Building Research Studies and Documentation) *CIB Report 66*, CIB Structural Timber Design Code, Luxembourg

Concrete Society and Institution of Structural Engineers *Standard Methods of Detailing Structural Concrete*, London

Higgins, J.B. and Rogers, B.R. *Design and Detailing (BS 8110: 1985)*, British Cement Association, Crowthorne

Rowe, R.E. *et al. Handbook to BS8110: 1985: Structural Use of Concrete*, Spon, London

Further Reading

Allen, A.H. *Reinforced Concrete Design to BS 8110 Simply Explained*, Spon, London

Baird, J.E. and Ozelton, E.C. *Timber Designers' Manual*, Crosby Lockwood Staples, London

Curtin, W.G., Shaw, G., Beck, J.K. and Bray, W.A. *Structural Masonry Designers' Manual*, BSP Professional Books, Oxford

Institution of Structural Engineers, Institution of Civil Engineers *Manual for the Design of Reinforced Concrete Building Structures*, London

Institution of Structural Engineers, Institution of Civil Engineers *Manual for the Design of Steel Building Structures*, London

MacGinley, T.J. and Choo, B.S. *Reinforced Concrete: Design Theory and Examples*, Spon, London

MacGinley, T.J. and Ang, T.C. *Structural Steelwork Design to Limit State Theory*, Butterworths, London

Mosley, W.H. and Bungey, J.H. *Reinforced Concrete Design*, Macmillan, London

Pask, J.W. *Manual on Connections*, BCSA Ltd, London

Index

Index

P ✓ ✓ F·details — scale — ~~7/8~~ 5/4
I   ✗ F·layout   "   ~~8/24~~
P ✓ ✓ Siteplan   "   no scales.
P ✓ ✓ Site   "   7/8
P   DRW 2   "